Lecture Notes in Computer Science 5831

Commenced Publication in 1973
Founding and Former Series Editors:
Gerhard Goos, Juris Hartmanis, and Jan van Leeuwen

Axel Polleres   Terrance Swift (Eds.)

Web Reasoning and Rule Systems

Third International Conference, RR 2009
Chantilly, VA, USA, October 25-26, 2009
Proceedings

 Springer

Volume Editors

Axel Polleres
National University of Ireland
Digital Enterprise Research Institute (DERI)
IDA Business Park, Lower Dangan - Galway, Ireland
E-mail: axel.polleres@deri.org

Terrance Swift
Centre for Artificial Intelligence (CENTRIA)
Departamento de Informatica, FCT/UNL
Quinta da Torre, 2829-516 Caparica, Portugal
E-mail: tswift@cs.sunysb.edu

Library of Congress Control Number: 2009936484

CR Subject Classification (1998): D.4, H.2.4, I.2.4, H.3.5, H.5

LNCS Sublibrary: SL 3 – Information Systems and Application, incl. Internet/Web and HCI

ISSN 0302-9743

ISBN 978-3-642-05081-7 Springer Berlin Heidelberg New York

springer.com

© Springer-Verlag Berlin Heidelberg 2009

Typesetting: Camera-ready by author, data conversion by Scientific Publishing Services, Chennai, India
Printed on acid-free paper   SPIN: 12778508   06/3180   5 4 3 2 1 0

Preface

The promise of the Semantic Web, at its most expansive, is to allow knowledge to be freely accessed and exchanged by software. It is now recognized that if the Semantic Web is to contain deep knowledge, the need for new representation and reasoning techniques is going to be critical. These techniques need to find the right trade-off between expressiveness, scalability and robustness to deal with the inherently incomplete, contradictory and uncertain nature of knowledge on the Web. The International Conference on Web Reasoning and Rule Systems (RR) was founded to address these needs and has grown into a major international forum in this area. The third RR conference was held during October 25–26, 2009 in Chantilly, Virginia, co-located with the International Semantic Web Conference (ISWC 2009).

This year 41 papers were submitted from authors in 21 countries. The Program Committee performed outstandingly to ensure that each paper submitted to RR 2009 was thoroughly reviewed by at least three referees in a short period of time. The resulting conference presented papers of high quality on many of the key issues for reasoning on the Semantic Web. RR 2009 was fortunate to have two distinguished invited speakers. Robert Kowalski, in his talk "Integrating Logic Programming and Production Systems with Abductive Logic Programming Agents" addressed some of the fundamental considerations behind reasoning about evolving systems. Benjamin Grossof's talk "SILK: Higher Level Rules with Defaults and Semantic Scalability" described the design of a major next-generation rule system. The invited tutorial "Uncertainty Reasoning for the Semantic Web" by Thomas Lukasiewicz provided perspectives on a central issue in this area.

Regular papers addressed fundamental issues of reasoning with topics including deduction procedures for ontologies with defaults and for conceptual logic programs, evaluation procedures for path query languages, analysis of production systems using fixed-point logic, and general perspectives on control in rule engines. The importance of scalability was reflected by papers on distributed resolution for ontologies, parallel logic programming techniques for Abox querying, and the separation of terminological from assertional data. The topic of knowledge amalgamation was studied by papers on alignment, modularity and paraconsistency for ontologies. Uncertainty was explored by papers on semantics and inference procedures for fuzzy reasoning, and a paper on inference procedures for a logic of belief.

The results of Web Reasoning and Rule Systems are not confined only to foundational issues, but are being applied to Web standards and real-world applications. Papers considered paraconsistent and fuzzy extensions to the RDF standard, which just celebrated its 10th anniversary and is becoming widely used. Another paper explored scalability of the OWL-2 standard, which is soon to be published by W3C. Finally, one paper described how Semantic Web techniques were applied in a risk-assessment system for elective surgery.

We thank the following sponsors for supporting RR 2009:

- BBN TECHNOLOGIES, USA
- LOGIC PROGRAMMING ASSOCIATES LTD., UK
- KNO.E.SIS CENTER, USA
- IOS PRESS
- THE LARKC - LARGE KNOWLEDGE COLLIDER PROJECT
- CLARK&PARSIA LLC, USA
- DIGITAL ENTERPRISE RESEARCH INSTITUTE (DERI) AT THE NATIONAL UNIVERSITY OF IRELAND, GALWAY
- VULCAN INC., USA
- THE NEON PROJECT

We are particularly thankful also to the authors, the invited speakers, and attendees for contributing and discussing the latest results in relevant areas to this conference, as well as to all members of the Program Committee, and the external reviewers for their critical reviews of submissions.

October 2009   Terrance Swift
                                                        Axel Polleres

Organization

General Chair

Michael Kifer   SUNY Stony Brook, NY, USA

Program Co-chairs

Terrance Swift   CENTRIA, Universidade Nova de Lisboa,
                                 Portugal
Axel Polleres   Digital Enterprise Research Institute
                                 National University of Ireland, Galway

Local Arrangements Chair

Guilin Qir   Universität Karlsruhe, Germany

Sponsorship Chair

Pascal Hitzler   Universität Karlsruhe, Germany

Program Committee

Grigoris Antoniou   Wolfgang May
Marcelo Arenas   David Pearce
Leopoldo Bertossi   Enrico Pontelli
Piero Bonatti   Guilin Qi
Carlos Damásio   Marie-Christine Rousset
Wlodek Drabent   Sebastian Rudolph
Bernardo Cuenca Grau   Sebastian Schaffert
Volker Haarslev   Michael Sintek
Giovambattista Ianni   Heiner Stuckenschmidt
Manolis Koubarakis   Péter Szeredi
Domenico Lembo   Sergio Tessaris
Thomas Lukasiewicz   Hans Tompits
Francesca Alessandra Lisi   Dirk Vermeir

Additional Reviewers

Anne Schlicht   Francesco Ricca
Johannes Oetsch   Gergely Lukácsy
Luigi Sauro   Zsolt Zombori

Arash Shaban-Nejad
Jörg Pührer
Héctor Pérez-Urbina
Christian Meilicke
Jun Fang
Antoine Zimmermann

Ratnesh Sahay
Jianfeng Du
Jeroen Janssen
Irini Fundulaki
Manuel Möller

Table of Contents

Knowledge Amalgamation and Querying

Rules for Decision Support and Production Systems

Integrating Logic Programming and Production Systems in Abductive Logic Programming Agents

Robert Kowalski and Fariba Sadri

Department of Computing, Imperial College London, 180 Queens Gate, London SW7 2AZ
{rak,fs}@doc.ic.ac.uk

Abstract. In this paper we argue the case for integrating the distinctive functionalities of logic programs and production systems within an abductive logic programming agent framework. In this framework, logic programs function as an agent's beliefs and production rules function as the agent's goals. The semantics and proof procedures are based on abductive logic programming, in which logic programs are integrated with integrity constraints that behave like production rules.

Similarly to production systems, the proof procedure is an operational semantics, which manipulates the current state of a database, which is modified by actions implemented by destructive assignment. The semantics can be viewed as generating a model, based on the sequence of database states and logic program, which makes the production rules true.

Keywords: Abductive logic programming, Production systems, Integrity constraints, agents.

1 Introduction

Rules are the basic form of knowledge representation in many areas of Artificial Intelligence, including both production systems and logic programming, and more recently in BDI (Belief Desire Intentions) agent languages. Despite their wide-spread use, there is a great deal of confusion between the different kinds of rules, and little agreement about the relationship between them.

In this paper we argue that production rules and logic programming rules have complementary characters and that one cannot usefully be reduced to the other. We show how abductive logic programming (ALP) combines the two kinds of rules in a single unified framework. The ALP framework gives a model-theoretic semantics to both kinds of rules and provides them with powerful proof procedures, combining both backward and forward reasoning. We present evidence from BDI agents, deductive databases and psychological experiments to support the distinct nature of the two kinds of rules.

We discuss the impact of including the two kinds of rules in a production system or agent cycle, which embeds the rules in a destructively changing environment, which is like a production system working memory. In this embedding, the environment can be viewed as the semantic structure that gives meaning to the two kinds of rules. Although the rules themselves need to be understood as explicitly or implicitly representing

A. Polleres and T. Swift (Eds.): RR 2009, LNCS 5837, pp. 1–23, 2009.

change of state, the fact that the environment changes destructively and exists at any given time only in its current state gives an efficient solution to the frame problem.

We assume the reader is familiar with the basic concepts of logic programming and production systems, although not necessarily with all of their technicalities.

1.1 Confusions

The most popular textbook on Artificial Intelligence Russell and Norvig [45] views production rules as just logical conditionals used to reason forward (page 286). In contrast, in one of the main textbooks on Cognitive Science, Thagard [50] argues that "Rules are if-then structures ...very similar to the conditionals..., but they have different representational and computational properties." (page 43). "Unlike logic, rule-based systems can also easily represent strategic information about what to do. Rules often contain actions that represent goals, such as *IF you want to go home for the weekend and you have bus fare, THEN you can catch a bus*." (page 45).

Thagard [50] characterizes Prolog as "a programming language that uses logic representations and deductive techniques". Simon [48], on the other hand, includes Prolog "among the production systems widely used in cognitive simulation."

There is a similar confusion in the field of agents. Rao [41], for example, characterises AgentSpeak as similar to logic programming. But in his comparison, he considers only the similarities between the operational semantics of plans in AgentSpeak and the execution of clauses in logic programming. He ignores the declarative semantics of logic programs (LPs).

1.2 Production Systems and Logic Programs in Practice

There have been many theoretical studies of the relationship between production rules and logic programs, which we discuss below in Section 2. Most of this work has been focussed on giving a declarative semantics to production systems by translating them into logic programs. However, there seems to have been little attention paid to the way in which logic programs and production rules are used in practice, and consequently little attempt to use this practice to guide the theoretical analysis. We argue that in practice, the two kinds of rules have both distinct and overlapping functionalities, and that the distinct functionalities are lost by translating one kind of rule into the other. We will show that abductive logic programming (ALP) both capitalises on the distinct functionalities and eliminates the overlap.

We argue that, in addition to the production rule (PR) cycle and destructively changing database, which are absent in LP, PRs offer three distinct functionalities: reactive rules that implement stimulus-response associations; forward chaining logic rules; and goal-reduction rules.

Reactive rules are, arguably, the most distinctive type of production rules, which are responsible for their general characterisation as *condition-action* rules. This kind of rule typically has implicit or *emergent* goals. For example, the rule *if a car coming towards you then get out of its way* has the implicit goal *to stay safe*. Reactive rules provide a functionality that is not directly available in logic programming.

The second kind of rule, for example *if X is a cat then X is an animal* uses forward chaining to implement forward reasoning with a logical conditional. It is probably this kind of rule that gives the impression that production rules are just conditionals used to reason forward.

It is the third kind of rule, exemplified by Thagard's example of the goal-reduction rule *"IF you want to go home for the weekend and you have bus fare, THEN you can catch a bus."*, that overlaps the most with logic programming. In logic programming, such strategic rules would be obtained by reasoning backward with the clause *you go home for the weekend if you have bus fare and you catch a bus.* The two best known cognitive models of human thinking, SOAR [34] and ACT-R [3], are based on production systems and focus on the use of production rules for goal-reduction.

Logic programming has its own confusions, mostly about whether clauses are to be understood declaratively or procedurally. The purely declarative interpretation of clauses, which is neutral about reasoning method, is probably the one that is most attractive to its admirers. It is well-suited for high-level program specifications and for certain applications where efficiency is not a major concern.

However, it is probably the procedural interpretation, in which clauses are used to obtain goal-reduction by backward reasoning, that is the main way in which logic programs are used in practice. This is also where there is the greatest overlap with production rules. Arguably, logic programs with backward reasoning are more suitable for this purpose than production rules with forward reasoning, because logic programs can also be interpreted declaratively. The declarative interpretation of logic programs makes it possible to give goal-reduction procedures the declarative semantics that is missing with production rules.

The confusion between the declarative and procedural uses of logic programs and how best to combine them is well-known even though it is not very well solved. However, there is another use of logic programs that has received less attention, and is perhaps even more confusing. It is the use of logic programs for forward reasoning. This use is not very common in practice, but is prevalent in theoretical investigations of logic programming. We will see that in ALP, clauses can be used to reason both backward and forward.

1.3   Combining Production Systems and Logic Programs

Broadly speaking, there are four motivations for combining PRs and LPs:

1. To eliminate the overlap between forward logic rules in PRs and forward/declarative clauses in LP. For example, one very simple combination of PRs and LPs is to use LPs to define ramifications of the working memory/database. Then existing PRs could simply query a deductive database rather than a relational database. This would hand over the forward reasoning logic rules from the PR to the LP component. Moreover, it would allow the decision about how to execute the ramifications to be taken by the implementation. The declarative semantics of ramifications would be compatible with executing them forward, backward, or any combination of the two.

2. To eliminate the overlap for goal-reduction, by using LPs for this purpose. Using LP for goal-reduction provides system support for managing goals as and-or trees, which is missing in production systems. Whereas most production systems just treat goals as ordinary facts in the working memory, SOAR and ACT-R manipulate them in goal stacks. Using LP for goal-reduction allows the declarative nature of LP

clauses to be exploited, so they can be used either for goal-reduction or for forward reasoning, as the context requires.

3. To provide a declarative semantics for PRs and for the combination of LP and PRs. Declarative semantics provides an independent specification for implementations, as illustrated by the discussion above about the implementation of ramifications. Declarative semantics also clarifies the nature of PRs as a representation language. Without a declarative semantics, which establishes a relationship between syntactic expressions and semantic structures, of the kind provided by the model theory of logic, the term "representation" has no meaning.

Our proposal is to combine LP and PR in the same way that ALP combines LP and integrity constraints (ICs), and to use the model-theoretic semantics of ALP to give a model-theoretic semantics to the combination of LP and PR. We will show that integrity constraints in ALP, especially when embedded in ALP agents, generalise production rules to include *condition-goal* rules, where the *goal* is like the body of a plan in BDI agents.

4. To provide a cycle and destructive database of facts, missing in LP. Without a cycle, LP is both closed and passive – closed because logic programs cannot be updated by the environment, and passive because they cannot perform updates on the environment. Without a destructive database of facts, LP suffers from the inefficiencies of the frame problem.

2   Other Approaches

Typically PRs, as well as event-condition-action (ECA) rules and active integrity constraints are defined by means of an operational semantics based on state transitions. However several authors have studied the relationship between these various kinds of rules and LP, with the aim of providing the rules with a declarative LP-based semantics. In the majority of these approaches PRs, ECA rules or active integrity constraints are mapped into LP to provide them with LP-based semantics. To our knowledge, there has been no proposal that would accommodate both LP and PR (or ECA rules or active integrity constraints) side-by-side with an integrated semantics or proof procedure that would exploit the strengths of both paradigms.

Raschid [52] combines LPs and ICs, but focuses on only two functionalities of PRs, namely on their use as reactive rules and as forward logic rules. She represents rules that add facts as LPs, and rules that delete facts as ICs. She then transforms their combination into LP, and uses the fixed point semantics of LP to chain forward and thereby simulate the production system cycle.

Ceri and Widom [7] and Ceri *et al.* [6] implement ICs by PRs. Dung and Mancarella [14] use an argumentation theoretic framework to provide semantics for PRs with negation as failure.

Caroprese *et al.* [5] transform active integrity constraints into LPs. They characterise the set of "founded" repairs for the database as the stable model of the database augmented by the LP representation of the active integrity constraints. Fraternali and Tanca [17] also consider active databases but provide a logic-based *core* syntax for representing low-level, procedural features of active database rules. They provide procedural semantics for core rules and show how this can capture the procedural semantics of known active database systems.

Most other work regards the cycle and actions in condition-action rules and ECA rules declaratively as performing a change of state. Zaniolo [54], for example, uses a situation calculus-like representation with frame axioms, and reduces PRs and ECA rules to LPs. Statelog [53] also uses a situation-calculus-like representation for the succession of database states. Like Zaniolo, Statelog represents PRs and ECAs as LPs, and gives them LP-based semantics. Neither is concerned with the role of ICs or with the use of LPs and PRs for goal-reduction.

Fernandes *et al.* [16] also view ECAs in terms of change of state, but use the event calculus as the basis for an ECA language coupled with a deductive database. The event calculus is used to evaluate the condition part of the ECA rules and to provide a specification for the effects of executing the action part. The ECA language also allows the recognition of complex events from an event history.

ERA (Evolving Reactive Algebraic Programs), developed by Alferes *et al.* [1], extends the dynamic logic programming system EVOLP [2] by adding complex events and actions as well as external actions. ERA combines ECA and LP rules, and the firing of the ECA rules can generate actions that add or delete ECA or LP rules, as well as external actions. In the operational semantics the ECA and LP rules maintain their distinct characteristics, but in the declarative semantics the ECA rules are translated to LP. The declarative semantics is based on a variant of stable models developed for EVOLP.

3 The Selection Task

Psychological evidence from the selection task suggests that people reason differently with two kinds of conditionals. One school of thought is that the difference depends, at least in part, on whether conditionals are interpreted descriptively or deontically. We will argue that descriptive conditionals are like logic programs, and deontic conditionals are like integrity constraints in abductive logic programming.

In Wason's original selection task, there are four cards, with letters on one side and numbers on the other. The cards are lying on a table, and only one side of each card is visible, showing the letters D and F, and the numbers 3 and 7.

The task is to select those and only those cards that need to be turned over, to determine whether the following conditional is true:

> *If there is a D on one side,*
> *then there is a 3 on the other side.*

Variations of this experiment have been performed numerous times. The surprising result is that only about 10% of the subjects give the correct answer according to the norms of classical logic.

Almost everyone recognizes, correctly, that the card showing D needs to be turned over, to make sure there is a 3 on the other side. Most people also recognize, correctly, that the card showing F does not need to be turned over. But many subjects also think, incorrectly, that it is necessary to turn over the card showing 3, to make sure there is a D on the other side. This is logically incorrect, because the implication does not claim that *conversely:*

> *If there is a 3 on one side,*
> *then there is a D on the other side.*

Only a few subjects realise that it is necessary to turn over the card showing 7, to make sure that D is not on the other side. It is necessary to turn over the 7, because the original implication is logically equivalent to its *contrapositive*:

> *If the number on one side is not 3 (e.g. 7),*
> *then the letter on the other side is not D.*

It has been shown that people perform far better, according to the norms of classical logic, when the selection task experiment is conducted with certain other formulations of the problem that are formally equivalent to the card version of the task. The classic experiment of this kind considers the situation in which people are drinking in a bar, and the subject is asked to check whether the following conditional holds:

> *If a person is drinking alcohol in a bar,*
> *then the person is at least eighteen years old.*

Again the subject is presented with four cases to consider, but this time instead of four cards there are four people. We can see what two people are drinking, but cannot see how old they are; and we can see how old two people are, but not what they are drinking. In contrast with the card version of the selection task, most people solve the bar version correctly. They realise that it is necessary to check the person drinking alcohol to make sure that he is at least eighteen years old, and to check the person under eighteen to make sure that she is not drinking alcohol. They also realise that it is not necessary to check the person who is eighteen years old or older, nor the person who is drinking a non-alcoholic beverage.

Cognitive psychologists have proposed a bewildering number of theories to explain why people are so much better at solving such versions of the selection task compared with the original card version. One of the most influential of these is the theory put forward by Cheng and Holyoak [8] that people tend to reason in accordance with classical logic when conditionals involve *deontic* notions concerned with permission, obligation and prohibition. However, except for [49] and [25], there has been little attempt to explain why people reason as they do with descriptive variants of the selection task, such as the card version.

Stenning and van Lambalgen [49] propose that understanding and solving the selection task is a two stage process: interpreting the conditional and then reasoning with the interpretation. They argue that people interpret conditionals of the kind involved in the card version of the task as logic programs. Interpreted as a logic programming clause, a conditional is understood, according to the *completion semantics*, as the if-half of a definition in if-and-only-if form [9]. The completion semantics entails the converse of conditionals in the selection task and inhibits the application of reasoning with contrapositives. This is exactly the kind of reasoning most people display in the card version of the selection task.

Stenning and vanLambalgen also argue that it is natural to interpret conditionals of the kind involved in the bar version of the task in deontic logic. However, Kowalski [25] has argued that the deontic interpretation can be obtained more simply by interpreting conditionals as integrity constraints.

It is curious that, although production systems, such as SOAR and ACT-R, are widely used in cognitive psychology as a model of human thinking, it seems that conditionals in the form of condition-action rules have not been studied in relation to the selection task.

4 Intelligent Agents

The psychological evidence that people reason differently with descriptive and deontic conditionals is mirrored by the notion that the mental state of an intelligent agent is best understood as having separate goals and beliefs. An agent's beliefs represent the way things are, and its goals represent the way the agent would like them to be. Thus beliefs have a descriptive character, whereas goals have a prescriptive or deontic character. The BDI (Belief, Desire, Intention) [4] model of agents adds to beliefs and desires (or goals) the notion of intention, which is an agent's plan of actions for achieving its goals. Intentions are derived from goals using beliefs to reduce goals to subgoals.

Arguably, the most influential of the BDI agent models is that of Rao and Georgeff [42] and its successors dMARS [13] and AgentSpeak [41]. The abstract agent intermediate language AIL of Dennis et al. [12] is an abstraction of these languages, based mainly on AgentSpeak and its successors.

The earliest BDI agent systems were specified in multi-modal logics, with separate modal operators for goals, beliefs and intentions. However, their procedural implementations bore little resemblance to their logical specifications. AgentSpeak abandoned the attempt to relate the modal logic specifications with their procedural implementations, observing instead that "...one can view agent programs as multi-threaded interruptible logic programming clauses". This abandonment of modal logic specifications is inherited by AgentSpeak's successors and their abstraction AIL.

However, this view of AgentSpeak in logic programming terms applies only to the procedural interpretation of clauses. In fact, programs in AgentSpeak are better viewed as a generalisation of production rules than as variant of logic programming. AgentSpeak programs, also called plans, have the form:

Event E: conditions C goals G and actions A.

AgentSpeak plans manipulate a "declarative" database, like the working memory in production systems. The database contains both *belief literals* (atoms and negations of atoms) and *goal atoms*. The belief literals represent the current state of the environment and are added and deleted destructively, simulating the execution of atomic actions. Goal atoms are added when they are generated as sub-goals, and deleted when they are solved.

The event E in the head of a plan can be the addition or deletion of a belief or of a goal. Plans are embedded in a cycle similar to the production system cycle, and are executed in the direction in which they are written. With the arrow written backwards, the execution of plans can be viewed as backward chaining. However, if the arrow is reversed, their execution can be viewed as forward chaining. No matter how their execution is viewed, plans have only an operational semantics.

The following are examples of possible AgentSpeak plans:

+ *there is a fire: true*   +*there is an emergency.*
+ *?there is an emergency: true   ? there is a fire.*
+ *! there is an emergency: true   ! there is a fire.*

Observations and actions do not have associated times, and the database provides only a snapshot of the current state of the world. To compensate for this lack of a temporal representation, the prefixes +,-, !, and ? are used to stand for *add, delete, achieve,* and *test*, respectively.

Notice that the first plan behaves like a logical conditional used to reason forwards, but in the opposite direction of the arrow, to conclude there is an emergency if it is observed that there is a fire. The other two plans are goal-reduction rules, one for testing whether there is an emergency and the other for creating an emergency.

In general, in the case where a plan has the form:

Goal E: conditions C goals G and actions A

and the triggering event E is the addition of a goal, the plan can be reformulated as a logic programming clause:

E' if C' and G' and A'and temporal constraints

where the prefixed predicates of AgentSpeak are replaced by predicates with explicit associated times. The corresponding clause subsumes the behaviour of the plan, but also has a declarative reading.

In the simple example of the three plans above, the corresponding clause is:

there is an emergency at time T if there is a fire at time T.

Represented in this way, the clause can be viewed as defining a ramification, which views fires more abstractly as emergencies.

Thus, although BDI agent models were inspired by the modal logic representation of goals and beliefs, this inspiration has largely been lost in recent years. Most agent systems today represent goals as facts, mixed with belief facts in database or represented in a separate stack, as in ACT-R and SOAR. Belief facts and goal facts are manipulated uniformly by procedures, often called plans, which generalise production rules.

The only kind of goal that can easily be represented as a fact in a database or in a goal stack, in this way, is an *achievement goal*, which is a one-off problem to be solved, including the problem of achieving some desired future state of the world. The higher-level notion of *maintenance goal*, which persists over all states of the world, is lost in the process.

In ALP agents, as we will see below, maintenance goals are integrity constraints, which have the form of universally quantified conditionals with existentially quantified conclusions. Thus maintenance goals are higher-level than achievement goals in ALP, because an achievement goal is derived as an instance of the conclusion of a maintenance goal, whenever an instance of the conditions of the maintenance goal are satisfied. For example, given the maintenance goal:

For all times T_1
If there is an emergency at time T_1 then there exists a time T_2 such that
I get help at time T_2 and $T_1 < T_2$

and an emergency at some specific time t_1, forward reasoning in ALP would derive the achievement goal of getting help at some later time. The later time could be bounded by an additional conjunct in the conclusion, or it could be left to the decision-making component of the agent cycle to take into account how urgently the achievement goal needs to be accomplished.

In AgentSpeak and its successors, maintenance goals and goal-reduction rules are just different kinds of plans. Our aim is to restore the high level distinction between goals and beliefs, to recognise the importance of maintenance goals in particular, to combine the distinctive forms of reasoning appropriate to the distinction between goals and beliefs, and to give their combination a logical, model-theoretic semantics. For this purpose we interpret beliefs as logic programs, goals as integrity constraints, and combine goals and beliefs in the way that logic programs and integrity constraints are combined in abductive logic programming.

5 Deductive Databases

The combination of logic programs and integrity constraints in ALP evolved from their relationship in deductive databases. The semantics of integrity constraints and the development of proof procedures for constraint satisfaction were active research areas in deductive databases in the 1980s.

The distinction between a database and its integrity constraints is intuitively clear in database systems, where integrity constraints have the same semantics as database queries. But, whereas *ad hoc* queries are concerned with properties that hold in a given state of the database, integrity constraints are persistent queries that are intended to hold in all states of the database. *Ad hoc* queries can be viewed as achievement goals, and integrity constraints can be viewed as maintenance goals and include prohibitions as a special case. The database itself can be thought of as a set of beliefs. Thus conventional database systems can be viewed as passive agents, which are open to updates from the environment, but are unable to perform actions themselves.

In relational databases, there is a clear distinction between the syntax of beliefs in the database and the syntax of goals. Beliefs are simple, ground, atomic sentences. Goals, both ad hoc queries and persistent integrity constraints, are sentences of first-order logic. However, the syntactic distinction is less clear in deductive databases, where the database consists of both ground facts and more general logic programs (also called deduction rules). The distinction is complicated by the fact that it is often natural to express both deduction rules and integrity constraints in a similar conditional form. Informal criteria for distinguishing between deduction rules and integrity constraints were proposed by Nicolas and Gallaire [37]. Consider, for example, the two conditionals:

The bus leaves at time X:00 if X is an integer and $9 \leq X \leq 18$.

If the bus leaves at time X:00, then for some integer Y,
the bus arrives at its destination at time X:Y and $20 \leq Y \leq 30$

The first conditional defines bus departure times constructively and therefore can function as a general rule in a deductive database. However, the second conditional has an existential quantifier in the conclusion, which means that it cannot be used to define data, but can only be used to constrain data, as an integrity constraint. In a passive database, the integrity constraint can be used to check updates to the database. But in an agent, the integrity constraint can be used as a maintenance goal, to generate achievement goals, which can then be reduced to plans of action. Thus an agent can be thought of as an active database, and its maintenance goals can be regarded as active database rules.

Several competing views of the semantics of integrity constraints were intensively investigated in the 1980s. The two main views, to begin with, were the consistency view and the theorem-hood view, both of which were defined relative to the completion of the database. In the *consistency view*, an integrity constraint is satisfied if it is consistent with the completion of the database. In the *theorem-hood view*, it is satisfied if it is a theorem, logically entailed by the completion.

Reiter [44] proposed an *epistemic view* of integrity constraints, according to which integrity constraints are statements about what the database knows. For example, the integrity constraint:

If X is an employee then for some integer Y
X has social security number Y

would be interpreted as:

If the database knows that X is an employee then for some integer Y
the database knows that X has social security number Y

However, Reiter [44] also showed that all three views are equivalent in many cases for databases augmented with the *closed world assumption* [43] which is the set of all the negations of atomic sentences that are not entailed by the database. For relational databases, the three views are also equivalent to the standard view in relational databases that a database satisfies an integrity constraint if it is *true* in the database regarded as a Herband model.

More generally, the four views of integrity satisfaction (consistency, theorem-hood, epistemic and truth-theoretic) coincide for any database whose closure has a single model, In the case of Horn clause databases, the four views are equivalent to the view that an integrity constraint is satisfied if (and only if) it is true in the unique minimal model of the set of Horn clauses.

However, whether or not the different views of integrity satisfaction are equivalent for a given database, it is generally accepted that queries and integrity constraints have the same semantics. Therefore, the most obvious way to check integrity satisfaction is to treat each integrity constraint as a query, using the same procedure for integrity checking as for query evaluation. The problem with this approach, is that in a dynamic setting, where the current database state is largely identical to the previous state, much of the work involved in processing the constraints in a new state duplicates the work performed in the previous state.

To alleviate this problem, the vast majority of integrity checking procedures developed in the 1980s incrementally checked the integrity of a new state of the database, assuming that the integrity constraints already hold in the previous state. As a consequence any new violation of integrity must be due to the update itself, and integrity checking can be focused on the update and its consequences. This focus can be achieved by using resolution to perform forward reasoning.

The combination of a sequence of updates to a database and forward reasoning triggered by updates can be viewed in terms that resemble the production system cycle. Early work on such an approach for integrity checking in deductive databases includes the integrity checking procedure of Sadri and Kowalski [46, 47, 30]. In this approach integrity constraints are formalised as denials (clauses with conclusion *false*), for example:

>   *If X is an employee and not X has a social security number then false*

Transactions are sets of literals, in which the positive atoms represent requested additions, and the negated atoms represent requested deletions. In the simple case where all the requested updates are additions:

>   If an update matches one of the conditions of a clause or integrity constraint,
>   forward reasoning (via resolution) is performed to generate the resolvent.
>   SLDNF is used to try to verify the remaining conditions of the resolvent.
>   If the conditions are verified, then
>   the instantiated conclusion is added as a new update.
>   If the new update is *false*, the procedure terminates,
>   and integrity has been violated.
>   Otherwise, the procedure is repeated, and
>   the new update is treated as an (implicit) update in the same transaction.
>   If the procedure terminates without generating false
>   then the transaction satisfies the integrity constraints.

If any of the updates is a deletion, a similar procedure is applied with an extended resolution step that allows resolution with negative literals. Here is an example without deletions:

ICs:	*If P and not R then false*		*If S and Q then false*
Database:	*S if M*	*Q if T*	*R if T*
Updates:	*{P, T}*		

Forward reasoning from the two updates produces two resolvents:

>   *if not R then false*
>   *if S then false.*

Forward reasoning from the update *P* produces the first resolvent, using the first IC. Forward reasoning from the update *T* produces the second resolvent, using the second database clause to derive *Q* and then using the second IC.

The SLDNF evaluation of the two conditions *not R* and *S* terminates unsuccessfully, and therefore the integrity checking procedure terminates successfully.

In the Sadri Kowalski (SK) integrity checking procedure sketched above, ICs are represented as denials, which conceptually represent prohibitions. However, the procedure can be modified easily to deal with ICs in conditional form. Integrity checking with ICs in conditional form highlights their relationship with production rules, and was introduced in the IFF proof procedure for ALP.

6 Abductive Logic Programming

Deductive databases and the closely related Datalog were sidelined in the late 1980s with the arrival of object-oriented and active databases. As we have seen in section 2, some of the attention of the Datalog community turned to the problem of providing a declarative semantics for active database rules. In the meanwhile, some of the work on integrity checking in deductive databases contributed to the development of abductive logic programming (ALP) [19].

ALP can be viewed as a variant of deductive databases in which integrity constraints are used actively to generate new facts that are candidates for addition to the database. However, ALP is normally viewed as an extension of logic programming, combining *closed predicates* that are defined in the conclusions of clauses with *abducible* (or *open*) *predicates* that occur in the conditions, but not in the conclusions of clauses. Abducible predicates are like extensional predicates in deductive databases, and closed predicates are like intensional predicates.

The classical application of abduction is to generate hypotheses, which are atomic sentences in the abducible predicates, to explain observations. The best known example of this is the use of abduction to explain the observation that *the grass is wet*, using the clauses:

> *the grass is wet if it rained.*
> *the grass is wet if the sprinkler was on.*

Abduction generates the two alternative hypotheses, *it rained* and *the sprinkler was on*. Assuming that the state of the environment is stored in a database, the hypotheses are alternative candidates for adding to the database. We will argue in the next section that deciding which of the alternatives to add is like conflict resolution in production systems.

ALP extends classical abduction in two ways: First, it not only generates abductive hypotheses to explain observations, but also it generates hypothetical actions to achieve goals; and second it constrains hypothetical explanations so that they do not violate integrity constraints.

Suppose that, in addition to the two clauses in the example above, we have:

> Fact: *the sun was shining.*
> Integrity constraint: *not (it rained and the sun was shining)*
> or equivalently: *if it rained and the sun was shining then false.*

Then the hypothesis *it rained* violates the integrity constraint, leaving the alternative hypothesis *the sprinkler was on* as the only acceptable explanation of the observation that *the grass is wet*.

In general, given a logic program *LP*, integrity constraints *IC* and a problem *G*, which is either an achievement goal or an observation to be explained, a set *Δ* of atomic sentences in the abducible predicates is an *abductive explanation* or a *solution* of *G* if and only if both *IC* and *G* hold with respect to the extended logic program *LP* *Δ*.

This characterisation of ALP is compatible with different semantics specifying what it means for a goal or integrity constraint to hold with respect to a logic program. It is also compatible with different proof procedures.

The most straight-forward proof procedures for ALP [15, 11, 20] are simple extensions of SLD or SLDNF in which the integrity constraints *IC* are represented as denials. The constraints *IC* and problem *G* are conjoined together and transformed into the form of a normal logic programming goal clause. The set *Δ* is constructed incrementally, adding new hypotheses to *Δ* to solve sub-goals in the abducible predicates that would otherwise fail. The incremental construction of *Δ* is interleaved with checking that the new hypotheses added to *Δ* satisfy the integrity constraints. Checking the integrity constraints may cause further updates to *Δ*.

For the sake of simplicity, these proof procedures avoid general-purpose integrity checking, and instead perform only one step of forward reasoning to match a newly added abductive hypothesis with a condition of an integrity constraint. For this purpose, the integrity constraints are preprocessed into a form in which at least one condition is abducible. The resulting preprocessed integrity constraints are similar to *event-condition-action* rules where the *event* and *action* predicates are abducible.

These approaches all use backward reasoning with SLD or SLDNF resolution, to reduce goals to sub-goals and to generate abductive hypotheses. An alternative approach, developed originally by Console, Dupre and Torasso [10] and extended by Fung and Kowalski [18] is to reason instead with the completions of the logic programming clauses defining the non-abducible predicates.

The completions in [28] all have the form of if-and-only-if (IFF) definitions:

atomic formula iff disjunction of conjunctions

where the *conjunctions* (also called *disjuncts*) are conjunctions of atomic formulae or conditionals. Negative literals *not p* are written as conditionals *if p then false*. The *atomic formula* is called the *head*, and the *disjunction of conjunctions* is called the *body* of the definition. Integrity constraints are represented as conditionals of the form:

if conditions then disjunction of conjunctions

where the *conditions* are a conjunction of atomic formulae and the conclusion has the same form as the body of a definition.

Because the conclusions of integrity constraints have the same form as the bodies of definitions, they can contain existentially quantified variables, and the quantification of the constraints can be implicit, as in the example:

If X is an employee then X has social security number Y.

Here X is universally quantified and Y is existentially quantified.

Because disjuncts in the bodies of definitions can contain conditionals, these conditionals can have the same form as integrity constraints with the same implicit quantification of variables, for example:

The banking department gets a 5 % bonus starting tomorrow iff
if X is an employee in the banking department today and X has salary S today
then X has salary S + .05S tomorrow.

Given an initial problem *G* and integrity constraints *IC*, IFF conjoins *G* and *IC* in an initial goal, and rewrites it as a *disjunction of conjunctions*. Starting from the initial goal, equivalence-preserving inferences transform one state of the goal into another, equivalent state, also represented in the form of a *disjunction of conjunctions*. Each disjunct corresponds to a branch of the search space of abductive proof procedures based on SLD and SLDNF.

The inference rules of IFF include both backward and forward reasoning. Backward reasoning (also called *unfolding*) replaces an atomic formulae by the body of its definition. Forward reasoning (also called *propagation*) performs resolution between an atomic formula and a conditional in the same disjunct. The resolvent is added to the disjunct. An abductive explanation is generated, when a disjunct contains only atoms or negative literals in the abducible predicates (or equalities) and no further inferences can be performed within that disjunct.

The fact that every disjunct in the search space contains a copy of all the ICs is a potential source of inefficiency. This inefficiency can be avoided, simply by factoring out the ICs, representing them explicitly only once, but treating them as though they belong to every disjunct.

The IFF proof procedure was developed with the theorem-hood view of integrity satisfaction in mind. It is sound and, with certain modest restrictions on the form of clauses and integrity constraints, complete with respect to the completion of the program in the Kunen [33] three-valued semantics. The SLP proof procedure [32] is a refinement of IFF, in which integrity constraints are also used for constraint handling rules in constraint logic programming, but with the consistency view of integrity satisfaction. SLP is not complete in the general case, because consistency is not semi-decidable. CIFF [36], which is a successor of SLP, reverts to the theorem-hood semantics, and is complete in certain cases.

Compared with proof procedures that extend SLD or SLDNF, which represent integrity constraints as denials, the main attraction of the IFF proof procedure and its successors is that they include forward reasoning with integrity constraints represented explicitly as conditionals. Forward reasoning can also be used to derive consequences of abducible hypotheses, to help in choosing between them. This is important in ALP agents, as we will discuss in the next section.

Another useful feature of these proof procedures is that their representation of negative literals in the conditional form *if p then false* means that negative conditions can be unfolded even if they contain variables, and the IFF selection function, which is the analogue of the SLDNF selection function, can be much more liberal compared with SLDNF.

The completeness of IFF means that ICs can be satisfied, not only, like condition-action rules, by satisfying their conclusions when their conditions are satisfied, but also by making their conditions fail. For example, the integrity constraint *if you commit a mortal sin and don't go to confession then you will go to hell* can be satisfied either by not committing a mortal sin, committing a mortal sin and going to confession, or committing a mortal sin, not going to confession and going to hell. Here is a symbolic example:

G: p
IC: *if p and not q then a*
LP: *q if b*

where p is an observation, and a and b represent actions. IFF derives *not q then a* by forward reasoning, which it then rewrites as *q or a*. It unfolds *q* to obtain *b or a*. Thus the problem has two solutions, either do *b* or do *a*.

In the informally described abductive agent proof procedures of [23, 24], the forward reasoning integrity checking method of SK is combined with the IFF representation of integrity constraints as conditionals within an SLD-style framework. A variant of this is formalised in LPS [29]. Here is an example from [24, 26].

Integrity constraint: *If there is an emergency then I get help.*

Logic program:
        A person gets help if the person alerts the driver.
        A person alerts the driver if the person presses the alarm signal button.
        There is an emergency if there is a fire.
        There is an emergency if one person attacks another.
        There is an emergency if someone becomes suddenly ill.
        There is an emergency if there is an accident.
        There is a fire if there are flames.
        There is a fire if there is smoke.

Abducible predicates: *there are flames, there is smoke, a person presses the alarm signal button*

Observation:	*There is smoke.*
Forward reasoning:	*There is a fire.*
Forward reasoning:	*There is an emergency.*
Forward reasoning, Goal:	*I get help*
Backward reasoning, Goal:	*I alert the driver*
Backward reasoning, Solution:	*I press the alarm signal button.*

Notice that the example integrates the behaviour of the three kinds of production rules identified in section 1.2. The first two steps of forward reasoning use clauses of the logic program as forward reasoning logic rules; the third step of forward reasoning uses the integrity constraint as a reactive rule, and the two steps of backward reasoning use clauses of the logic program as goal-reduction rules.

An alternative to using forward reasoning with logic programs in the first two steps in this example is to preprocess the integrity constraint (by unfolding the definitions of *there is an emergency* and *there is a fire*) into several separate integrity constraints:

If there are flames then I get help.
If there is smoke then I get help.
If one person attacks another then I get help.
If someone becomes suddenly ill then I get help.
If there is an accident then I get help.

The use of a single integrity constraint and forward reasoning with logic programs simulates forward chaining with production rules, and is arguably more natural. (Notice that these integrity constraints could be further pre-processed into condition-action rules, by unfolding the conclusion *I get help,* replacing it by the action *I press the alarm signal button.*)

Following the lead of other abductive proof procedures, the proof procedure illustrated in this example can be given a variety of logical semantics. To achieve our desired combination of logic programs, production systems and agents, it remains only to:

- represent actions by abducible predicates,
- embed the abductive proof procedure in an agent cycle, and
- justify the use of a destructively changing database of facts.

7 ALP Agents

The notion of ALP agent, in which ALP is embedded as the thinking component of an observation-thought-decision-action cycle, was introduced in [22] and developed in [27, 28, 23, 24]. It is the basis of the KGP (Knowledge, Goals and Plans) agent model of [21]. The ALP agent proof procedure of [28] is the IFF proof procedure. The proof procedure of KGP is the CIFF proof procedure [36] which extends IFF with constraint handling procedures.

The observation-thought-decision-action cycle of an ALP agent is similar to other agent cycles. However, in ALP agents, an agent's beliefs are represented by logic programs, its goals are represented by integrity constraints, and its actions are represented by abducible predicates.

7.1 Observations

In the original ALP agent model, observations are added incrementally to the goals, and the IFF proof procedure generates explanations of observations as well as actions to achieve goals. However, in production systems and BDI agents, observations contribute to the current state of the database of facts. In the KGP agent model, observations are added to a separate database of facts, similar to the production system/BDI agent database. But whereas in PR/BDI systems the database represents only the current state of the environment, in KGP the database is a monotonically increasing representation of both the current state and all past states.

The ALP agent and KGP beliefs include the event calculus axioms [31] to derive new time-stamped facts from previous observations. However, the agent can bypass reasoning with the event calculus, by directly observing the environment instead. In both the ALP agent and KGP proof procedures, facts, whether directly observed or derived by the event calculus, are used both to solve atomic goals and to propagate with integrity constraints.

In both the ALP and the KGP agent models, in the routine use of the agent cycle, forward reasoning derives consequences of observations and triggers integrity constraints; and backward reasoning reduces goals to plans of action. Reasoning can be interrupted both by incoming observations and by outgoing actions. An incoming observation, in particular, might trigger an integrity constraint and derive an action that needs to be performed immediately, interrupting the derivation of plans and the execution of actions that need to be performed only later in the future.

However, in more demanding situations, backward reasoning can also be used to generate explanations of observations, and forward reasoning can also be used to derive consequences of explanations, to help in choosing between alternative explanations. For example, if *it rained* and *the sprinkler was on* are alternative explanations for the observation *the grass is wet*, then forward reasoning might derive *the street is wet* from the hypothesis that *it rained*, and derive *the water meter reading is high* from the hypothesis that *the sprinkler was on*. Checking these consequences by attempting to observe them in the environment can help to prefer one explanation to the other.

A similar kind of reasoning forward from alternative candidate actions, to derive and evaluate their possible outcomes, can also help with decision-making and conflict resolution.

7.2   Decision-Making and Conflict Resolution

The easy part of the extension of ALP to ALP agents is making ALP open to observations, and embedding ALP in a down-sized observation-thought cycle, in which the agent passively thinks about alternative plans of actions, but doesn't make any commitments to perform any actions. The hard part is to decide among the alternatives, and actually do something. To choose between different actions, an extended cycle can incorporate a decision making component, which generalises conflict resolution in production systems.

Consider the following "rules":

> If someone attacks me then I attack them back
> If someone attacks me then I run away

and suppose I observe *someone attacks me*. Treated as condition-action rules in a production system, both rules would be triggered, and conflict resolution would be needed to fire only one of them. However, in ALP and ALP agents, both rules and both conclusions *I attack them back* and *I run away* would need to be satisfied, which might not be possible. To avoid this problem and to have a logical semantics, the rules need to be re-formulated:

Maintenance goal:	*If X attacks me then I protect myself against X*
Beliefs:	*I protect myself against X if I attack X back*
	I protect myself against X if I run away.

In ALP, given this re-formulation and the observation *someone attacks me*, it suffices to satisfy only one of *I attack them back* or *I run away*. But this still involves making a choice between the two alternatives, and then successfully performing one or more actions to make that choice succeed, and thereby to make the given instance of the goal become true. In the IFF syntax and proof procedure, the goal and the beliefs could be pre-processed, by unfolding the conclusion *I protect myself* into the still simpler form of a single maintenance goal:

Maintenance goal: *If someone attacks me*
                     then (I attack them back or I run away).

Viewed in this way, conflict resolution in production systems is partly a compensation for the restricted syntax of production rules (no disjunction in conclusions of rules) and partly a form of Decision Theory. ALP agents do not suffer from the same restrictions on syntax; but, like production systems, they need to be augmented with a decision-making component.

In classical Decision Theory, actions are chosen to optimise expected outcomes. ALP can help with this, because it can be used, not only to generate alternative actions, but also to reason forward from candidate actions to derive their likely outcomes – for example to reason forward from the candidate action *we go for a picnic*, to derive the following outcome from the following belief:

Belief: *We will have a whale of a good time*
             if we go for a picnic and the weather is good.
Outcome: *We will have a whale of a good time if the weather is good.*

To fit into a full-fledged decision-making procedure, outcomes (e.g. *We will have a whale of a good time*) need to be evaluated for their utility, and abducible conditions beyond the agent's control (e.g. *the weather is good*) need to be assigned a probability. ALP is particularly well-suited to this, as Poole [39, 40] has shown, because it is very natural and very easy to associate probabilities with abducible predicates. The resulting Decision Theoretic analysis, or some computationally less expensive approximation to it, fits comfortably into an ALP agent cycle [24, 38]

Having done the analysis and made the decision, the agent must still commit to performing an action. The action may be part of a plan; and even if the action succeeds the plan might fail, because for some reason or other the agent might not be able to perform the other actions in the plan. For example the preconditions of later actions might not hold or the actions might get timed out. The upshot of all these complications is that it may not be possible to commit to only one alternative plan for achieving a higher level goal. It might be necessary to embark upon one plan, and then switch to another plan if the first plan fails. It may also be necessary to re-perform an action in the same plan. The ALP agent cycle allows all these options.

7.3 Semantics

The semantics of ALP agents can be understood in ALP terms. The fact that observations and actions occur in a temporal order can be dealt with simply by including the time or state of observations and actions as explicit parameters of predicates. The effect of executing actions can be taken into account by including an action theory, such as the situation calculus or event calculus in the agent's set of beliefs B. With these assumptions, and the further assumption that the observations do not include any post-conditions (effects) of the agent's own actions, the semantics of the ALP agent cycle is a special case of the ALP semantics.

> Given beliefs B, goals G, initial database state S_0 and possibly an infinite set $O = \{O_1, O_2, ..., O_n, ...\}$ of input (external) observations, an *ALP agent solution* is a possibly infinite set $\Delta = \{A_1, A_2, ..., A_m, ...\}$ of actions such that G is satisfied by the logic program $B \cup O \cup \Delta$.

Theorem 7.1 of [28] establishes a correspondence between the static behaviour of the IFF proof procedure and the dynamic behaviour of ALP agents with observations. The soundness of the ALP agent cycle follows from that theorem.

8   The Representation of State Change and the Frame Problem

For the ALP agent semantics to work, observations and actions need to have an explicit representation of time or state. This gives rise to the *frame problem* of how to represent and reason about change of state correctly and efficiency. It is generally held that the frame problem can be solved by means of an appropriate action theory such as situation calculus or event calculus. These and similar calculi all include some form of frame axiom, such as:

> *fact F holds in state S+1 if fact F holds in state S*
> *and S+1 is obtained from S by action/event A*
> *and A does not terminate F.*

Given an appropriate semantics, including the various semantics developed for logic programming, it has been argued that these formulations of the frame axiom solve the frame problem.

    However, all of these solutions reason explicitly, whether forward or backward, that a fact holds in a state $S+1$ if it held in state S and was not terminated in the state transition. Reasoning in the forward direction, for example, every unterminated fact holding in state S needs to be explicitly copied-to the new state $S+1$. Moreover, without sophisticated optimisations and garbage collection, the same unterminated facts need to be duplicated in both states S and $S+1$, and indeed in all states from the initial state to the current state. Backward reasoning does not store unterminated facts redundantly, but requires a potentially expensive calculation instead. Thus it can be argued that that aspect of the frame problem that is concerned with efficiently reasoning about change of state has no solution within a purely logical representation.

    It is probably for such reasons of efficiency that production systems and BDI agents store only the current state, use destructive assignment to generate a new current state from the previous state, and do not employ an explicit representation of states or time at all.

It can be argued that the ALP agent model with its event calculus representation of change also suffers from the inefficiencies associated with the frame problem. However, in ALP agents, to determine whether a fact holds in the current state, these inefficiencies can be avoided, by directly observing whether the fact holds in the environment instead. In such a case, the environment serves as an auxiliary, external database, which contains a complete record of the current state. To test these ideas, we have developed LPS (Logic-based Production System) [29], which is a variant of ALP agents without frame axioms, but with an operational semantics that incorporates a destructively changing database. For simplicity, as is common in production systems, we assume that there are no external observations, once the initial state of the database has been given as input.

The LPS operational semantics maintains the current state of a deductive database, with facts (atomic sentences) defining the extensional predicates and logic programs (or deduction rules) defining intensional predicates. The actions that change the current state affect only the extensional predicates, which are represented without explicit state parameters. As a consequence, to execute an action and to change the current state, it suffices to delete the facts terminated by the action and to add the facts initiated by the action. Facts that are not affected by the action persist without the need to reason that they persist, simply because they do not change. Intentional predicates change implicitly as ramifications of changes to the extensional predicates.

In contrast with the operational semantics, which maintains only the current state, the LPS declarative semantics is based on the sequence of database states with the implicit state parameter of the operational semantics made explicit. The semantics also requires that the sequence of states conforms to a logical representation of change, such as the event calculus EC represented in logic programming form:

> Given beliefs B in the form of a logic program, goals G in the form of a set of integrity constraints, and an initial database state S_0, a possibly infinite set $\Delta = \{A_1, A_2, ..., A_m, ...\}$ of actions is an *LPS solution* if and only if G is satisfied by the logic program $B \cup S_0 \cup \Delta \cup EC$.

As shown in [29], LPS is sound, but is not complete for the same reason that condition-action rules are not complete (because they cannot make their conditions false).

9   Conclusions

We have argued for combining the functionalities of production systems and logic programs, while retaining their individual contributions and eliminating their overlap. To defend our thesis that the two kinds of rules provide distinct functionalities and need to be combined, we have appealed to the distinctions

- in psychology between descriptive and deontic conditionals,
- in intelligent agents between beliefs and goals,
- in deductive databases between deduction rules and integrity constraints, and
- in ALP between logic programs and integrity constraints.

We have advocated an approach that combines the two kinds of rules in an observation-thought-decision-action cycle, along the lines of ALP agents, and sketched an approach that combines a declarative, logic-based semantics with an operational semantics that operates with the current state of a destructively changing database.

References

1. Alferes, J.J., Banti, F., Brogi, A.: An Event-Condition-Action Logic Programming Language. In: Fisher, M., van der Hoek, W., Konev, B., Lisitsa, A. (eds.) JELIA 2006. LNCS (LNAI), vol. 4160, pp. 29–42. Springer, Heidelberg (2006)
2. Alferes, J.J., Brogi, A., Leite, J.A., Pereira, L.M.: Evolving Logic Programs. In: Flesca, S., Greco, S., Leone, N., Ianni, G. (eds.) JELIA 2002. LNCS (LNAI), vol. 2424, pp. 50–61. Springer, Heidelberg (2002)
3. Anderson, J., Bower, G.: Human Associative Memory. Winston, Washington (1973)
4. Bratman, M.E., Israel, D.J., Pollack, M.E.: Plans and Resource-bounded Practical Reasoning. Computational Intelligence 4, 349–355 (1988)
5. Caroprese, L., Greco, S., Sirangelo, C., Zumpano, E.: Declarative Semantics of Production Rules for Integrity Maintenance. In: Etalle, S., Truszczyński, M. (eds.) ICLP 2006. LNCS, vol. 4079, pp. 26–40. Springer, Heidelberg (2006)
6. Ceri, S., Fraternali, P., Paraboschi, S., Tanca, L.: Automatic Generation of Production Rules for Integrity Maintenance. ACM Transactions on Database Systems (TODS) 19(3), 367–422 (1994)
7. Ceri, S., Widom, J.: Deriving production rules for constant maintenance. In: 16th International Conference on Very Large Data Bases, pp. 566–577 (1990)
8. Cheng, P.W., Holyoak, K.J.: Pragmatic Reasoning Schemas. Cognitive Psychology 17, 391–416 (1985)
9. Clark, K.: Negation as Failure. In: Readings in Nonmonotonic Reasoning, pp. 311–325. Morgan Kaufmann, San Francisco (1978)
10. Console, L., Theseider Dupre, D., Torasso, P.: On the Relationship Between Abduction and Deduction. Journal of Logic and Computation 1(5), 661–690 (1991)
11. Denecker, M., De Schreye, D.: SLDNFA: An Abductive Procedure for Normal Abductive Programs. J. Logic Programming 34(2), 111–167 (1998)
12. Dennis, L.A., Bordini, R.H., Farwer, B., Fisher, M.: A Common Semantic Basis for BDI Languages. In: Dastani, M.M., El Fallah Seghrouchni, A., Ricci, A., Winikoff, M. (eds.) PROMAS 2007. LNCS (LNAI), vol. 4908, pp. 124–139. Springer, Heidelberg (2008)
13. d'Inverno, M., Luck, M., Georgeff, M.P., Kinny, D., Wooldridge, M.: A Formal Specification of dMARS. In: Rao, A., Singh, M.P., Wooldridge, M.J. (eds.) ATAL 1997. LNCS (LNAI), vol. 1365, pp. 155–176. Springer, Heidelberg (1998)
14. Dung, P.M., Mancarella, P.: Production Systems with Negation as Failure. IEEE Transactions on Knowledge and Data Engineering 14(2), 336–352 (2002)
15. Eshghi, K., Kowalski, R.: Abduction Compared with Negation by Failure. In: Levi, G., Martelli, M. (eds.) 6th International Conference on Logic Programming, pp. 234–254. MIT Press, Cambridge (1989)
16. Fernandes, A.A.A., Williams, M.H., Paton, N.: A Logic-Based Integration of Active and Deductive Databases. New Generation Computing 15(2), 205–244 (1997)
17. Fraternali, P., Tanca, L.: A Structured Approach for the Definition of the Semantics of Active Databases. ACM Transactions on Database Systems (TODS) 20(4), 414–471 (1995)

18. Fung, T.H., Kowalski, R.: The IFF Proof Procedure for Abductive Logic Programming. Journal of Logic Programming 33(2), 151–164 (1997)

19. Kakas, A., Kowalski, R., Toni, F.: The Role of Logic Programming in Abduction. In: Gabbay, D., Hogger, C.J., Robinson, J.A. (eds.) Handbook of Logic in Artificial Intelligence and Programming, vol. 5, pp. 235–324. Oxford University Press, Oxford (1998)

20. Kakas, A.C., Mancarella, P.: Abduction and Abductive Logic Programming. In: 11th International Conference on Logic Programming, pp. 18–19 (1994)

21. Kakas, A., Mancarella, P., Sadri, S., Stathis, K., Toni, F.: Computational Logic Foundations of KGP Agents. Journal of Artificial Intelligence Research (2009)

22. Kowalski, R.: Using Metalogic to Reconcile Reactive with Rational Agents. In: Apt, K., Turini, F. (eds.) Meta-Logics and Logic Programming. MIT Press, Cambridge (1995)

23. Kowalski, R.: Artificial Intelligence and the Natural World. Cognitive Processing 4, 547–573 (2001)

24. Kowalski, R.: The Logical Way to be Artificially Intelligent. In: Toni, F., Torroni, P. (eds.) CLIMA 2005. LNCS (LNAI), vol. 3900, pp. 1–22. Springer, Heidelberg (2006)

25. Kowalski, R.: Reasoning with Conditionals in Artificial Intelligence. In: Oaksford, M. (ed.) The Psychology of Conditionals. Oxford University Press, Oxford (to appear, 2009)

26. Kowalski, R.: How to be Artificially Intelligence. Cambridge University Press, Cambridge (2010) (to be published), http://www-lp.doc.ic.ac.uk/UserPages/staff/rak/rak.html

27. Kowalski, R., Sadri, F.: Towards a Unified Agent Architecture that Combines Rationality with Reactivity. In: Pedreschi, D., Zaniolo, C. (eds.) LID 1996. LNCS, vol. 1154, pp. 131–150. Springer, Heidelberg (1996)

28. Kowalski, R., Sadri, F.: From Logic Programming towards Multi-agent Systems. Annals of Mathematics and Artificial Intelligence 25, 391–419 (1999)

29. Kowalski, R., Sadri, F.: LPS - a Logic-Based Production System Framework. Department of Computing, Imperial College (2009)

30. Kowalski, R., Sadri, F., Soper, P.: Integrity Checking in Deductive Databases. In: 13th VLDB, pp. 61–69. Morgan Kaufmann, Los Altos (1987)

31. Kowalski, R., Sergot, M.: A Logic-based Calculus of Events. New Generation Computing 4(1), 67–95 (1986); also In: Mani, I., Pustejovsky, J., Gaizauskas, R. (eds.): The Language of Time: A Reader. Oxford University Press, Oxford (2005)

32. Kowalski, R., Toni, F., Wetzel, G.: Executing Suspended Logic Programs. Fundamenta Informatica 34(3), 1–22 (1998)

33. Kunen, K.: Negation in Logic Programming. Journal of Logic Programming 4(4), 289–308 (1987)

34. Laird, J.E., Newell, A., Rosenblum, P.S.: SOAR: an Architecture for General Intelligence. Artificial Intelligence 33(1), 1–64 (1987)

35. Lloyd, J.W., Topor, R.W.: A Basis for Deductive Database Systems. J. Logic Programming 2, 93–109 (1985)

36. Mancarella, P., Terreni, G., Sadri, F., Toni, F., Endriss, U.: The CIFF Proof Procedure for Abductive Logic Programming with Constraints: Theory, Implementation and Experiments. Theory and Parctice of Logic Programming (to appear, 2009)

37. Nicolas, J.M., Gallaire, H.: Database: Theory vs. Interpretation. In: Gallaire, H., Minker, J. (eds.) Logic and Databases. Plenum, New York (1978)

38. Pereira, L.M., Anh, H.T.: Evolution Prospection. In: Nakamatsu, K., et al. (eds.) KES-IDT 2009. SCI, vol. 199, pp. 51–64. Springer, Heidelberg (2009)

39. Poole, D.: Probabilistic Horn Abduction and Bayesian Networks. Artificial Intelligence 64(1), 81–129 (1993)

40. Poole, D.: The Independent Choice Logic for Modeling Multiple Agents Under Uncertainty. Artificial Intelligence 94, 7–56 (1997)
41. Rao, A.S.: Agents Breaking Away. In: Perram, J., Van de Velde, W. (eds.) MAAMAW 1996. LNCS, vol. 1038. Springer, Heidelberg (1996)
42. Rao, A.S., Georgeff, M.P.: Modeling Rational Agents with a BDI-Architecture. In: 2nd International Conference on Principles of Knowledge Representation and Reasoning, pp. 473–484 (1991)
43. Reiter, R.: On Closed World Data Bases. In: Gallaire, H., Minker, J. (eds.) Logic and Data Bases, pp. 55–76. Plenum Press, New York (1978)
44. Reiter, R.: On Integrity Constraints. In: 2nd Conference on Theoretical Aspects of Reasoning about Knowledge, pp. 97–111 (1988)
45. Russell, S.J., Norvig, P.: Artificial Intelligence: A Modern Approach, 2nd edn. Prentice Hall, Upper Saddle River (2003)
46. Sadri, F.: A Theorem-Proving Approach to Database Integrity. PhD Thesis, Imperial College (1988)
47. Sadri, F., Kowalski, R.: A Theorem-Proving Approach to Database Integrity. In: Minker, J. (ed.) Foundations of Deductive Databases and Logic Programming, pp. 313–362. Morgan Kaufmann, San Francisco (1988)
48. Simon, H.: Production Systems. In: Wilson, R., Keil, F. (eds.) The MIT Encyclopedia of the Cognitive Sciences, pp. 676–677. The MIT Press, Cambridge (1999)
49. Stenning, K., van Lambalgen, M.: Human Reasoning and Cognitive Science. MIT Press, Cambridge (2008)
50. Thagard, P.: Mind: Introduction to Cognitive Science, 2nd edn. MIT Press, Cambridge (2005)
51. Wason, P.C.: Reasoning About a Rule. The Quarterly Journal of Experimental Psychology 20(3), 273–281 (1968)
52. Raschid, L.: A Semantics for a Class of Stratified Production System Programs. J. Log. Program. 21(1), 31–57 (1994)
53. Lausen, G., Ludäscher, B., May, W.: On Active Deductive Databases: The Statelog Approach. In: Kifer, M., Voronkov, A., Freitag, B., Decker, H. (eds.) Dagstuhl Seminar 1997, DYNAMICS 1997, and ILPS-WS 1997. LNCS, vol. 1472, pp. 69–106. Springer, Heidelberg (1998)
54. Zaniolo, C.: On the Unification of Active Databases and Deductive databases. In: Worboys, M.F., Grundy, A.F. (eds.) BNCOD 1993. LNCS, vol. 696, pp. 23–39. Springer, Heidelberg (1993)

SILK: Higher Level Rules
with Defaults and Semantic Scalability
(Abstract of Invited Talk)

Benjamin N. Grosof

Vulcan Inc., 505 Fifth Ave. S., Suite 900, Seattle, WA 98104, USA
BenjaminG@vulcan.com,
http://www.mit.edu/~bgrosof

1   SILK and Its KR Overall

We overview the technical approach and motivations of the SILK system for semantic rules and ontologies, that radically extends the knowledge representation (KR) power of currently commercially important business rule systems, including not only Prologs but also production rules and event-condition-action rules, database systems, and semantic web.

The newest part of Vulcan Inc.'s Project Halo, SILK is a new, highly ambitious effort that aims to provide key infrastructure for widely-authored VLKBs (Very Large Knowledge Bases) for business and science that answer questions, proactively supply information, and reason powerfully.

Practical semantic rules and ontologies KR today is based primarily on declarative logic programs.

SILK's KR is *hyper* logic programs, which adds:

- prioritized defaults and robust conflict handling;
- higher-order and flexible meta-reasoning;
- sound interchange with classical logic (including OWL, Common Logic, and SBVR); and
- actions and events, cf. production rules and process models.

SILK thus provides a significantly higher expressive abstraction level than previous approaches to semantic rules.

The SILK system includes components for:

- large-scale reasoning;
- web knowledge interchange; and
- (in future) collaborative knowledge acquisition.

We survey use cases for SILK in business and science. We discuss prospects for the SILK approach to effectively interchange and integrate a high percentage of the world's structured knowledge starting from today's legacy forms.

A. Polleres and T. Swift (Eds.): RR 2009, LNCS 5837, pp. 24–25, 2009.

2   Defaults for Semantic Scalability

We focus particularly on how SILK can overcome previous fundamental obstacles to semantic scalability, not just inferencing performance scalability, of semantic rules and ontologies on web scale. To do so, SILK newly combines Courteous style defaults, HiLog style higher-order, well founded semantics, and sound interchange with first-order logic via a hypermonotonic mapping from Courteous.

"SILK" stands for "Semantic Inferencing on Large Knowledge". It hopes to be what much of the next generation Web will be spun from.

For More Info

For more info about SILK, please see: `http://silk.semwebcentral.org`.
For more info about Project Halo and Vulcan Inc., please see
`http://www.projecthalo.com` and `http://www.vulcan.com`, respectively.

Uncertainty Reasoning for the Semantic Web

Thomas Lukasiewicz[*]

Computing Laboratory, University of Oxford
Wolfson Building, Parks Road, Oxford OX1 3QD, UK
thomas.lukasiewicz@comlab.ox.ac.uk

Abstract. Significant research activities have recently been directed towards the Semantic Web as a potential future substitute of the current World Wide Web. Many experts predict that the next huge step forward in Web information technology will be achieved by adding semantics to Web data. An important role in research towards the Semantic Web is played by formalisms and technologies for handling uncertainty and/or vagueness. In this paper, I first provide some motivating examples for handling uncertainty and/or vagueness in the Semantic Web. I then give an overview of some own recent formalisms for handling uncertainty and/or vagueness in the Semantic Web.

1 Introduction

During the recent decade, the *Semantic Web* [1,2,3,4] has attracted much attention, both from academia and industry, and is commonly regarded as the next step in the evolution of the World Wide Web. It aims at an extension of the current Web by standards and technologies that help machines to understand the information on the Web so that they can support richer discovery, data integration, navigation, and automation of tasks. The main ideas behind it are to add a machine-understandable "meaning" to Web pages, to use ontologies for a precise definition of shared terms in Web resources, to use KR technology for automated reasoning from Web resources, and to apply cooperative agent technology for processing the information of the Web.

The Semantic Web is divided into several hierarchical layers, where the *Ontology layer*, in the form of the *OWL Web Ontology Language* [5], is currently the highest layer of sufficient maturity. OWL consists of three increasingly expressive sublanguages, namely, *OWL Lite*, *OWL DL*, and *OWL Full*. OWL Lite and OWL DL are essentially very expressive description logics with an RDF syntax. As shown in [6], ontology entailment in OWL Lite (resp., OWL DL) reduces to knowledge base (un)satisfiability in the description logic $\mathcal{SHIF}(\mathbf{D})$ (resp., $\mathcal{SHOIN}(\mathbf{D})$). On top of the Ontology layer, sophisticated representation and reasoning capabilities for the *Rules*, *Logic*, and *Proof layers* of the Semantic Web are currently being developed next.

A key requirement of the layered architecture of the Semantic Web is in particular to integrate the Rules and the Ontology layer. Here, it is crucial to allow for building rules on top of ontologies, that is, for rule-based systems that use vocabulary from

---

[*] Alternative address: Institut für Informationssysteme, Technische Universität Wien, Favoritenstraße 9-11, 1040 Wien, Austria; e-mail: lukasiewicz@kr.tuwien.ac.at

A. Polleres and T. Swift (Eds.): RR 2009, LNCS 5837, pp. 26–39, 2009.
© Springer-Verlag Berlin Heidelberg 2009

ontology knowledge bases. Another type of combination is to build ontologies on top of rules, where ontological definitions are supplemented by rules or imported from rules. Both types of integration have been realized in recent hybrid integrations of rules and ontologies, called *description logic programs* (or *dl-programs*), which are of the form $KB = (L, P)$, where L is a description logic knowledge base, and P is a finite set of rules involving either queries to L in a loose integration (see, e.g., [7,8]) or concepts and roles from L as unary resp. binary predicates in a tight integration (see, e.g., [9]).

However, classical ontology languages and description logics as well as formalisms integrating rules and ontologies are less suitable in all those domains where the information to be represented comes along with (*quantitative*) *uncertainty* and/or *vagueness* (or *imprecision*). For this reason, during the recent years, handling uncertainty and vagueness has started to play an important role in research towards the Semantic Web. A recent forum for approaches to uncertainty reasoning in the Semantic Web is the annual *International Workshop on Uncertainty Reasoning for the Semantic Web (URSW)* at the *International Semantic Web Conference (ISWC)*. There has also been a W3C Incubator Group on *Uncertainty Reasoning for the World Wide Web*. The research focuses especially on probabilistic and fuzzy extensions of description logics, ontology languages, and formalisms integrating rules and ontologies. Note that probabilistic formalisms allow to encode ambiguous information, such as "John is a student with the probability 0.7 and a teacher with the probability 0.3" (roughly, John is either a teacher or a student, but more likely a student), while fuzzy approaches allow to encode vague or imprecise information, such as "John is tall with the degree of truth 0.7" (roughly, John is quite tall). Formalisms for dealing with uncertainty and vagueness are especially applied in ontology mapping, data integration, information retrieval, and database querying. For example, some of the most prominent technologies for dealing with uncertainty are probably the ranking algorithms standing behind Web search engines. Other important applications are belief fusion and opinion pooling, recommendation systems, user preference modeling, trust and reputation modeling, and shopping agents. Vagueness and imprecision also abound in multimedia information processing and retrieval, and are an important aspect of natural language interfaces to the Web.

In this paper, I give an overview of some own recent extensions of description logics and description logic programs by probabilistic uncertainty and fuzzy vagueness. The rest of this paper is organized as follows. Section 2 provides some motivating examples. In Section 3, I describe an approach to probabilistic description logics for the Semantic Web. Sections 4 and 5 focus on approaches to probabilistic and fuzzy description logic programs for the Semantic Web, respectively, while Section 6 describes an approach to description logic programs for handling both uncertainty and vagueness in a uniform framework for the Semantic Web. For a more detailed overview of extensions of description logics for handling uncertainty and vagueness in the Semantic Web, I also refer the reader to the recent survey [10].

2   Motivating Examples

We now provide some examples for the use of probabilistic ontologies and of probabilistic and vague extensions of formalisms integrating rules and ontologies.

In order to illustrate probabilistic ontologies, consider some medical knowledge about patients. In such knowledge, we often encounter terminological probabilistic and terminological default knowledge about classes of individuals, as well as assertional probabilistic knowledge about individuals. It is often advantageous to share such medical knowledge between hospitals and/or medical centers, for example, to follow up patients, to track medical history, for case studies research, and to get information on rare diseases and/or rare cures to diseases. The need for sharing medical knowledge is also at the core of the *W3C Semantic Web Health Care and Life Sciences Interest Group*, who state that the "key to the success of Life Science Research and Health Care is the implementation of new informatics models that will unite many forms of biological and medical information across all institutions" (see http://www.w3.org/2001/sw/hcls/).

Example 2.1 (Medical Example [11]). Consider patient records related to cardiological illnesses. We distinguish between heart patients (who have any kind of cardiological illness), pacemaker patients, male pacemaker patients, and female pacemaker patients, who all are associated with illnesses, illness statuses, symptoms of illnesses, and health insurances. Furthermore, we have the patients Tom, John, and Mary, where Tom is a heart patient, while John and Mary are male and female pacemaker patients, respectively, and John has the symptoms arrhythmia (abnormal heart beat), chest pain, and breathing difficulties, and the illness status advanced.

Then, *terminological default knowledge* is of the form "generally (or typically / in nearly all cases), heart patients suffer from high blood pressure" and "generally, pacemaker patients do not suffer from high blood pressure", while *terminological probabilistic knowledge* has the form "generally, pacemaker patients are male with a probability of at least 0.4" (that is, "generally, a randomly chosen pacemaker patient is male with a probability of at least 0.4"), "generally, heart patients have a private insurance with a probability of at least 0.9", and "generally, pacemaker patients have the symptoms arrhythmia, chest pain, and breathing difficulties with probabilities of at least 0.98, 0.9, and 0.6, respectively". Finally, *assertional probabilistic knowledge* is of the form "Tom is a pacemaker patient with a probability of at least 0.8", "Mary has the symptom breathing difficulties with a probability of at least 0.6", "Mary has the symptom chest pain with a probability of at least 0.9", and "Mary's illness status is final with a probability between 0.2 and 0.8".

Uncertain medical knowledge may also be collected by a medical company from own databases and public sources (e.g., client data, web pages, web inquiries, blogs, and mailing lists) and be used in an advertising campaign for a new product.

Example 2.2 (Medical Example cont'd [11]). Suppose that a medical company wants to carry out a targeted advertising campaign about a new pacemaker product. The company may then first collect all potential addressees of such a campaign (e.g., pharmacies, hospitals, doctors, and heart patients) by probabilistic data integration from different data and web sources (e.g., own databases with data of clients and their shopping histories; and web listings of pharmacies, hospitals, and doctors along with their product portfolio resp. fields of expertise). The result of this process is a collection of individuals with probabilistic memberships to a collection of concepts in a medical ontology as the one above. The terminological probabilistic and terminological default knowledge of this ontology can then be used to derive probabilistic concept memberships that are relevant for a potential addressee of the advertising campaign. For example, for persons that are known to be heart patients with certain probabilities, we may derive the probabilities with which they are also pacemaker patients.

The next example illustrates the use of probabilistic ontologies in information retrieval for an increased recall (which has especially been explored in [12,13]).

Example 2.3 (Literature Search [11]). Suppose that we want to obtain a list of research papers in the area of "logic programming". Then, we should not only collect those papers that are classified as "logic programming" papers, but we should also search for papers in closely related areas, such as "rule-based systems" or "deductive databases", as well as in more general areas, such as "knowledge representation and reasoning" or "artificial intelligence" (since a paper may very well belong to the area of "logic programming", but is classified only with a closely related or a more general area). This expansion of the search can be done automatically using a probabilistic ontology, which has the papers as individuals, the areas as concepts, and the explicit paper classifications as concept memberships. The probabilistic degrees of overlap between the concepts in such a probabilistic ontology then provide a means of deriving a probabilistic membership to the concept "logic programming" and so a probabilistic estimation for the relevance to our search query.

We finally describe a shopping agent example, where we encounter both probabilistic uncertainty (in resource selection, ontology mapping / query transformation, and data integration) and fuzzy vagueness (in query matching with vague concepts).

Example 2.4 (Shopping Agent [37,38]). Suppose a person would like to buy "a sports car that costs at most about 22 000 € and that has a power of around 150 HP".

In todays Web, the buyer has to manually (i) search for car selling sites, e.g., using Google, (ii) select the most promising sites (e.g., http://www.autos.com), (iii) browse through them, query them to see the cars that they sell, and match the cars with our requirements, (iv) select the offers in each web site that match our requirements, and (v) eventually merge all the best offers from each site and select the best ones.

It is obvious that the whole process is rather tedious and time consuming, since, e.g., (i) the buyer has to visit many sites, (ii) the browsing in each site is very time consuming, (iii) finding the right information in a site (which has to match the requirements) is not simple, and (iv) the way of browsing and querying may differ from site to site.

A shopping agent may now support us as follows, automatizing the whole selection process once it receives the request / query q from the buyer:

– *Probabilistic Resource Selection.* The agent selects some sites / resources S that it considers as promising for the buyer's request. The agent has to select a subset of some relevant resources, since it is not reasonable to assume that it will access and query all the resources known to him. The relevance of a resource S to a query is usually (automatically) estimated as the probability $Pr(q|S)$ (the probability that the information need represented by the query q is satisfied by the searching resource S; see, e.g., [14,15]). It is not difficult to see that such probabilities can be represented by probabilistic rules.

- *Probabilistic Ontology Mapping / Query Reformulation.* For the top-k selected sites, the agent has to reformulate the buyer's query using the terminology / ontology of the specific car selling site. For this task, the agent relies on so-called transformation rules, which say how to translate a concept or property of the agent's ontology into the ontology of the information resource (which is called ontology mapping in the Semantic Web). To relate a concept B of the buyer's ontology to a concept S of the seller's ontology, one often automatically estimates the probability $P(B|S)$ that an instance of S is also an instance of B, which can then be represented as a probabilistic rule [16,17].
- *Vague Query Matching.* Once the agent has translated the buyer's request for the specific site's terminology, the agent submits the query. But the buyer's request often contains many so-called vague / fuzzy concepts such as "the price is around 22 000 € or less", rather than strict conditions, and thus a car may match the buyer's condition to a degree. As a consequence, a site / resource / web service may return a ranked list of cars, where the ranks depend on the degrees to which the sold items match the buyer's requests q.
- *Probabilistic Data Integration.* Eventually, the agent has to combine the ranked lists by considering the involved matching (or truth) degrees (vagueness) and probability degrees (uncertainty) and show the top-n items to the buyer.

3   Probabilistic Description Logics

In this section, we briefly describe the probabilistic description logic P-$\mathcal{SHOIN}(\mathbf{D})$, which is a probabilistic generalization of the description logic $\mathcal{SHOIN}(\mathbf{D})$ behind OWL DL towards sophisticated formalisms for reasoning under probabilistic uncertainty in the Semantic Web [11]. Closely related probabilistic generalizations of the *DL-Lite* family of tractable description logics (which lies between the Semantic Web languages RDFS and OWL Lite) and the description logics $\mathcal{SHIF}(\mathbf{D})$ and $\mathcal{SHOQ}(\mathbf{D})$ (which stand behind OWL Lite and DAML+OIL, respectively) have been introduced in [11,18]. A companion paper [19] combines *DL-Lite* with Bayesian networks.

Probabilistic description logics allow for representing probabilistic ontologies and for reasoning about them. There is a plethora of applications with an urgent need for handling probabilistic knowledge in ontologies, especially in areas like medicine, biology, defense, and astronomy. Moreover, probabilistic ontologies allow for quantifying the degrees of overlap between the ontological concepts in the Semantic Web, reasoning about them, and using them in Semantic Web applications and systems, such as information retrieval, personalization tasks, and recommender systems. Furthermore, probabilistic ontologies can be used to align the concepts of different ontologies (called ontology mapping) and for handling inconsistencies in Semantic Web data.

The syntax of P-$\mathcal{SHOIN}(\mathbf{D})$ uses the notion of a conditional constraint from [20] to express probabilistic knowledge in addition to the axioms of $\mathcal{SHOIN}(\mathbf{D})$. Its semantics is based on the notion of lexicographic entailment in probabilistic default reasoning [21,22], which is a probabilistic generalization of the sophisticated notion of lexicographic entailment by Lehmann [23] in default reasoning from conditional knowledge bases. Due to this semantics, P-$\mathcal{SHOIN}(\mathbf{D})$ allows for expressing both terminological probabilistic knowledge about concepts and roles, and also assertional probabilistic knowledge about instances of concepts and roles. It naturally interprets terminological and assertional probabilistic knowledge as statistical knowledge about concepts and roles, and as degrees of belief about instances of concepts and roles, respectively, and allows for deriving both statistical knowledge and degrees of belief. As an important additional feature, it also allows for expressing default knowledge about concepts (as a special case of terminological probabilistic knowledge), which is semantically interpreted as in Lehmann's lexicographic default entailment [23].

Example 3.1. Suppose a classical description logic knowledge base T is used to encode knowledge about cars and their properties (e.g., that sports cars and roadsters are cars). A probabilistic knowledge base $KB = (T, P, (P_o)_{o \in \mathbf{I}_P})$ in P-$\mathcal{SHOIN}(\mathbf{D})$ then extends T by terminological default and terminological probabilistic knowledge in P as well as by assertional probabilistic knowledge in P_o for certain objects $o \in \mathbf{I}_P$. For example, the terminological default knowledge (1) "generally, cars do not have a red color" and (2) "generally, sports cars have a red color", and the terminological probabilistic knowledge (3) "cars have four wheels with a probability of at least 0.9", can be expressed by the following conditional constraints in P:

(1) $(\neg \exists HasColor.\{red\} \mid Car)[1, 1]$,
(2) $(\exists HasColor.\{red\} \mid SportsCar)[1, 1]$,
(3) $(HasFourWheels \mid Car)[0.9, 1]$.

Suppose we want to encode some probabilistic information about John's car (which we have not seen so far). Then, the set of probabilistic individuals $\mathbf{I}_P$ contains the individual *John's car*, and the assertional probabilistic knowledge (4) "John's car is a sports car with a probability of at least 0.8" (we know that John likes sports cars) can be expressed by the following conditional constraint in $P_{John's car}$:

(4) $(SportsCar \mid \top)[0.8, 1]$.

Then, the following are some (terminological default and terminological probabilistic) tight lexicographic consequences of $PT = (T, P)$:

$(\neg \exists HasColor.\{red\} \mid Car)[1, 1]$,
$(\exists HasColor.\{red\} \mid SportsCar)[1, 1]$,
$(HasFourWheels \mid Car)[0.9, 1]$,
$(\neg \exists HasColor.\{red\} \mid Roadster)[1, 1]$,
$(HasFourWheels \mid SportsCar)[0.9, 1]$,
$(HasFourWheels \mid Roadster)[0.9, 1]$.

Hence, in addition to the sentences (1) to (3) directly encoded in P, we also conclude "generally, roadsters do not have a red color", "sports cars have four wheels with a probability of at least 0.9", and "roadsters have four wheels with a probability of at least 0.9". Observe here that the default property of not having a red color and the probabilistic property of having four wheels with a probability of at least 0.9 are inherited from cars down to roadsters. Roughly, the tight lexicographic consequences of $PT = (T, P)$ are given by all those conditional constraints that (a) are either in P, or (b) can be constructed by inheritance along subconcept relationships from the ones in P and are not overridden by more specific pieces of knowledge in P.

The following conditional constraints for the probabilistic individual *John's car* are some (assertional probabilistic) tight lexicographic consequences of KB, which informally say that John's car is a sports car, has a red color, and has four wheels with probabilities of at least 0.8, 0.8, and 0.72, respectively:

$$(SportsCar \mid \top)[0.8, 1],$$
$$(\exists HasColor.\{red\} \mid \top)[0.8, 1],$$
$$(HasFourWheels \mid \top)[0.72, 1].$$

4 Probabilistic Description Logic Programs

We now summarize the main ideas behind loosely and tightly coupled probabilistic dl-programs, introduced in [24,25,26,27] and [28,29,30,31,32], respectively. For further details on the syntax and semantics of these programs, their background, and their semantic and computational properties, we refer to the above works.

Loosely coupled probabilistic dl-programs [24,25,26] are a combination of loosely coupled dl-programs under the answer set and the well-founded semantics with probabilistic uncertainty as in Bayesian networks. Roughly, they consist of a loosely coupled dl-program (L, P) under different "total choices" B (they are the full joint instantiations of a set of random variables, and they serve as pairwise exclusive and exhaustive possible worlds), and a probability distribution μ over the set of total choices B. One then obtains a probability distribution over Herbrand models, since every total choice B along with the loosely coupled dl-program produces a set of Herbrand models of which the probabilities sum up to $\mu(B)$. As in the classical case, the answer set semantics of loosely coupled probabilistic dl-programs is a refinement of the well-founded semantics of loosely coupled probabilistic dl-programs. Consistency checking and tight query processing (i.e., computing the entailed tight interval for the probability of a conditional or unconditional event) in such probabilistic dl-programs under the answer set semantics can be reduced to consistency checking and query processing in loosely coupled dl-programs under the answer set semantics, while tight query processing under the well-founded semantics can be done in an anytime fashion by reduction to loosely coupled dl-programs under the well-founded semantics. For suitably restricted description logic components, the latter can be done in polynomial time in the data complexity. Query processing for stratified loosely coupled probabilistic dl-programs can be reduced to computing the canonical model of stratified loosely coupled dl-programs. Loosely coupled probabilistic dl-programs can especially be used for (database-oriented) probabilistic data integration in the Semantic Web, where probabilistic uncertainty is used to handle inconsistencies between different data sources [27].

Example 4.1. A university database may use a loosely coupled dl-program (L, P) to encode ontological and rule-based knowledge about students and exams. A probabilistic dl-program $KB = (L, P', C, \mu)$ then additionally allows for encoding probabilistic knowledge. For example, the following two probabilistic rules in P' along with a probability distribution on a set of random variables may express that if two master (resp., bachelor) students have given the same exam, then there is a probability of .0.9 (resp., 0.7) that they are friends:

$$friends(X,Y) \leftarrow given\_same\_exam(X,Y), DL[master\_student(X)],$$
$$DL[master\_student(Y)], choice_m ;$$
$$friends(X,Y) \leftarrow given\_same\_exam(X,Y), DL[bachelor\_student(X)],$$
$$DL[bachelor\_student(Y)], choice_b .$$

Here, we assume the set $C = \{V_m, V_b\}$ of value sets $V_m = \{choice_m, not\_choice_m\}$ and $V_b = \{choice_b, not\_choice_b\}$ of two random variables X_m resp. X_b and the probability distribution μ on all their joint instantiations, given by μ: $choice_m, not\_choice_m,$ $choice_b, not\_choice_b \mapsto 0.9, 0.1, 0.7, 0.3$ under probabilistic independence. For example, the joint instantiation $choice_m, choice_b$ is associated with the probability $0.9 \times 0.7 = 0.63$. Asking about the entailed tight interval for the probability that $john$ and $bill$ are friends can then be expressed by a probabilistic query $\exists(friends(john, bill))[R, S]$, whose answer depends on the available concrete knowledge about $john$ and $bill$ (namely, whether they have given the same exams, and are both master or bachelor students).

Tightly coupled probabilistic dl-programs [28,29] are a tight combination of disjunctive logic programs under the answer set semantics with description logics and Bayesian probabilities. They are a logic-based representation formalism that naturally fits into the landscape of Semantic Web languages. Tightly coupled probabilistic dl-programs can especially be used for representing mappings between ontologies [30,31], which are a common way of approaching the semantic heterogeneity problem on the Semantic Web. Here, they allow in particular for resolving inconsistencies and for merging mappings from different matchers based on the level of confidence assigned to different rules (see below). Furthermore, tightly coupled probabilistic description logic programs also provide a natural integration of ontologies, action languages, and Bayesian probabilities towards Web Services. Consistency checking and query processing in tightly coupled probabilistic dl-programs can be reduced to consistency checking and cautious/brave reasoning, respectively, in tightly coupled disjunctive dl-programs. Under certain restrictions, these problems have a polynomial data complexity.

Example 4.2. The two correspondences between two ontologies O_1 and O_2 that (i) an element of *Collection* in O_1 is an element of *Book* in O_2 with the probability 0.62, and (ii) an element of *Proceedings* in O_1 is an element of *Proceedings* in O_2 with the probability 0.73 (found by the matching system hmatch) can be expressed by the following two probabilistic rules:

$$O_2: Book(X) \leftarrow O_1: Collection(X) \land hmatch_1;$$
$$O_2: Proceedings(X) \leftarrow O_1: Proceedings(X) \land hmatch_2.$$

Here, we assume the set $C = \{\{hmatch_i, not\_hmatch_i\} \mid i \in \{1, 2\}\}$ of values of two random variables and the probability distribution μ on all joint instantiations of these variables, given by μ: $hmatch_1, not\_hmatch_1, hmatch_2, not\_hmatch_2 \mapsto 0.62, 0.38,$ $0.73, 0.27$ under probabilistic independence.

Similarly, two other correspondences between O_1 and O_2 (found by the matching system falcon) are expressed by the following two probabilistic rules:

$$O_2: InCollection(X) \leftarrow O_1: Collection(X) \wedge falcon_1;$$
$$O_2: Proceedings(X) \leftarrow O_1: Proceedings(X) \wedge falcon_2,$$

where we assume the set $C' = \{\{falcon_i, not\_falcon_i\} \mid i \in \{1,2\}\}$ of values of two random variables and the probability distribution μ' on all joint instantiations of these variables, given by $\mu': falcon_1, not\_falcon_1, falcon_2, not\_falcon_2 \mapsto 0.94, 0.06, 0.96,$ 0.04 under probabilistic independence.

Using the trust probabilities 0.55 and 0.45 for hmatch and falcon, respectively, for resolving inconsistencies between rules, we can now define a merged mapping set that consists of the following probabilistic rules:

$$O_2: Book(X) \leftarrow O_1: Collection(X) \wedge hmatch_1 \wedge sel\_hmatch_1;$$
$$O_2: InCollection(X) \leftarrow O_1: Collection(X) \wedge falcon_1 \wedge sel\_falcon_1;$$
$$O_2: Proceedings(X) \leftarrow O_1: Proceedings(X) \wedge hmatch_2;$$
$$O_2: Proceedings(X) \leftarrow O_1: Proceedings(X) \wedge falcon_2.$$

Here, we assume the set C'' of values of random variables and the probability distribution μ'' on all joint instantiations of these variables, which are obtained from $C \cup C'$ and $\mu \cdot \mu'$ (defined as $(\mu \cdot \mu')(B\, B') = \mu(B) \cdot \mu'(B')$, for all joint instantiations B of C and B' of C'), respectively, by adding the values $\{sel\_hmatch_1, sel\_falcon_1\}$ of a new random variable, with the probabilities $sel\_hmatch_1, sel\_falcon_1 \mapsto 0.55, 0.45$ under probabilistic independence, for resolving the inconsistency between the first two rules.

A companion approach to probabilistic description logic programs [32] combines probabilistic logic programs, probabilistic default theories, and the description logics behind OWL Lite and OWL DL. It is based on new notions of entailment for reasoning with conditional constraints, which realize the principle of inheritance with overriding for both classical and purely probabilistic knowledge. They are obtained by generalizing previous formalisms for probabilistic default reasoning with conditional constraints (similarly as for P-$\mathcal{SHOIN}$(**D**) in Section 3). In addition to dealing with probabilistic knowledge, these notions of entailment thus also allow for handling default knowledge.

5   Fuzzy Description Logic Programs

We next briefly describe loosely and tightly coupled fuzzy dl-programs, which have been introduced in [33,34] and [35,36], respectively, and extended by a top-k retrieval technique in [39]. All these fuzzy dl-programs have natural special cases where query processing can be done in polynomial time in the data complexity. For further details on their syntax and semantics, background, and properties, we refer to the above works.

Towards dealing with vagueness and imprecision in the reasoning layers of the Semantic Web, loosely coupled (normal) fuzzy dl-programs under the answer set semantics [33,34] generalize normal dl-programs under the answer set semantics by fuzzy vagueness and imprecision in both the description logic and the logic program component. This is the first approach to fuzzy dl-programs that may contain default negations in rule bodies. Query processing in such fuzzy dl-programs can be done by reduction to normal dl-programs under the answer set semantics. In the special cases of positive and stratified loosely coupled fuzzy dl-programs, the answer set semantics coincides with a canonical least model and an iterative least model semantics, respectively, and has a characterization in terms of a fixpoint and an iterative fixpoint semantics, respectively.

Example 5.1. Consider the fuzzy description logic knowledge base L of a car shopping Web site, which defines especially (i) the fuzzy concepts of sports cars (*SportsCar*), "at most 22 000 €" (*LeqAbout22000*), and "around 150 horse power" (*Around150HP*), (ii) the attributes of the price and of the horse power of a car (*hasInvoice* resp. *hasHP*), and (iii) the properties of some concrete cars (such as a *MazdaMX5Miata* and a *MitsubishiES*). Then, a loosely coupled fuzzy dl-program $KB = (L, P)$ is given by the set of fuzzy dl-rules P, which contains only the following fuzzy dl-rule encoding the request of a buyer (asking for a sports car costing at most 22 000 € and having around 150 horse power), where $\otimes$ may be the conjunction strategy of, e.g., Gödel Logic (that is, $x \otimes y = \min(x, y)$, for all $x, y \in [0, 1]$, is used to evaluate $\wedge$ and $\leftarrow$ on truth values):

$$query(x) \leftarrow_\otimes DL[SportsCar](x) \wedge_\otimes DL[\exists hasInvoice.LeqAbout22000](x) \wedge_\otimes$$
$$DL[\exists hasHP.Around150HP](x) \geqslant 1 .$$

The above fuzzy dl-program $KB = (L, P)$ is positive, and has a minimal model M_{KB}, which defines the degree to which some concrete cars in the description logic knowledge base L match the buyer's request, for example,

$$M_{KB}(query(MazdaMX5Miata)) = 0.36 , \quad M_{KB}(query(MitsubishiES)) = 0.32 .$$

That is, the car *MazdaMX5Miata* is ranked top with the degree 0.36, while the car *MitsubishiES* is ranked second with the degree 0.32.

Tightly coupled fuzzy dl-programs under the answer set semantics [35,36] are a tight integration of fuzzy disjunctive logic programs under the answer set semantics with fuzzy description logics. They are also a generalization of tightly coupled disjunctive dl-programs by fuzzy vagueness in both the description logic and the logic program component. This is the first approach to fuzzy dl-programs that may contain disjunctions in rule heads. Query processing in such programs can essentially be done by a reduction to tightly coupled disjunctive dl-programs. A closely related work [39] explores the evaluation of ranked top-k queries. It shows in particular how to compute the top-k answers in data-complexity tractable tightly coupled fuzzy dl-programs.

Example 5.2. A tightly coupled fuzzy dl-program $KB = (L, P)$ is given by a suitable fuzzy description logic knowledge base L and the set of fuzzy rules P, which contains only the following fuzzy rule (where $x \otimes y = \min(x, y)$):

$$query(x) \leftarrow_\otimes SportyCar(x) \wedge_\otimes hasInvoice(x, y_1) \wedge_\otimes hasHorsePower(x, y_2) \wedge_\otimes$$
$$LeqAbout22000(y_1) \wedge_\otimes Around150(y_2) \geqslant 1 .$$

Informally, *query* collects all sports cars, and ranks them according to whether they cost at most around 22 000 € and have around 150 HP. Another fuzzy rule involving also a negation in its body and a disjunction in its head is given as follows (where $\ominus x = 1 - x$ and $x \oplus y = \max(x, y)$):

$$Small(x) \vee_\oplus Old(x) \leftarrow_\otimes Car(x) \wedge_\otimes hasInvoice(x, y) \wedge_\otimes$$
$$not_\ominus GeqAbout15000(y) \geqslant 0.7.$$

This rule says that a car costing at most around $15\,000\,€$ is either small or old. Notice here that $Small$ and Old may be two concepts in the fuzzy description logic knowledge base L. That is, the tightly coupled approach to fuzzy dl-programs under the answer set semantics also allows for using the rules in P to express relationships between the concepts and roles in L. This is not possible in the loosely coupled approach to fuzzy dl-programs under the answer set semantics in [33,34], since the dl-queries there can only occur in rule bodies, but not in rule heads.

6   Probabilistic Fuzzy Description Logic Programs

We finally describe (loosely coupled) probabilistic fuzzy dl-programs [37,38], which combine fuzzy description logics, fuzzy logic programs (with stratified default-negation), and probabilistic uncertainty in a uniform framework for the Semantic Web. Intuitively, they allow for defining several rankings on ground atoms using fuzzy vagueness, and then for merging these rankings using probabilistic uncertainty (by associating with each ranking a probabilistic weight and building the weighted sum of all rankings). Such programs also give rise to important concepts dealing with both probabilistic uncertainty and fuzzy vagueness, such as the expected truth value of a crisp sentence and the probability of a vague sentence. Probabilistic fuzzy dl-programs can be used to model a shopping agent as described in Example 2.4.

Example 6.1. A (loosely coupled) probabilistic fuzzy dl-program is given by a suitable fuzzy description logic knowledge base L and the following set of fuzzy dl-rules P, modeling some query reformulation / retrieval steps using ontology mapping rules:

$$query(x) \leftarrow_\otimes SportyCar(x) \wedge_\otimes hasPrice(x, y_1) \wedge_\otimes hasPower(x, y_2) \wedge_\otimes$$
$$DL[LeqAbout22000](y_1) \wedge_\otimes DL[Around150HP](y_2) \geqslant 1, \quad (1)$$
$$SportyCar(x) \leftarrow_\otimes DL[SportsCar](x) \wedge_\otimes sc_{pos} \geqslant 0.9, \quad (2)$$
$$hasPrice(x, y) \leftarrow_\otimes DL[hasInvoice](x, y) \wedge_\otimes hi_{pos} \geqslant 0.8, \quad (3)$$
$$hasPower(x, y) \leftarrow_\otimes DL[hasHP](x, y) \wedge_\otimes hhp_{pos} \geqslant 0.8, \quad (4)$$

where we assume the set $C = \{\{sc_{pos}, sc_{neg}\}, \{hi_{pos}, hi_{neg}\}, \{hhp_{pos}, hhp_{neg}\}\}$ of values of random variables and the probability distribution μ on all joint instantiations of these variables, given by μ: $sc_{pos}, sc_{neg}, hi_{pos}, hi_{neg}, hhp_{pos}, hhp_{neg} \mapsto 0.91, 0.09, 0.78, 0.22, 0.83, 0.17$ under probabilistic independence. Here, rule (1) is the buyer's request, but in a "different" terminology than the one of the car selling site. Rules (2)–(4) are so-called ontology alignment mapping rules. For example, rule (2) states that the predicate "SportyCar" of the buyer's terminology refers to the concept "SportsCar" of the selected site with probability 0.91.

The following may be some tight consequences of the above probabilistic fuzzy dl-program (where for ground atoms q, we use $(\mathbf{E}[q])[L,U]$ to denote that the expected truth value of q lies in the interval $[L,U]$):

$$(\mathbf{E}[query(MazdaMX5Miata)])[0.21, 0.21], (\mathbf{E}[query(MitsubishiES)])[0.19, 0.19].$$

That is, the car $MazdaMX5Miata$ is ranked first with the degree 0.21, while the car $MitsubishiES$ is ranked second with the degree 0.19.

Acknowledgments. This work has been supported by the German Research Foundation (DFG) under the Heisenberg Programme.

References

1. Berners-Lee, T.: Weaving the Web. Harper, San Francisco (1999)
2. Berners-Lee, T., Hendler, J., Lassila, O.: The Semantic Web. Sci. Amer. 284(5), 34–43 (2001)
3. Fensel, D., Wahlster, W., Lieberman, H., Hendler, J. (eds.): Spinning the Semantic Web: Bringing the World Wide Web to Its Full Potential. MIT Press, Cambridge (2002)
4. Horrocks, I., Patel-Schneider, P.F., van Harmelen, F.: From $\mathcal{SHIQ}$ and RDF to OWL: The making of a web ontology language. J. Web Sem. 1(1), 7–26 (2003)
5. W3C: OWL web ontology language overview. W3C Recommendation, February 10 (2004), http://www.w3.org/TR/2004/REC-owl-features-20040210/
6. Horrocks, I., Patel-Schneider, P.F.: Reducing OWL entailment to description logic satisfiability. In: Fensel, D., Sycara, K., Mylopoulos, J. (eds.) ISWC 2003. LNCS, vol. 2870, pp. 17–29. Springer, Heidelberg (2003)
7. Eiter, T., Ianni, G., Lukasiewicz, T., Schindlauer, R., Tompits, H.: Combining answer set programming with description logics for the Semantic Web. Artif. Intell. 172(12/13), 1495–1539 (2008)
8. Eiter, T., Lukasiewicz, T., Schindlauer, R., Tompits, H.: Well-founded semantics for description logic programs in the Semantic Web. In: Antoniou, G., Boley, H. (eds.) RuleML 2004. LNCS, vol. 3323, pp. 81–97. Springer, Heidelberg (2004)
9. Lukasiewicz, T.: A novel combination of answer set programming with description logics for the Semantic Web. In: Franconi, E., Kifer, M., May, W. (eds.) ESWC 2007. LNCS, vol. 4519, pp. 384–398. Springer, Heidelberg (2007)
10. Lukasiewicz, T., Straccia, U.: Managing uncertainty and vagueness in description logics for the Semantic Web. J. Web Sem. 6(4), 291–308 (2008)
11. Lukasiewicz, T.: Expressive probabilistic description logics. Artif. Intell. 172(6/7), 852–883 (2008)
12. Udrea, O., Deng, Y., Hung, E., Subrahmanian, V.S.: Probabilistic ontologies and relational databases. In: Meersman, R., Tari, Z. (eds.) OTM 2005. LNCS, vol. 3760, pp. 1–17. Springer, Heidelberg (2005)
13. Hung, E., Deng, Y., Subrahmanian, V.S.: TOSS: An extension of TAX with ontologies and similarity queries. In: Proceedings ACM SIGMOD 2004, pp. 719–730. ACM Press, New York (2004)
14. Callan, J.: Distributed information retrieval. In: Croft, W.B. (ed.) Advances in Information Retrieval, pp. 127–150. Kluwer, Dordrecht (2000)
15. Fuhr, N.: A decision-theoretic approach to database selection in networked IR. ACM Trans. Inf. Syst. 3(17), 229–249 (1999)

16. Straccia, U., Troncy, R.: Towards distributed information retrieval in the Semantic Web. In: Sure, Y., Domingue, J. (eds.) ESWC 2006. LNCS, vol. 4011, pp. 378–392. Springer, Heidelberg (2006)
17. Nottelmann, H., Straccia, U.: Information retrieval and machine learning for probabilistic schema matching. Inf. Process. Manage. 43(3), 552–576 (2007)
18. Giugno, R., Lukasiewicz, T.: P-$\mathcal{SHOQ}$(**D**): A probabilistic extension of $\mathcal{SHOQ}$(**D**) for probabilistic ontologies in the Semantic Web. In: Flesca, S., Greco, S., Leone, N., Ianni, G. (eds.) JELIA 2002. LNCS (LNAI), vol. 2424, pp. 86–97. Springer, Heidelberg (2002)
19. d'Amato, C., Fanizzi, N., Lukasiewicz, T.: Tractable reasoning with Bayesian description logics. In: Greco, S., Lukasiewicz, T. (eds.) SUM 2008. LNCS (LNAI), vol. 5291, pp. 146–159. Springer, Heidelberg (2008)
20. Lukasiewicz, T.: Probabilistic deduction with conditional constraints over basic events. J. Artif. Intell. Res. 10, 199–241 (1999)
21. Lukasiewicz, T.: Probabilistic logic programming under inheritance with overriding. In: Proceedings UAI 2001, pp. 329–336. Morgan Kaufmann, San Francisco (2001)
22. Lukasiewicz, T.: Probabilistic default reasoning with conditional constraints. Ann. Math. Artif. Intell. 34(1–3), 35–88 (2002)
23. Lehmann, D.: Another perspective on default reasoning. Ann. Math. Artif. Intell. 15(1), 61–82 (1995)
24. Lukasiewicz, T.: Probabilistic description logic programs. In: Godo, L. (ed.) ECSQARU 2005. LNCS (LNAI), vol. 3571, pp. 737–749. Springer, Heidelberg (2005)
25. Lukasiewicz, T.: Probabilistic description logic programs. Int. J. Approx. Reasoning 45(2), 288–307 (2007)
26. Lukasiewicz, T.: Tractable probabilistic description logic programs. In: Prade, H., Subrahmanian, V.S. (eds.) SUM 2007. LNCS (LNAI), vol. 4772, pp. 143–156. Springer, Heidelberg (2007)
27. Calì, A., Lukasiewicz, T.: An approach to probabilistic data integration for the Semantic Web. In: da Costa, P.C.G., d'Amato, C., Fanizzi, N., Laskey, K.B., Laskey, K.J., Lukasiewicz, T., Nickles, M., Pool, M. (eds.) URSW 2005 - 2007. LNCS (LNAI), vol. 5327, pp. 52–65. Springer, Heidelberg (2008)
28. Calì, A., Lukasiewicz, T.: Tightly integrated probabilistic description logic programs for the Semantic Web. In: Dahl, V., Niemelä, I. (eds.) ICLP 2007. LNCS, vol. 4670, pp. 428–429. Springer, Heidelberg (2007)
29. Calì, A., Lukasiewicz, T., Predoiu, L., Stuckenschmidt, H.: Tightly coupled probabilistic description logic programs for the Semantic Web. J. Data Sem. 12, 95–130 (2009)
30. Calì, A., Lukasiewicz, T., Predoiu, L., Stuckenschmidt, H.: Rule-based approaches for representing probabilistic ontology mappings. In: da Costa, P.C.G., d'Amato, C., Fanizzi, N., Laskey, K.B., Laskey, K.J., Lukasiewicz, T., Nickles, M., Pool, M. (eds.) URSW 2005 - 2007. LNCS (LNAI), vol. 5327, pp. 66–87. Springer, Heidelberg (2008)
31. Calì, A., Lukasiewicz, T., Predoiu, L., Stuckenschmidt, H.: Tightly integrated probabilistic description logic programs for representing ontology mappings. In: Hartmann, S., Kern-Isberner, G. (eds.) FoIKS 2008. LNCS, vol. 4932, pp. 178–198. Springer, Heidelberg (2008)
32. Lukasiewicz, T.: Probabilistic description logic programs under inheritance with overriding for the Semantic Web. Int. J. Approx. Reasoning 49(1), 18–34 (2008)
33. Lukasiewicz, T.: Fuzzy description logic programs under the answer set semantics for the Semantic Web. In: Proceedings RuleML 2006, pp. 89–96. IEEE Computer Society, Los Alamitos (2006)
34. Lukasiewicz, T.: Fuzzy description logic programs under the answer set semantics for the Semantic Web. Fundam. Inform. 82(3), 289–310 (2008)

35. Lukasiewicz, T., Straccia, U.: Tightly integrated fuzzy description logic programs under the answer set semantics for the Semantic Web. In: Marchiori, M., Pan, J.Z., de Marie, C.S. (eds.) RR 2007. LNCS, vol. 4524, pp. 289–298. Springer, Heidelberg (2007)
36. Lukasiewicz, T., Straccia, U.: Tightly coupled fuzzy description logic programs under the answer set semantics for the Semantic Web. Int. J. Semantic Web Inf. Syst. 4(3), 68–89 (2008)
37. Lukasiewicz, T., Straccia, U.: Description logic programs under probabilistic uncertainty and fuzzy vagueness. In: Mellouli, K. (ed.) ECSQARU 2007. LNCS (LNAI), vol. 4724, pp. 187–198. Springer, Heidelberg (2007)
38. Lukasiewicz, T., Straccia, U.: Description logic programs under probabilistic uncertainty and fuzzy vagueness. Int. J. Approx. Reasoning 50(6), 837–853 (2009)
39. Lukasiewicz, T., Straccia, U.: Top-k retrieval in description logic programs under vagueness for the Semantic Web. In: Prade, H., Subrahmanian, V.S. (eds.) SUM 2007. LNCS (LNAI), vol. 4772, pp. 16–30. Springer, Heidelberg (2007)

A Preferential Tableaux Calculus for Circumscriptive $\mathcal{ALCO}^\star$

Stephan Grimm[1] and Pascal Hitzler[2]

[1] FZI Research Center for Information Technologies, Univ. of Karlsruhe, Germany
[2] Institute AIFB, University of Karlsruhe, Germany

Abstract. Nonmonotonic extensions of description logics (DLs) allow for default and local closed-world reasoning and are an acknowledged desired feature for applications, e.g. in the Semantic Web. A recent approach to such an extension is based on McCarthy's circumscription, which rests on the principle of minimising the extension of selected predicates to close off dedicated parts of a domain model. While decidability and complexity results have been established in the literature, no practical algorithmisation for circumscriptive DLs has been proposed so far. In this paper, we present a tableaux calculus that can be used as a decision procedure for concept satisfiability with respect to concept-circumscribed $\mathcal{ALCO}$ knowledge bases. The calculus builds on existing tableaux for classical DLs, extended by the notion of a preference clash to detect the non-minimality of constructed models.

1 Introduction

Modern description logics (DLs) [10] are formalisations of semantic networks and frame-based knowledge representation systems that build on classical logic and are the foundation of the W3C Web Ontology Language OWL [18]. To also capture non-classical features, such as default and local closed-world reasoning, nonmonotonic extensions to DLs have been investigated. While in the past such extensions were primarily devised using autoepistemic operators [5,14,12] and default inclusions [1], a recent proposal [2] is to extend DLs by circumscription and to perform nonmonotonic reasoning on circumscribed DL knowledge bases. In circumscription, the extension of selected predicates – i.e. concepts or roles in the DL case – can be explicitly minimised to close off dedicated parts of a domain model, resulting in a default reasoning behaviour. In contrast to the former approaches, nonmonotonic reasoning in circumscriptive DLs also applies to "unknown individuals" that are not explicitly mentioned in a knowledge base, but whose existence is guaranteed due to existential quantification (see also [8]).

The proposal in [2] presents a semantics for circumscriptive DLs together with decidability and complexity results, in particular for fragments of the logic $\mathcal{ALCQIO}$. However, a practical algorithmisation for reasoning in circumscriptive DLs has not been addressed so far. In this paper, we present an algorithm that builds on existing DL tableaux methods for guided model construction.

---

* This work is partially supported by the German Federal Ministry of Economics (BMWi) under the project THESEUS (number 01MQ07019).

A. Polleres and T. Swift (Eds.): RR 2009, LNCS 5837, pp. 40–54, 2009.
© Springer-Verlag Berlin Heidelberg 2009

In particular, we present a tableaux calculus that supports reasoning with concept-circumscribed knowledge bases in the logic $\mathcal{ALCO}$. We focus on the reasoning task of concept satisfiability, which is motivated by an application of nonmonotonic reasoning in a Semantic Web setting, described in [9]. While typical examples in the circumscription literature deal with defeasible conclusions of circumscriptive abnormality theories, in this setting we use minimisation of concepts to realise a local closed-world assumption for the matchmaking of semantically annotated resources.

The reason for our choice of $\mathcal{ALCO}$ as the underlying DL is twofold. First, we want to present the circumscriptive extensions for the simplest expressive DL $\mathcal{ALC}$ for sake of a clear and concise description of the tableaux modifications. Second, there is the necessity to deal with nominals within the calculus in order to keep track of extensions of minimised concepts, so we include $\mathcal{O}$.

The basic idea behind our calculus is to detect the non-minimality of candidate models, produced by a tableaux procedure for classical DLs, via the notion of a preference clash, and based on the construction of a classical DL knowledge base that has a model if and only if the original candidate model produced is not minimal. This check can be realised by reasoning in classical DLs with nominals. We formally prove this calculus to be sound and complete. A similar idea has been applied in [15] for circumscriptive reasoning in first-order logic. As presented in the pure first-order setting, however, that calculus does not directly yield a decision procedure for DL reasoning as it is only decidable if function symbols are disallowed, which correspond to existential restrictions in DLs.

The paper is structured as follows. In Section 2 we recall circumscriptive DLs from [2] for the case of $\mathcal{ALCO}$. In Section 3, we present our tableaux calculus and prove it to be a decision procedure for circumscriptive concept satisfiability. We conclude in Section 4. Full proofs, partly omitted here, can be found in [7].

2 Description Logics and Circumscription

Description Logics (DLs) [10] are typically fragments of first-order predicate logic that provide a well-studied formalisation for knowledge representation systems. Circumscription [13], on the other hand, is an approach to nonmonotonic reasoning based on the explicit minimisation of selected predicates. In this section, we present the description logic $\mathcal{ALCO}$ extended with circumscription according to [2], which allows for nonmonotonic reasoning with DL knowledge bases.

2.1 Circumscriptive $\mathcal{ALCO}$

The basic elements to represent knowledge in DLs are *individuals*, which represent objects in a domain of discourse, *concepts*, which group together individuals with common properties, and *roles*, which put individuals in relation. The countably infinite sets N_I, N_C and N_r of individual, concept and role names, respectively, form the basis to construct the syntactic elements of $\mathcal{ALCO}$ according to the following grammar, in which $A \in N_C$ denotes an atomic concept, $C_{(i)}$ denote complex concepts, $r \in N_r$ denotes a role and $a_i \in N_I$ denote individuals.

$$C_{(i)} \longrightarrow \bot \mid \top \mid A \mid \neg C \mid C_1 \sqcap C_2 \mid C_1 \sqcup C_2 \mid \exists r.C \mid \forall r.C \mid \{a_1, \ldots, a_n\}$$

The *negation normal form* of a concept C, which we denote by $\|C\|$, is obtained from pushing negation symbols $\neg$ into concept expressions to occur in front of atomic concepts only, as described in [17].

The semantics of the syntactic elements of $\mathcal{ALCO}$ is defined in terms of an *interpretation* $\mathcal{I} = (\Delta^{\mathcal{I}}, \cdot^{\mathcal{I}})$ with a non-empty set $\Delta^{\mathcal{I}}$ as the *domain* and an *interpretation function* $\cdot^{\mathcal{I}}$ that maps each individual $a \in N_I$ to a distinct element $a^{\mathcal{I}} \in \Delta^{\mathcal{I}}$ and that interprets (possibly) complex concepts and roles as follows.

$$\top^{\mathcal{I}} = \Delta^{\mathcal{I}} \ , \ \bot^{\mathcal{I}} = \emptyset \ , \ A^{\mathcal{I}} \subseteq \Delta^{\mathcal{I}} \ , \ r^{\mathcal{I}} \subseteq \Delta^{\mathcal{I}} \times \Delta^{\mathcal{I}}$$
$$(C_1 \sqcap C_2)^{\mathcal{I}} = C_1^{\mathcal{I}} \cap C_2^{\mathcal{I}}$$
$$(C_1 \sqcup C_2)^{\mathcal{I}} = C_1^{\mathcal{I}} \cup C_2^{\mathcal{I}}$$
$$(\neg C)^{\mathcal{I}} = \Delta^{\mathcal{I}} \setminus C^{\mathcal{I}}$$
$$(\forall r . C)^{\mathcal{I}} = \{x \in \Delta^{\mathcal{I}} \mid \forall y.(x,y) \in r^{\mathcal{I}} \rightarrow y \in C^{\mathcal{I}}\}$$
$$(\exists r . C)^{\mathcal{I}} = \{x \in \Delta^{\mathcal{I}} \mid \exists y.(x,y) \in r^{\mathcal{I}} \wedge y \in C^{\mathcal{I}}\}$$
$$(\{a_1, \ldots, a_n\})^{\mathcal{I}} = \{a_1^{\mathcal{I}}, \ldots, a_n^{\mathcal{I}}\}$$

Notice that we assume unique names for individuals, i.e. $a_1^{\mathcal{I}} \neq a_2^{\mathcal{I}}$ for any interpretation $\mathcal{I}$ and any pair $a_1, a_2 \in N_I$.

An $\mathcal{ALCO}$ knowledge base *KB* is a finite set of *axioms* formed by concepts, roles and individuals. A *concept assertion* is an axiom of the form $C(a)$ that assigns membership of an individual a to a concept C. A *role assertion* is an axiom of the form $r(a_1, a_2)$ that assigns a directed relation between two individuals a_1, a_2 by the role r. A *concept inclusion* is an axiom of the form $C_1 \sqsubseteq C_2$ that states the subsumption of the concept C_1 by the concept C_2, while a *concept equivalence* axiom $C_1 \equiv C_2$ is a shortcut for two inclusions $C_1 \sqsubseteq C_2$ and $C_2 \sqsubseteq C_1$. An interpretation $\mathcal{I}$ satisfies a concept assertion $C(a)$ if $a^{\mathcal{I}} \in C^{\mathcal{I}}$, a role assertion $r(a_1, a_2)$ if $(a_1^{\mathcal{I}}, a_2^{\mathcal{I}}) \in r^{\mathcal{I}}$, a concept inclusion $C_1 \sqsubseteq C_2$ if $C_1^{\mathcal{I}} \subseteq C_2^{\mathcal{I}}$ and a concept equivalence $C_1 \equiv C_2$ if $C_1^{\mathcal{I}} = C_2^{\mathcal{I}}$. An interpretation that satisfies all axioms of a knowledge base *KB* is called a *model* of *KB*. A concept C is called *satisfiable with respect to KB* if *KB* has a model in which $C^{\mathcal{I}} \neq \emptyset$ holds.

We now turn to the circumscription part of the formalism, which allows for nonmonotonic reasoning by explicit minimisation of selected $\mathcal{ALCO}$ concepts. We adopt a slightly simplified form of the circumscriptive DLs presented in [2] by restricting our formalism to parallel concept circumscription (without prioritisation among minimised concepts). For this purpose we define the notion of a *circumscription pattern* as follows.

Definition 1 (circumscription pattern, $<_{CP}$). *A circumscription pattern[1] CP is a tuple (M, F, V) of sets of atomic concepts called the* minimised, fixed *and* varying *concepts. Based on CP, a preference relation on interpretations is defined by setting $\mathcal{J} <_{CP} \mathcal{I}$ if and only if the following conditions hold:*

---

[1] The notion of circumscription pattern introduced in [2] is more general and allows the sets M, F and V to also contain roles. There, a circumscription pattern according to Definition 1 is called a *concept* circumscription pattern. However, in the general case role circumscription leads to undecidability, which was also shown in [2]. As our calculus does not allow for role circumscription, we use the term circumscription pattern to denote a concept circumscription pattern in the sense of [2].

(i) $\Delta^{\mathcal{J}} = \Delta^{\mathcal{I}}$ and $a^{\mathcal{J}} = a^{\mathcal{I}}$ for all $a^{\mathcal{J}} \in \Delta^{\mathcal{J}}$
(ii) $\bar{A}^{\mathcal{J}} = \bar{A}^{\mathcal{I}}$ for all $\bar{A} \in F$
(iii) $\tilde{A}^{\mathcal{J}} \subseteq \tilde{A}^{\mathcal{I}}$ for all $\tilde{A} \in M$
(iv) there is an $\tilde{A} \in M$ such that $\tilde{A}^{\mathcal{J}} \subset \tilde{A}^{\mathcal{I}}$

For nonmonotonic reasoning, a classical $\mathcal{ALCO}$ knowledge base is circumscribed with a circumscription pattern and reasoning is performed by means of the resulting *circumscribed knowledge base*, defined as follows.

Definition 2 (circumscribed knowledge base). *A circumscribed knowledge base $circ_{CP}(KB)$ is a knowledge base KB together with a circumscription pattern $CP = (M, F, V)$, such that the sets M, F and V partition the atomic concepts that occur in KB. An interpretation $\mathcal{I}$ is a model of $circ_{CP}(KB)$ if $\mathcal{I}$ is a model of KB and there exists no model $\mathcal{J}$ of KB with $\mathcal{J} <_{CP} \mathcal{I}$.*

The intuition behind the preference relation is to identify interpretations that are "smaller" in the extensions of minimised concepts than others, to select only the "smallest" ones as the *preferred* models. Fixed concepts can be used to restrict this selection and to prevent certain models from being preferred.

2.2 Reasoning with Circumscribed Knowledge Bases

The typical DL reasoning tasks are defined as expected (see [2]) with respect to the models of a circumscribed knowledge base $circ_{CP}(KB)$, which are just the preferred models of KB with respect to CP. For our calculus, we focus on concept satisfiability, which we define next. Other reasoning tasks can be reduced to concept satisfiability, as described in [2].

Definition 3 (circumscriptive concept satisfiability). *A concept C is satisfiable with respect to a circumscribed knowledge base $circ_{CP}(KB)$ if some model $\mathcal{I}$ of $circ_{CP}(KB)$ satisfies $C^{\mathcal{I}} \neq \emptyset$.*

Observe that in classical DLs an atomic concept A is satisfiable with respect to a knowledge base KB "by default" if there is no evidence for its unsatisfiability in KB, i.e. any A is satisfiable with respect to the empty knowledge base. Now suppose that A is a minimised concept in a circumscription pattern CP by which KB is circumscribed. Then, A is unsatisfiable with respect to $circ_{CP}(KB)$ for $KB = \emptyset$. Only if we explicitly assure that the extension of A is non-empty, e.g. by setting $KB = \{A(a)\}$, A becomes satisfiable.

A known result in circumscription is that there is a close relation between fixed and minimised predicates. Namely, fixed predicates can be simulated by minimising them together with their complements. In case of concept circumscription this is achieved by introducing additional concept names and respective equivalence axioms, as reflected by the following proposition (see [2,4,7]).

Proposition 1 (simulation of concept fixation). *Let C be a concept, let KB be a knowledge base and let $CP = (M, F, V)$ be a circumscription pattern with $F = \{\bar{A}_1, \ldots, \bar{A}_n\}$. Furthermore, let*

$$KB' = KB \cup \{\tilde{A}_i \equiv \cdot \bar{A}_i \mid 1 \le i \le n\}$$
$$and \; let \quad \mathrm{CP}' = (M \cup \{\tilde{A}_1, \ldots, \tilde{A}_n, \bar{A}_1, \ldots, \bar{A}_n\}, \emptyset, V) \,,$$

where $\tilde{A}_1, \ldots, \tilde{A}_n$ are atomic concepts that do not occur in KB, CP or C. Then, C is satisfiable with respect to $circ_{CP}(KB)$ if and only if it is satisfiable with respect to $circ_{CP'}(KB')$.

To illustrate the reasoning task of checking concept satisfiability with respect to circumscribed knowledge bases we present the following example.

Example 1. The following knowledge base describes species of the arctic sea.

$$KB_1 = \{ \; Bears(PolarBear), \; \neg Bears(BlueWhale), \; EndangeredSpecies(BlueWhale) \; \}$$

According to KB_1, the polar bear is a kind of bear, while the blue whale is not. Moreover, the blue whale is explicitly listed to be an endangered species, while the polar bear does not occur on this list. The following circumscription pattern allows to "switch off" the open-world assumption for the list of endangered species by minimising the extension of the concept *EndangeredSpecies*.

$$\mathrm{CP} = (M = \{EndangeredSpecies\}, F = \emptyset, V = \{Bears\})$$

The concept $Bears \sqcap EndangeredSpecies$ is unsatisfiable with respect to the circumscribed knowledge base $circ_{CP}(KB_1)$, reflecting that there cannot be an individual that is both an endangered species and a kind of bear. The only endangered species in the preferred models of KB_1 is the blue whale, which is explicitly said to be no kind of bear.

Recently, however, the polar bear unfortunately had to be included in the list of endangered species, which is reflected by the following update of KB_1.

$$KB_2 = KB_1 \cup \{ \; EndangeredSpecies(PolarBear) \; \}$$

With respect to $circ_{CP}(KB_2)$, the concept $Bears \sqcap EndangeredSpecies$ is satisfiable, as the polar bear is a kind of bear and at the same time an endangered species in the preferred models of KB_2.

Instead of using a concept assertion for the explicitly mentioned individual *PolarBear*, we could alternatively update KB_1 by introducing an existentially quantified object through an inclusion axiom stating that the arctic sea is a habitat for an endangered bear species, as follows.

$$KB_3 = KB_1 \cup \{ \; \exists isHabitatFor \,.(Bears \sqcap EndangeredSpecies)(ArcticSea) \; \}$$

The concept $Bears \sqcap EndangeredSpecies$ is also satisfiable with respect to $circ_{CP}(KB_3)$. Observe that in any preferred model of KB_3 the extension of *EndangeredSpecies* contains an unknown individual whose existence is propagated from the known individual *ArcticSea* via the role *isHabitatFor*. Alternative approaches to nonmonotonic reasoning in DL, such as [6,1], typically treat unknown objects differently and do not allow for their defeating of conclusions (see also [8]).

3   Tableaux Calculus for Circumscriptive $\mathcal{ALCO}$

In this section, we introduce a preferential tableaux calculus that decides the
satisfiability of a concept with respect to a circumscribed knowledge base. We
build on the notion of constraint systems, which map to tableaux branches in
tableaux calculi, and we keep the presentation similar to that in [3,5].

3.1   Constraint Systems and Their Solvability

In addition to the alphabet of individuals N_I, we introduce a set N_V of variable
symbols. We denote elements of N_I by a, elements of N_V by x and elements of
$N_I \cup N_V$ by o, all possibly with an index. A *constraint* is a syntactic entity of
one of the forms $o : C$ or $(o_1, o_2) : r$ or $\forall x.x : C$, where C is an $\mathcal{ALCO}$ concept,
r is a role and the o's are objects in $N_I \cup N_V$. A *constraint system*, denoted by
S, is a finite set of constraints. By N_I^S we denote the individuals and by N_V^S the
variables that occur in a constraint system S.

Given an interpretation $\mathcal{I}$, we define an $\mathcal{I}$-*assignment* as a function $\alpha^{\mathcal{I}}$:
$N_I \cup N_V \mapsto \Delta^{\mathcal{I}}$, that maps every variable of N_V to an element of $\Delta^{\mathcal{I}}$ and every
individual a to $a^{\mathcal{I}}$, i.e. $\alpha^{\mathcal{I}}(a) = a^{\mathcal{I}}$ for all $a \in N_I$.

A pair $(\mathcal{I}, \alpha^{\mathcal{I}})$ of an interpretation $\mathcal{I}$ and an $\mathcal{I}$-assignment $\alpha^{\mathcal{I}}$ *satisfies* a
constraint $o : C$ if $\alpha^{\mathcal{I}}(o) \in C^{\mathcal{I}}$, a constraint $(o_1, o_2) : r$ if $(\alpha^{\mathcal{I}}(o_1), \alpha^{\mathcal{I}}(o_2)) \in r^{\mathcal{I}}$
and a constraint $\forall x.x : C$ if $C^{\mathcal{I}} = \Delta^{\mathcal{I}}$. A *solution* for a constraint system S is a
pair $(\mathcal{I}, \alpha^{\mathcal{I}})$ that satisfies all constraints in S.

We denote by $S[o_1/o_2]$ the constraint system that is obtained by replacing any
occurrence of object o_1 by object o_2 in every constraint in S. Furthermore, we
define the constraint system S_{KB} to be obtained from an $\mathcal{ALCO}$ knowledge base
KB by including one constraint of the form $a : \|C\|$ for each concept assertion
$C(a) \in KB$, one constraint $(a_1, a_2) : r$ for each role assertion $r(a_1, a_2) \in KB$ and
one constraint $\forall x.x : \|\neg C_1 \sqcup C_2\|$ for each concept inclusion $C_1 \sqsubseteq C_2 \in KB$, such
that S_{KB} captures all the information in KB.

To ensure termination of our calculus in the presence of general inclusion axioms, we need to introduce the notion of blocking (see e.g. [11]). Given constraint
systems S and $S^* \subseteq S$, we say that an object o_1 is a *direct predecessor* of an object o_2, if S^* contains a role constraint $(o_1, o_2) : r$ for some role r. We denote by
predecessor the transitive closure in S^* of the direct predecessor relation. Moreover, we say that, in a constraint system S with $S^* \subseteq S$, an object o_2 is *blocked by*
an object o_1 if o_1 is a predecessor of o_2 and if $\{C \mid o_2 : C \in S\} \subseteq \{C \mid o_1 : C \in S\}$
holds. The set S^* is maintained by the tableaux calculus and used to control
which role constraints in S shall be taken into consideration for blocking.

Due to the analogy between a constraint system and a knowledge base the
following Lemma holds.

Lemma 1. *Let KB be an $\mathcal{ALCO}$ knowledge base, S be a constraint system with
$S_{KB} \subseteq S$ and $\mathcal{I}$ be an interpretation. If $\mathcal{I}$ is a model of KB then, for any $\mathcal{I}$-
assignment $\alpha^{\mathcal{I}}$, $(\mathcal{I}, \alpha^{\mathcal{I}})$ is a solution for S_{KB}. Furthermore, for any solution
$(\mathcal{I}, \alpha^{\mathcal{I}})$ for S, $\mathcal{I}$ is a model of KB.*

Our calculus is based on finding a solution for constraint systems the interpretation of which is a preferred model of an initial knowledge base with respect to a circumscription pattern. For this purpose we define the notion of solvability.

Definition 4 (CP-solvability). *A constraint system S is* CP-*solvable with respect to* KB *if there is a model $\mathcal{I}$ of* KB *and an $\mathcal{I}$-assignment $\alpha^{\mathcal{I}}$ such that $(\mathcal{I}, \alpha^{\mathcal{I}})$ is a solution for S and there is no model $\mathcal{J}$ of* KB *with $\mathcal{J} <_{CP} \mathcal{I}$.*

By the next proposition, we reduce circumscriptive concept satisfiability to checking a constraint system for its solvability.

Proposition 2 (satisfiability reduction). *Let* KB *be a knowledge base,* CP *be a circumscription pattern and C be a concept. C is satisfiable with respect to* $circ_{CP}(KB)$ *if and only if $S_{KB} \cup \{x : C\}$ is* CP-*solvable with respect to* KB.

Proof. $\Rightarrow$: Since C is satisfiable with respect to $circ_{CP}(KB)$, there is a model $\mathcal{I}$ of $circ_{CP}(KB)$ in which $C^{\mathcal{I}}$ is nonempty. Let a be an individual with $a^{\mathcal{I}} \in C^{\mathcal{I}}$. Since $\mathcal{I}$ is also a model of KB and due to Lemma 1, $(\mathcal{I}, \alpha^{\mathcal{I}})$ is a solution for S_{KB} for any $\mathcal{I}$-assignment $\alpha^{\mathcal{I}}$. Let $\alpha^{\mathcal{I}}_{x,a}$ be an $\mathcal{I}$-assignment with $\alpha^{\mathcal{I}}_{x,a}(x) = a^{\mathcal{I}}$. Then, $(\mathcal{I}, \alpha^{\mathcal{I}}_{x,a})$ satisfies, besides the constraints in S_{KB}, also the constraint $x : C$, because of $\alpha^{\mathcal{I}}_{x,a}(x) \in C^{\mathcal{I}}$, and is therefore a solution for $S_{KB} \cup \{x : C\}$. Since there is no other model $\mathcal{J}$ of KB with $\mathcal{J} <_{CP} \mathcal{I}$, $S_{KB} \cup \{x : C\}$ is CP-solvable with respect to KB.
$\Leftarrow$: Since $S_{KB} \cup \{x : C\}$ is CP-solvable with respect to KB, there is a model $\mathcal{I}$ of KB and an $\mathcal{I}$-assignment $\alpha^{\mathcal{I}}$ such that $(\mathcal{I}, \alpha^{\mathcal{I}})$ is a solution for $S_{KB} \cup \{x : C\}$. Moreover, there exists an element $a^{\mathcal{I}} \in \Delta^{\mathcal{I}}$ with $\alpha^{\mathcal{I}}(x) = a^{\mathcal{I}} \in C^{\mathcal{I}}$ because $(\mathcal{I}, \alpha^{\mathcal{I}})$ satisfies the constraint $x : C$. By definition of CP-solvability, there is no model $\mathcal{J}$ of KB with $\mathcal{J} <_{CP} \mathcal{I}$, and thus, $\mathcal{I}$ is a model of $circ_{CP}(KB)$ in which $C^{\mathcal{I}}$ is non-empty. Hence, C is satisfiable with respect to $circ_{CP}(KB)$. $\square$

3.2 Tableaux Expansion Rules

Constraint systems are manipulated by tableaux expansion rules, which decompose the structure of complex logical constructs or replace variables by concrete individuals. By expanding a constraint system with the resulting constraints, our calculus tries to build a model for the initial knowledge base that is represented by the constraint system. To decide the satisfiability of a concept C with respect to a circumscribed knowledge base $circ_{CP}(KB)$ according to Proposition 2, we initialise the calculus with the constraint system $S = S_{KB} \cup \{x : \|C\|\}$ and $S^* = \emptyset$. Without loss of generality, we assume all fixed predicates to be simulated according to Proposition 1, and thus, the set F in CP to be empty. The algorithm exhaustively performs the tableau rules given in Table 1, however, the $\longrightarrow_{<_{CP}}$-rule must not be applied if any of the other rules is applicable, i.e. the $\longrightarrow_{<_{CP}}$-rule has a lower precedence than the other rules. The notion of predecessor is evaluated with respect to S^*.

Observe that the rules are parametric with respect to KB and CP. The rules $\longrightarrow_{\forall_x}$, $\longrightarrow_{\sqcap}$, $\longrightarrow_{\exists}$ and $\longrightarrow_{\forall}$ are *deterministic* and their application yields a

Table 1. Tableau Expansion Rules for Circumscriptive $\mathcal{ALCO}$. The $\longrightarrow_{<_{\mathsf{CP}}}$-rule must not be executed if any of the other rules is applicable. Blocking is evaluated with respect to S^*.

$\longrightarrow_{\forall_x}$	:	**if** $\forall x.x : C \in S$ and $o : C \notin S$
		then $S \leftarrow S \cup \{o : C\}$
$\longrightarrow_{\sqcap}$	:	**if** $o : C_1 \sqcap C_2 \in S$ and $\{o : C_1, o : C_2\} \not\subseteq S$
		then $S \leftarrow S \cup \{o : C_1, o : C_2\}$
$\longrightarrow_{\sqcup}$	:	**if** $o : C_1 \sqcup C_2 \in S$ and $\{o : C_1, o : C_2\} \cap S = \emptyset$
		then $S \leftarrow S \cup \{o : C_1\}$ or $S \leftarrow \{o : C_2\}$
$\longrightarrow_{\exists}$	:	**if** $o_1 : \exists r.C \in S$ and $\{(o_1, o_2) : r, o_2 : C\} \not\subseteq S$ and o_1 is not blocked
		then $S \leftarrow S \cup \{(o_1, x) : r, x : C\}$, with x a new variable
		and $(o_1, x) : r$ is added to S^*
$\longrightarrow_{\forall}$	:	**if** $o_1 : \forall r.C \in S$ and $(o_1, o_2) : r \in S$ and $o_2 : C \notin S$
		then $S \leftarrow S \cup \{o_2 : C\}$
$\longrightarrow_{\mathcal{O}}$	:	**if** $x : \{a_1, \ldots, a_k\} \in S$
		then $S \leftarrow S[x/a_i]$ for any $i \in \{1, \ldots, k\} \subset \mathbb{N}$
		and all $(o, x) : r$ are removed from S^*
$\longrightarrow_{<_{\mathsf{CP}}}$	:	**if** $x : \tilde{A} \in S$ and $\tilde{A} \in M$
		then $S \leftarrow S[x/a]$ for any $a \in \mathsf{N}_I^S \cup \{\iota\}$, with ι a new individual
		and $S^* \leftarrow S^*[x/\iota]$ if $a = \iota$

single constraint system. Contrarily, the rules $\longrightarrow_{\sqcup}$, $\longrightarrow_{\mathcal{O}}$ and $\longrightarrow_{<_{\mathsf{CP}}}$ are *non-deterministic*, meaning that they can be applied in multiple ways and yield different constraint systems. Any such non-deterministic choice produces a branching point for backtracking. In the $\longrightarrow_{\sqcup}$-rule, the disjunction leads to the choice of expanding on either of the disjuncts, while in the $\longrightarrow_{\mathcal{O}}$- and $\longrightarrow_{<_{\mathsf{CP}}}$-rules the presence of several individuals leads to a choice of selecting one for replacement of the variable x. Moreover, the $\longrightarrow_{<_{\mathsf{CP}}}$-rule introduces new individuals into the constraint system whenever ι is selected for replacement,[2] while the $\longrightarrow_{\exists}$-rule introduces new variables whenever an object lacks a role filler.

Definition 5 (completion). *A completion of a constraint system S with regard to CP and KB is any constraint system that results from the application of the algorithm to S, using CP and KB, and to which none of the rules is applicable.*

The algorithm finally leads to a completion of the initial constraint system that contains the exhaustive decomposition of complex constraints, which is established by the following lemma.

Lemma 2 (termination). *For any constraint system S, the algorithm always terminates, and yields a completion of S.*

Proof (Sketch). Note that the top part of Table 1 (without the $\longrightarrow_{<_{\mathsf{CP}}}$-rule) and corresponding algorithm coincides with that of [11] for $\mathcal{ALCO}$. In fact, the termination proof from [11], can easily be adapted to our setting.

---

[2] The idea of including a new individual ι as a representative for the infinitely many remaining objects in $\mathsf{N}_I \setminus \mathsf{N}_I^S$ in the domain is taken from [5].

Moreover, we establish the result that the tableaux expansion rules of our calculus preserve the solvability of constraint systems as follows.

Proposition 3 (solvability preservation). *Let KB be an $\mathcal{ALCO}$ knowledge base, CP be a circumscription pattern and S, S' be two constraint systems.*

1. *If S' results from S by application of a deterministic rule then S is CP-solvable with respect to KB if and only if S' is CP-solvable with respect to KB.*
2. *If S' results from S by application of a non-deterministic rule then S is CP-solvable with respect to KB if S' is CP-solvable with respect to KB. Furthermore, if S is CP-solvable with respect to KB and a non-deterministic rule applies to S then it can be applied in such a way that the resulting constraint system S' is also CP-solvable with respect to KB.*

Proof. The claim 1. for the rules $\longrightarrow_\sqcap$, $\longrightarrow_\exists$, $\longrightarrow_\forall$, $\longrightarrow_\sqcup$ and $\longrightarrow_\mathcal{O}$ follows from the results in [11]. (See also [7] for a full proof.) Therefore, we concentrate on the claim 2. for the $\longrightarrow_{<_{CP}}$-rule.

$\Leftarrow$: Assume that S' is obtained from S by application of the $\longrightarrow_{<_{CP}}$-rule and S' is CP-solvable with respect to KB. Let $(\mathcal{I}, \alpha^\mathcal{I})$ be a solution for S' such that $\mathcal{I}$ is a model of KB and there is no model $\mathcal{J}$ of KB with $\mathcal{J} <_{CP} \mathcal{I}$. As the $\longrightarrow_{<_{CP}}$-rule has been applied, $S' = S[x/a]$ for some individual $a \in \mathsf{N}_I$. As a solution for S', $(\mathcal{I}, \alpha^\mathcal{I})$ satisfies all the constraints in $S[x/a]$, in particular those in which x has been replaced by a. Let $\alpha^\mathcal{I}_{x,a}$ be the $\mathcal{I}$-assignment that coincides with $\alpha^\mathcal{I}$ except that $\alpha^\mathcal{I}_{x,a}(x) = a^\mathcal{I}$. Then, $(\mathcal{I}, \alpha^\mathcal{I}_{x,a})$ satisfies all the constraints in S in which x occurs, and since S and S' differ only by these, also all remaining constraints in S. Hence, $(\mathcal{I}, \alpha^\mathcal{I}_{x,a})$ is a solution for S, and since there is no model $\mathcal{J}$ of KB with $\mathcal{J} <_{CP} \mathcal{I}$ by assumption, S is CP-solvable with respect to KB.

$\Rightarrow$: Assume that S' is obtained from S by application of the $\longrightarrow_{<_{CP}}$-rule and that S is CP-solvable with respect to KB. Let $(\mathcal{I}, \alpha^\mathcal{I})$ be a solution for S such that $\mathcal{I}$ is a model of KB and there is no model $\mathcal{J}$ of KB with $\mathcal{J} <_{CP} \mathcal{I}$. As the $\longrightarrow_{<_{CP}}$-rule has been applied, S contains a constraint of the form $x : \tilde{A}$ with $\tilde{A} \in M$. As a solution for S, $(\mathcal{I}, \alpha^\mathcal{I})$ satisfies this constraint and there is some individual $a \in \mathsf{N}_I$ with $\alpha^\mathcal{I}(x) = a^\mathcal{I}$. We distinguish the two cases in which a) a is in N_I^S and b) a is a new individual not in N_I^S:

- a) In case $a \in \mathsf{N}_I^S$, a can be picked for the application of the $\longrightarrow_{<_{CP}}$-rule and it directly follows that $(\mathcal{I}, \alpha^\mathcal{I})$ is a solution for the resulting constraint system $S' = S[x/a]$.
- b) In case $a \in \mathsf{N}_I \setminus \mathsf{N}_I^S$, $\iota \in \mathsf{N}_I \setminus \mathsf{N}_I^S$ can be picked for the application of the $\longrightarrow_{<_{CP}}$-rule as a representative for any new individual. Then, $S[x/a]$ and $S[x/\iota]$ differ only by the naming of an individual new to S and are in this sense isomorphic[3]. Hence, as $(\mathcal{I}, \alpha^\mathcal{I})$ is a solution for $S[x/a]$ it is also a solution for the resulting constraint system $S' = S[x/\iota]$.

Finally, since $(\mathcal{I}, \alpha^\mathcal{I})$ is a solution for S' and there is no model $\mathcal{J}$ of KB with $\mathcal{J} <_{CP} \mathcal{I}$ by assumption, the $\longrightarrow_{<_{CP}}$-rule can be applied to S in such a way that S' is CP-solvable with respect to KB.   □

---

[3] See also the analogous argument in [5, Lemma 3.6].

Algorithm 1. Construct a knowledge base KB'

Require: a constraint system S produced for an initial $\mathcal{ALCO}$ knowledge base KB
circumscribed with a circumscription pattern $\mathsf{CP} = (M, F, V)$

$\quad KB' \leftarrow KB$, $\quad D \leftarrow \{\bot\}$
$\quad$**for all** $\tilde{A} \in M_{KB}$ **do**
$\quad\quad E_{\tilde{A}} := \{a \mid a : \tilde{A} \in S\}$
$\quad\quad$**if** $\#E_{\tilde{A}} > 0$ **then**
$\quad\quad\quad KB' \leftarrow KB' \cup \{\tilde{A} \sqsubseteq \{a_1, \ldots, a_n\}\}, a_1, \ldots a_n \in E_{\tilde{A}}$
$\quad\quad\quad D \leftarrow D \cup \{\{a_1, \ldots, a_n\} \sqcap \neg\tilde{A}\}, a_1, \ldots a_n \in E_{\tilde{A}}$
$\quad\quad$**else**
$\quad\quad\quad KB' \leftarrow KB' \cup \{\tilde{A} \sqsubseteq \bot\}$
$\quad\quad$**end if**
$\quad$**end for**
$\quad KB' \leftarrow KB' \cup \{(\bigsqcup_{D_{\tilde{A}} \in D} D_{\tilde{A}})(\iota)\}$, with ι a new individual

---

3.3 Notions of Clash and Detection of Inconsistencies

Once a completion of an initial constraint system has been produced, its solvability can be verified by using the notion of a *clash*. In addition to the clashes defined in [5,16], which represent obvious contradictions in a knowledge base, we introduce the notion of a *preference clash*, which reflects non-minimality of the respective model with regard to the preference relation $<_{\mathsf{CP}}$.

Definition 6 (Clashes). *Let S be a constraint system.*
$\quad$*S contains an* inconsistency clash *if at least one of the following holds:*
$\quad$*(i) S contains a constraint of the form $o : \bot$.*
$\quad$*(ii) S contains two constraints of the form $o : A$, $o : \neg A$.*

$\quad$*S contains an* individual clash *if at least one of the following holds:*
$\quad$*(iii) S contains a constraint of the form $a : \{a_1, \ldots, a_k\}$.*
$\quad\quad$*with $a \neq a_i$ for all $i \in \{1, \ldots, k\} \subset \mathbb{N}$.*
$\quad$*(vi) S contains a constraint of the form $a : \neg\{a_1, \ldots, a_k\}$.*
$\quad\quad$*with $a = a_i$ for some $i \in \{1, \ldots, k\} \subset \mathbb{N}$.*

$\quad$*S contains a* preference clash, *parameterised with a circumscription pattern CP and an $\mathcal{ALCO}$ knowledge base KB, if the following condition holds:*
$\quad$*(v) the constraint system $S_{KB'}[\iota/x]$ has a completion, with regard to*
$\quad\quad$*$\mathsf{CP}' = (\emptyset, \emptyset, F \cup M \cup V)$ and KB', that does neither contain an*
$\quad\quad$*inconsistency clash nor an individual clash, while the $\mathcal{ALCO}$*
$\quad\quad$*knowledge base KB' is constructed according to Algorithm 1.*

The idea behind the construction of KB' in Algorithm 1 is to freeze the instance situation for minimised concepts as asserted in the current constraint system perceived as reflecting some model $\mathcal{I}$ of the original knowledge base KB. Then, KB' is constructed such that for any of its models $\mathcal{J}$ it holds that $\mathcal{J} <_{\mathsf{CP}} \mathcal{I}$, and thus, checking KB' for unsatisfiability verifies minimality of $\mathcal{I}$. By inclusion axioms for minimised concepts $\tilde{A}$ the conditions (iii) (and indirectly also (ii)) of Definition 1 are assured to hold for each model of KB'. Moreover, by the disjunctive concept assertion condition (iv) of Definition 1 is assured to hold, such that any model of KB' is actually "smaller" than $\mathcal{I}$ in some minimised concept, which is achieved by mapping the not uniquely named individual ι to one that already occurs in the extension of a minimised concept. Although in general we assume unique names in the formalism, the replacement of the new individual ι by the variable x within $S_{KB'}[\iota/x]$ in condition (v) of Definition 6 allows ι to be (indirectly) identified with some other individual.

We illustrate the detection of clashes in our calculus by means of an example.

Example 2. Consider the circumscribed knowledge base $\mathrm{circ_{CP}}(KB)$ with the following $\mathcal{ALCO}$ knowledge base KB and circumscription pattern CP.

$$KB = \{\ \neg Bears(Blue\,Whale)\,,\ EndangeredSpecies(Blue\,Whale)\ \}$$
$$\mathrm{CP} = (M = \{EndangeredSpecies\}, F = \emptyset, V = \{Bears\})$$

We perform our calculus to check whether the concept $Bears \sqcap EndangeredSpecies$ is satisfiable with respect to $\mathrm{circ_{CP}}(KB)$.

We start with the constraint system initialised as follows.

$$S_{KB} \cup \{x : Bears \sqcap EndangeredSpecies\} = \{\ Blue\,Whale : \neg Bears\,,$$
$$Blue\,Whale : EndangeredSpecies\,,\ x : Bears \sqcap EndangeredSpecies\ \}$$

From the application of the $\longrightarrow_\sqcap$-rule and subsequently of the $\longrightarrow_{<_{CP}}$-rule, the following two resulting completions are produced.

$$S_1 = \{\ Blue\,Whale : \neg Bears\,,\ Blue\,Whale : EndangeredSpecies\,,\ Blue\,Whale : Bears\ \}$$
$$S_2 = \{\ Blue\,Whale : \neg Bears\,,\ Blue\,Whale : EndangeredSpecies\,,$$
$$\iota_0 : Bears\,,\ \iota_0 : EndangeredSpecies\ \}$$

The completion S_1 obviously contains an inconsistency clash, since it contains both the constraints $Blue\,Whale : Bears$ and $Blue\,Whale : \neg Bears$.

For the completion S_2, we construct KB' according to Algorithm 1 as follows.

$$KB' = \{\ \neg Bears(Blue\,Whale)\,,\ EndangeredSpecies(Blue\,Whale)\,,$$
$$EndangeredSpecies \sqsubseteq \{Blue\,Whale, \iota_0\}\,,$$
$$\neg EndangeredSpecies \sqcap \{Blue\,Whale, \iota_0\}(\iota)\ \}$$

It can be verified by classical reasoning techniques that KB' has a model when the new individual ι is not uniquely named serving as a variable, and thus, the completion S_2 contains a preference clash.

Since both S_1 and S_2 contain some clash, the initial constraint system $S_{KB} \cup \{x : Bears \sqcap EndangeredSpecies\}$ has no clash-free completion. Hence, the concept $Bears \sqcap EndangeredSpecies$ is unsatisfiable with respect to $\mathrm{circ_{CP}}(KB)$.

In the description logic literature, tableaux methods for sound and complete reasoning have been proposed for various DL variants including $\mathcal{ALCO}$. They detect inconsistencies in DL knowledge bases by checking completions of constraint systems for the occurrence of a clash. We include this result adapted to our setting in form of the following proposition.

Proposition 4 ($\mathcal{ALCO}$ correctness). *Let KB be an $\mathcal{ALCO}$ knowledge base and S be the completion of a constraint system containing at least the constraints of S_{KB}, with regard to any circumscription pattern and KB. Then S has a solution if and only if it contains neither an inconsistency clash nor an individual clash.*

Proof (Sketch). The top part of Table 1 (without the $\longrightarrow_{<\text{CP}}$-rule) captures the algorithm from [11], which is known to be correct. In fact, the proof from [11] essentially carries over.

Based on this correspondence between clash-free completions and their solutions, we can establish the correlation between solvability of constraint systems and the absence of preference clashes in their completions as the main result of this paper by the following proposition.

Proposition 5 (circumscriptive $\mathcal{ALCO}$ correctness). *Let KB be an $\mathcal{ALCO}$ knowledge base, CP be a circumscription pattern and S be the completion of a constraint system containing at least the constraints of S_{KB}, with regard to CP and KB. S is CP-solvable with respect to KB if and only if it contains no inconsistency clash, no individual clash and no preference clash with respect to CP and KB.*

Proof

$\Rightarrow$: Assume that S is CP-solvable with respect to *KB*. According to Definition 4 there is a solution $(\mathcal{I}, \alpha^{\mathcal{I}})$ for S, such that $\mathcal{I}$ is a model of *KB* and there is no model $\mathcal{J}$ of *KB* with $\mathcal{J} <_{\text{CP}} \mathcal{I}$. From Proposition 4, we know that S does neither contain an inconsistency clash nor an individual clash. We show by contradiction that S does also not contain a preference clash.

Assume that S contains a preference clash with respect to CP and *KB*. Then, $S_{KB'}[\iota/x]$ has a completion S' with regard to CP $= (\emptyset, \emptyset, M \cup F \cup V)$ and *KB'* that contains no inconsistency and no individual clash, where the knowledge base *KB'* is constructed based on CP and *KB* according to Algorithm 1. Observe that, by construction, $KB \subset KB'$ and that ι is a new individual in *KB'* that cannot occur in *KB*. Hence, we have that $S_{KB} \subset S_{KB'}[\iota/x] \subseteq S'$. Proposition 4($\Leftarrow$) implies that there is a solution $(\mathcal{J}, \alpha^{\mathcal{J}})$ for S', since S' is clash-free. Due to Lemma 1, and since $S_{KB} \subseteq S'$, it follows that $\mathcal{J}$ is a model of both *KB'* and *KB*. It remains to show that $\mathcal{J} <_{\text{CP}} \mathcal{I}$, to contradict the containment of a preference clash in S. Without loss of generality, we can assume that $\Delta^{\mathcal{I}} = \Delta^{\mathcal{J}}$ and that $a^{\mathcal{I}} = a^{\mathcal{J}}$ for all individuals $a \in \mathsf{N}_I$. Moreover, we assumed $F = \emptyset$ due to Proposition 1, such that $\bar{A}^{\mathcal{J}} = \bar{A}^{\mathcal{I}}$ for all $\bar{A} \in F$ vacuously holds. We prove the following claims: a) $\tilde{A}^{\mathcal{J}} \subseteq \tilde{A}^{\mathcal{I}}$ for all $\tilde{A} \in M$, and b) $\tilde{A}^{\mathcal{J}} \subset \tilde{A}^{\mathcal{I}}$ for some $\tilde{A} \in M$.

- a) Due to the inclusion axioms for minimised concepts inserted into *KB'* by Algorithm 1, and since $\mathcal{J}$ is a model of *KB'*, $\mathcal{J}$ has the property $\tilde{A}^{\mathcal{J}} \subseteq \{\alpha^{\mathcal{J}}(a) \mid a : \tilde{A} \in S\}$ for each $\tilde{A} \in M$. For every $\tilde{A} \in M$, all the constraints $a : \tilde{A} \in S$ are satisfied by $(\mathcal{I}, \alpha^{\mathcal{I}})$, i.e. $\alpha^{\mathcal{I}}(a) \in \tilde{A}^{\mathcal{I}}$, and therefore we have that $\{\alpha^{\mathcal{I}}(a) \mid a : \tilde{A} \in S\} \subseteq \tilde{A}^{\mathcal{I}}$. Since $\alpha^{\mathcal{I}}$ and $\alpha^{\mathcal{J}}$ coincide on individuals, it follows that $\tilde{A}^{\mathcal{J}} \subseteq \tilde{A}^{\mathcal{I}}$ for all $\tilde{A} \in M$.
- b) By construction of *KB'*, $S_{KB'}[\iota/x]$ contains a constraint $x : \bigsqcup_{\tilde{A}} D_{\tilde{A}}$, and for one of the disjuncts $D_{\tilde{A}}$ its completion S' contains a constraint of the form $x : \{a_1, \ldots, a_n\} \sqcap \neg \tilde{A}$ with $a_i : \tilde{A} \in S$ for $i = 1 \ldots n$. Since S' is a completion to which none of the tableaux rules apply, the $\longrightarrow_{\sqcap}$- and the $\longrightarrow_{\mathcal{O}}$-rule have produced the constraints $a : \{a_1, \ldots, a_n\}$ and $a : \neg \tilde{A}$ in S' in which the variable x has been replaced by an individual a. As a solution for S', $(\mathcal{J}, \alpha^{\mathcal{J}})$ satisfies these two constraints and we have that both $\alpha^{\mathcal{J}}(a) \in (\Delta^{\mathcal{J}} \setminus \tilde{A}^{\mathcal{J}})$

and $\alpha^{\mathcal{J}}(a) \in \{\alpha^{\mathcal{J}}(a) \mid a : \tilde{A} \in S\}$ hold. This implies that $\alpha^{\mathcal{J}}(a) \notin \tilde{A}^{\mathcal{J}}$ and, since $(\mathcal{I}, \alpha^{\mathcal{I}})$ satisfies the constraint $a : \tilde{A}$, that $\alpha^{\mathcal{J}}(x) = \alpha^{\mathcal{I}}(a) \in \tilde{A}^{\mathcal{I}}$. From the arguments under b) we already know that $\tilde{A}^{\mathcal{J}} \subseteq \tilde{A}^{\mathcal{I}}$, and since we have an element $a^{\mathcal{I}}$ which is in $\tilde{A}^{\mathcal{I}}$ but not in $\tilde{A}^{\mathcal{J}}$, it follows that $\tilde{A}^{\mathcal{J}} \subset \tilde{A}^{\mathcal{I}}$.

$\Leftarrow$: Let S contain no clash. From Proposition 4 we know that there is a solution $(\mathcal{I}, \alpha^{\mathcal{I}})$ for S. (We assume here that this solution is directly obtained from S by construction, in analogy to the notion of a canonical interpretation; see e.g. [3].) We show by contradiction that there is no model $\mathcal{J}$ of KB such that $\mathcal{J} <_{\mathsf{CP}} \mathcal{I}$.

Assume that there is a model $\mathcal{J}$ of KB with $\mathcal{J} <_{\mathsf{CP}} \mathcal{I}$. First we show that for some $\mathcal{J}$-assignment $\alpha^{\mathcal{J}}$, $(\mathcal{J}, \alpha^{\mathcal{J}})$ is a solution for $S_{KB'}[\iota/x]$, where the knowledge base KB' is constructed according to Algorithm 1. Due to $\mathcal{J} <_{\mathsf{CP}} \mathcal{I}$ we know that $\Delta^{\mathcal{J}} = \Delta^{\mathcal{I}}$ and $a^{\mathcal{J}} = a^{\mathcal{I}}$ for all individuals $a \in \Delta^{\mathcal{I}}$, and that for some $\tilde{A} \in M$ there is an element $\iota^{\mathcal{J}} \in \Delta^{\mathcal{J}}$ which is in $\tilde{A}^{\mathcal{I}}$ but not in $\tilde{A}^{\mathcal{J}}$. Let $\alpha^{\mathcal{J}}_{x,\iota}$ be a $\mathcal{J}$-assignment with $\alpha^{\mathcal{J}}_{x,\iota}(x) = \iota^{\mathcal{J}}$. Since $\mathcal{J}$ is a model of KB, $(\mathcal{J}, \alpha^{\mathcal{J}}_{x,\iota})$ is a solution for S_{KB} due to Lemma 1. Moreover, as the individual ι is new to KB' and $KB \subset KB'$ by construction of KB', the replacement of ι by x does not affect any constraint in S_{KB} and we have that $S_{KB} \subset S_{KB'}[\iota/x]$. Hence, it suffices to show that the constraints in $S_{KB'}[\iota/x] \setminus S_{KB}$ are satisfied by $(\mathcal{J}, \alpha^{\mathcal{J}}_{x,\iota})$. For this purpose, we consider the axioms in $KB' \setminus KB$ that are inserted into KB' by Algorithm 1, and that can be a) concept inclusion axioms of the form $\tilde{A} \sqsubseteq \{a_1, \ldots, a_n\}$, or b) the concept assertion axiom $(\bigsqcup_{\tilde{A}} D_{\tilde{A}})(\iota)$ with disjuncts $D_{\tilde{A}}$ of the form $\neg \tilde{A} \sqcap \{a_1, \ldots, a_n\}$, for individuals $\{a_i \mid a_i : \tilde{A} \in S\}$ with $i \in \{1, \ldots, n\}$.

- a) For every $\tilde{A} \in M$, KB' contains an axiom $\tilde{A} \sqsubseteq \{a_1, \ldots, a_n\}$ with individuals a_i that occur in concept constraints of the form $a_i : \tilde{A}$ within S. Since S is a completion, in any constraint of the form $x : \tilde{A}$ the variable x has been replaced by an individual $a \in N_I^S$ in S due to the $\longrightarrow_{<_{\mathsf{CP}}}$-rule, such that for any constraint $o : \tilde{A} \in S$ we have that $o = a_i$ for some $i \in \{1, \ldots, n\}$. Since $\mathcal{I}$ is obtained by construction from S, we have that $\tilde{A}^{\mathcal{I}} = \{\alpha^{\mathcal{I}}(a_1), \ldots, \alpha^{\mathcal{I}}(a_n)\} = \{a_1^{\mathcal{I}}, \ldots, a_n^{\mathcal{I}}\}$. Since $\tilde{A}^{\mathcal{J}} \subseteq \tilde{A}^{\mathcal{I}}$ holds by assumption and $a_i^{\mathcal{I}} = a_i^{\mathcal{J}}$ for all individuals a_i, $\mathcal{J}$ satisfies $\tilde{A}^{\mathcal{J}} \subseteq \{a_1^{\mathcal{J}}, \ldots, a_n^{\mathcal{J}}\}$, and thus, the axiom $\tilde{A} \sqsubseteq \{a_1, \ldots, a_n\}$ for every $\tilde{A} \in M$. If there are no constraints $a_i : \tilde{A}$ in S' then $\tilde{A}^{\mathcal{I}} = \emptyset$ and the respective axiom has the form $\tilde{A} \sqsubseteq \bot$. Hence, $(\mathcal{J}, \alpha^{\mathcal{J}}_{x,\iota})$ satisfies all the constraints $\forall x.x : C$ that result from these inclusion axioms in $S_{KB'}[\iota/x]$.
- b) Furthermore, due to the concept assertion $(\bigsqcup_{\tilde{A}} D_{\tilde{A}})(\iota)$ in KB', $S_{KB'}[\iota/x]$ contains the constraint $x : \bigsqcup_{\tilde{A}} D_{\tilde{A}}$ with disjuncts $D_{\tilde{A}}$ of the form $\neg \tilde{A} \sqcap \{a_1, \ldots, a_n\}$. As in a), we know that $\tilde{A}^{\mathcal{I}} = \{a_1^{\mathcal{J}}, \ldots, a_n^{\mathcal{J}}\}$. As for some $\tilde{A} \in M$ the element $\iota^{\mathcal{J}}$ is in $\tilde{A}^{\mathcal{I}}$ but not in $\tilde{A}^{\mathcal{J}}$, we have that $\alpha^{\mathcal{J}}_{x,\iota}(x) \in \tilde{A}^{\mathcal{I}} \setminus \tilde{A}^{\mathcal{J}}$, and thus, $\alpha^{\mathcal{J}}_{x,\iota}(x) \in (\{a_1^{\mathcal{J}}, \ldots, a_k^{\mathcal{J}}\} \setminus \tilde{A}^{\mathcal{J}}) = (\Delta^{\mathcal{J}} \setminus \tilde{A}^{\mathcal{J}}) \cap \{\alpha^{\mathcal{J}}_{x,\iota}(a_1), \ldots, \alpha^{\mathcal{J}}_{x,\iota}(a_n)\}$. Hence, the pair $(\mathcal{J}, \alpha^{\mathcal{J}}_{x,\iota})$ satisfies the constraint $x : \bigsqcup_{\tilde{A}} D_{\tilde{A}}$ for some $\tilde{A} \in M$ with $\tilde{A}^{\mathcal{J}} \subset \tilde{A}^{\mathcal{I}}$, as one of its disjuncts is satisfied.

Having shown that $(\mathcal{J}, \alpha^{\mathcal{J}}_{x,\iota})$ is a solution for $S_{KB'}[\iota/x]$, from Proposition 3($\Rightarrow$) and from Proposition 4($\Rightarrow$) it follows that there is a clash-free completion of

$S_{KB'}[\iota/x]$. Hence, S must contain a preference clash, which contradicts the existence of $\mathcal{J}$. □

As a direct result of the propositions 2, 3, 5 and Lemma 2, we obtain that the presented calculus provides an effective procedure for reasoning with circumscribed knowledge bases, reflected by the following theorem.

Theorem 1 (soundness/completeness). *Let KB be an $\mathcal{ALCO}$ knowledge base, cp be a circumscription pattern and C be an $\mathcal{ALCO}$ concept. C is satisfiable with respect to $circ_{CP}(KB)$ if and only if the algorithm based on Table 1 results in a clash-free completion of the constraint system $S_{KB} \cup \{x : C\}$.*

By Theorem 1, the proposed tableaux calculus is a decision procedure for reasoning in $\mathcal{ALCO}$ with concept circumscription.

Although we did not perform a complexity analysis of our calculus, we want to report that in [2] the theoretical runtime complexity for reasoning with concept-circumscribed knowledge bases was shown to be $\mathsf{NEXP}^{\mathsf{NP}}$ for the cases of $\mathcal{ALCIO}$ and $\mathcal{ALCQO}$.

4   Conclusion

We have presented a tableaux calculus for concept satisfiability with respect to circumscribed DL knowledge bases in the logic $\mathcal{ALCO}$. Building on tableaux procedures for classical DLs, the calculus checks a constraint system not only for clashes due to inconsistent concept assertion and individual naming, but also for preference clashes, which occur whenever the model associated with the produced constraint system is not minimal with respect to the preference relation $<_{\mathsf{CP}}$. This check is performed by testing a specifically constructed classical $\mathcal{ALCO}$ knowledge base for satisfiability, which requires reasoning in classical DL with nominals and equality between individuals.

We have proved that the presented calculus is sound and complete for verifying concept satisfiability in circumscriptive $\mathcal{ALCO}$. By this we have devised a first guided algorithmisation for description logic with circumscription that integrates well with state of the art tableaux methods for DL reasoning. This lays a basis for further investigations on optimisation of the calculus within the framework of tableaux procedures as a guided way for model construction. We have implemented a first prototype[4] of the calculus in Java that works together with ontology development tools, such as Protégé, via the DIG interface.

As future work we see the update of the calculus to support more expressive features, such as prioritisation between minimised concepts or the remaining constructs of the Web Ontology Language OWL [18]. Moreover, optimisation issues need to be addressed to obtain a more efficient reasoning procedure. First ideas for specific optimisations would be to employ model caching techniques for the inner classical tableaux step as KB' might be identical in multiple cases, to postpone assertions of individuals to minimised predicates in order to avoid constructing non-minimal models, and to exploit early closing of tableaux branches

---

[4] Available at http://www.fzi.de/downloads/wim/sgr/CircDL.zip

through preference clash detection. Besides these, it would be interesting to see how well preferential tableaux performs when included in optimised state-of-the-art DL reasoners.

References

1. Baader, F., Hollunder, B.: Embedding Defaults into Terminological Knowledge Representation Formalisms. Journal of Automated Reasoning 14(1), 149–180 (1995)
2. Bonatti, P., Lutz, C., Wolter, F.: Expressive Non-Monotonic Description Logics Based on Circumscription. In: Proc. of the Tenth Int. Conf. on Principles of Knowledge Representation and Reasoning (KR 2006), pp. 400–410 (2006)
3. Buchheit, M., Donini, F.M., Schaerf, A.: Decidable Reasoning in Terminological Knowledge Representation Systems. J. Artif. Intell. Res (JAIR) 1, 109–138 (1993)
4. de Kleer, J., Konolige, K.: Eliminating the Fixed Predicates from a Circumscription. Artif. Intell. 39(3), 391–398 (1989)
5. Donini, F.M., Lenzerini, M., Nardi, D., Nutt, W., Schaerf, A.: An Epistemic Operator for Description Logics. Artificial Intelligence 100(1-2), 225–274 (1998)
6. Donini, F.M., Nardi, D., Rosati, R.: Description Logics of Minimal Knowledge and Negation as Failure. ACM Trans. on Computational Logic 3(2), 177–225 (2002)
7. Grimm, S.: Semantic Matchmaking with Nonmonotonic Description Logics. IOS Press, Amsterdam (2008)
8. Grimm, S., Hitzler, P.: Defeasible Inference with OWL Ontologies. In: Proceedings of the ESWC 2007 Workshop on Advancing Reasoning on the Web: Scalability and Commonsense, ARea 2008 (June 2008)
9. Grimm, S., Hitzler, P.: Semantic Matchmaking of Resources with Local Closed-World Reasoning. Int. Journal of eCommerce (IJEC) 12(2), 89–126 (2008)
10. Hitzler, P., Krötzsch, M., Rudolph, S.: Foundations of Semantic Web Technologies. Chapman & Hall/CRC (2009)
11. Horrocks, I., Sattler, U.: Ontology Reasoning in the SHOQ(D) Description Logic. In: Proc. IJCAI 2001, pp. 199–204 (2001)
12. Knorr, M., Alferes, J.J., Hitzler, P.: A Coherent Well-founded Model for Hybrid MKNF Knowledge Bases. In: Proceedings of the 18th European Conference on Artificial Intelligence, ECAI 2008, pp. 99–103. IOS Press, Amsterdam (2008)
13. McCarthy, J.: Circumscription – A Form of Non-Monotonic Reasoning. Artificial Intelligence 13(1–2), 27–39 (1980)
14. Motik, B., Rosati, R.: A Faithful Integration of Description Logics with Logic Programming. In: Proc. of the 20th Intern. Joint Conference on Artificial Intelligence (IJCAI 2007), Hyderabad, India, January 2007, pp. 477–482. AAAI Press, Stanford (2007)
15. Niemelä, I.: Implementing Circumscription Using a Tableau Method. In: Proc. of the 12th Europ. Conf. on Artificial Intelligence (ECAI 1996). J. Wiley & Sons, Chichester (1996)
16. Schaerf, A.: Reasoning with Individuals in Concept Languages. Data Knowl. Eng. 13(2), 141–176 (1994)
17. Schmidt-Schauß, M., Smolka, G.: Attributive Concept Descriptions with Complements. Artif. Intell. 48(1), 1–26 (1991)
18. W3C OWL Working Group. OWL 2 Web Ontology Language: Document Overview (2009), http://www.w3.org/TR/owl2-overview/

A Reasoner for Simple Conceptual Logic Programs[*]

Stijn Heymans, Cristina Feier, and Thomas Eiter

Knowledge-Based Systems Group, Institute of Information Systems
Vienna University of Technology
Favoritenstrasse 9-11, A-1040 Vienna, Austria
{heymans,feier,eiter}@kr.tuwien.ac.at

Abstract. Open Answer Set Programming (OASP) can be seen as a
framework to represent tightly integrated combined knowledge bases of
ontologies and rules that are not necessarily DL-safe. The framework
makes the open-domain assumption and has a rule-based syntax supporting negation under a stable model semantics. Although decidability
of different fragments of OASP has been identified, reasoning and effective algorithms remained largely unexplored. In this paper, we describe
an algorithm for satisfiability checking of the fragment of *simple Conceptual Logic Programs* and provide a BProlog implementation. To the
best of our knowledge, this is the first implementation of a (fragment) of
a framework that can tightly integrate ontologies and non-DL-safe rules
under an expressive nonmonotonic semantics.

1 Introduction

Integrating Description Logics (DLs) with rules for the Semantic Web has received considerable attention over the past years with tightly-coupled approaches
such as *Description Logic Programs* [7,13][1], *DL-safe rules* [14], *r-hybrid knowledge bases* [16], $\mathcal{DL}{+}log$ [15], and *Description Logic Rules* [12], as well as loosely-
coupled approaches such as *dl-programs* [4]. In [8], we proposed a tightly-coupled
approach to combine knowledge bases using a similar semantics as Rosati's r-
hybrid knowledge bases. However, instead of syntactically restricting the rule set
to DL-safe[2] rules, we required the rule set to fall into a decidable fragment of
Open Answer Set Programming (OASP) [11].

OASP is a language that combines attractive features from both the DL and
the Logic Programming (LP) world: an open domain semantics from the DL
side allows for stating generic knowledge, without mentioning actual constants,

---

[*] A preliminary version of this work, without the implementation, was presented at
the 3rd Int. Workshop on Applications of Logic Programming to the (Semantic) Web
and Web Services (ALPSWS 2008) for a limited audience.

[1] Note that even though the approaches of [7] and [13] carry the same name, they are
different.

[2] A rule is DL-safe if each variable appears positively in a non-DL atom, where a
non-DL atom is an atom that is not formed using a DL concept as predicate.

A. Polleres and T. Swift (Eds.): RR 2009, LNCS 5837, pp. 55–70, 2009.
© Springer-Verlag Berlin Heidelberg 2009

and a rule-based syntax from the LP side supports nonmonotonic reasoning via *negation as failure*. Decidability of several fragments of OASP was identified by syntactically restricting the shape of logic programs, while carefully safe-guarding expressiveness, e.g., *Conceptual Logic Programs (CoLPs)* [9] and *Forest Logic Programs (FoLPs)* [10].

Decidability of combined knowledge bases $KB = \langle \Phi, P \rangle$, where Φ is a theory in a DL $\mathcal{DL}$ and P is a program in a decidable fragment $\mathcal{L}'$ of OASP, can then be shown whenever $\mathcal{DL}$ is reducible to $\mathcal{L}'$. For example, if Φ is a $\mathcal{SHIQ}$ DL theory and P is a CoLP, KB can be translated to a CoLP P' such that satisfiability is preserved. Relying on such OASP fragments, rules can then deduce results about anonymous individuals, in contrast with (weakly) DL-safe rules where deductions are only taking into account the instances in the knowledge base. OASP is thus a suitable alternative integrative formalism for both ontologies and rules. To make it a suitable implementation vehicle though, the lack of effective reasoning procedures for decidable fragments of OASP has to be overcome.

In this paper, we describe a terminating, sound and complete algorithm for satisfiability checking in a fragment of Conceptual Logic Programs, and report on a prototype implementation.

The major contributions of the paper can be summarized as follows:

- We identify a fragment of Conceptual Logic Programs (CoLPs), called *simple CoLPs*, that are expressive enough to simulate the DL $\mathcal{ALCH}$.
- We define an algorithm for deciding satisfiability, inspired by tableaux-based methods from DLs, that constructs a finite representation of an open answer set. We show that this algorithm is terminating, sound, complete, and runs in nondeterministic exponential time.
- We provide a prototypical implementation in BProlog [2]. Note that, to date, this provides the basis for the first implementation of a non-trivial tightly-coupled approach that supports a minimal model semantics and is not depending on (a variant of) DL-safeness.

Detailed proofs and discussion of related work can be found in [5].

2   Preliminaries

We recall the open answer set semantics from [11]. *Constants* $a, b, c, \ldots$, *variables* $X, Y, \ldots$, *terms* $s, t, \ldots$, and *atoms* $p(t)$ are defined as usual. A *literal* is an atom $p(t)$ or a *naf-atom* $not\ p(t)$. For a set α of literals or (possibly negated) predicates, $\alpha^+ = \{l \mid l \in \alpha, l$ an atom or a predicate$\}$ and $\alpha^- = \{l \mid not\ l \in \alpha, l$ an atom or a predicate$\}$. For a set X of atoms, $not\ X = \{not\ l \mid l \in X\}$. For a set of (possibly negated) predicates α, we will often write $\alpha(x)$ for $\{a(x) \mid a \in \alpha\}$ and $\alpha(x, y)$ for $\{a(x, y) \mid a \in \alpha\}$.

A *program* is a countable set of rules $\alpha \leftarrow \beta$, where α and β are finite sets of literals. The set α is the *head* of the rule and represents a disjunction, while β is called the *body* and represents a conjunction. If $\alpha = \emptyset$, the rule is called a *constraint*. *Free rules* are rules $q(\boldsymbol{X}) \vee not\ q(\boldsymbol{X}) \leftarrow$ for variables $\boldsymbol{X}$; they enable a choice for the inclusion of atoms. We call a predicate q *free* in a program if there is a *free rule* $q(\boldsymbol{X}) \vee not\ q(\boldsymbol{X}) \leftarrow$ in the program. Atoms, literals, rules, and programs that do not contain variables are *ground*. For a rule or a program χ, let $\mathcal{C}(\chi)$ be the constants in χ, $\mathcal{V}(\chi)$ its variables, and $\mathcal{P}(\chi)$ its predicates with $\mathcal{P}^1(\chi)$ the unary and $\mathcal{P}^2(\chi)$ the binary predicates. A *universe* U for a program P is a non-empty countable set $U \supseteq \mathcal{C}(P)$. We denote by P_U the ground program obtained from P by substituting every variable in P by every possible element in U. Let $\mathcal{B}_P$ ($\mathcal{L}_P$) be the set of atoms (literals) that can be formed from a ground program P. An *interpretation* I of a ground P is any subset of $\mathcal{B}_P$ and it is an *answer set* of P if the usual definition holds, see, e.g., [6].

In the following, programs are assumed to be finite; infinite programs only appear as byproducts of grounding a finite program with an infinite universe. An *open interpretation* of a program P is a pair (U, M) where U is a universe for P and M is an interpretation of P_U. An *open answer set* of P is an open interpretation (U, M) of P with M an answer set of P_U. For example, an open answer set of the rules $p \leftarrow not\ q(X)$, $p \leftarrow not\ p$, and $q(a) \leftarrow$ is $(\{a, b\}, \{q(a), p\})$. Note that if the universe U is constrained to $\{a\}$, the set of constants appearing in the rules, the set of rules is inconsistent, i.e., there is no answer set. An n-ary predicate p in P is *satisfiable*, if there is an open answer set (U, M) of P and a $\boldsymbol{x} \in U^n$ such that $p(\boldsymbol{x}) \in M$.

We define *trees* as tuples $T = (N_T, A_T)$, with N_T the set of *nodes* of the T, and A_T the set of edges of T. We denote the root of T with ε. For a node $x \in N_T$, we denote with $succ_T(x)$ the *successors* of x. The *arity* of T is the maximum number of successors any node has in T. For $x, y \in N_T$, $x \leq y$ iff x is an ancestor of y (possibly $x = y$). As usual, $x < y$ if $x \leq y$ and $y \nleq x$.

3   Simple Conceptual Logic Programs

In [9], *Conceptual Logic Programs (CoLPs)*, were defined as a syntactical fragment of logic programs for which satisfiability checking under the open answer set semantics is decidable. We restrict this fragment by disallowing the occurrence of inequalities and inverse predicates, and by restricting the dependencies between predicate symbols which appear in the program.

Definition 1. *A* simple conceptual logic program (simple CoLP) *is a program with only unary and binary predicates, without constants, and such that any rule is either (i) a* free rule, *(ii) a* unary rule

$$a(X) \leftarrow \beta(X), \left(\gamma_m(X, Y_m), \delta_m(Y_m)\right)_{1 \leq m \leq k} \tag{1}$$

where for all m, $\gamma_m^+ \neq \emptyset$, or (iii) a binary rule

$$f(X, Y) \leftarrow \beta(X), \gamma(X, Y), \delta(Y) \tag{2}$$

with $\gamma^+ \neq \emptyset$.

Furthermore, for such a set of rules P, *let* $D(P)$ *be the* marked predicate dependency graph: $D(P)$ *has as nodes the predicates from* P *and as edges tuples* (p, q) *if there is either a rule (1) or a rule (2) with a head predicate* p *and a positive body predicate* q. *An edge* (p, q) *is marked, if* q *is a predicate in some* δ_m *for rules (1), respectively* δ *for rules (2). In order for* P *to be a simple CoLP,* $D(P)$ *must not contain any cycle that has a marked edge.*

Intuitively, the free rules allow for a free introduction of atoms (in a first-order way) in answer sets; unary rules consist of a root atom $a(X)$ that is motivated by a syntactically tree-shaped body, and binary rules motivate $f(X, Y)$ for x and its 'successor' Y by a body that only considers literals involving Y and Y. The restriction on $D(P)$ ensures that there is no path from some $p(x)$ to some $p(y)$ in the positive atom dependency graph of P_U, where $p \in \mathcal{P}^1(P)$, and x, y are from an arbitrary universe U. Indeed, observe that any marked cycle in $D(P)$ contains a unary predicate and thus corresponds to a path from some $p(x)$ to some $p(y)$ in the atom dependency graph of P_U. Consider the program P:

$$r_1 : \quad a(X) \leftarrow b(X), f(X, Y), not\ a(Y)$$
$$r_2 : \quad b(X) \leftarrow a(X)$$
$$r_3 : f(X, Y) \leftarrow g(X, Y), b(Y)$$

The marked dependency graph is depicted in Figure 1.

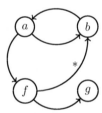

Fig. 1. Marked dependency graph of a CoLP P

There is a single marked edge in $D(P)$, viz. the edge (f, b). Although the cycle (a, b, a) does not contain any marked edge, the cycle (a, f, b, a) contains the marked edge (f, b) and thus P is not a simple CoLP. If we leave r_3 out, the remaining program is a simple *CoLP*. Intuitively, simple CoLPs allow for local recursion (through X) and non-local negative recursion (through Y, see r_1), but not for non-local positive recursion.

As satisfiability checking of CoLPs is EXPTIME-complete [9], checking satisfiability of simple CoLPs is in EXPTIME. Moreover, using a similar simulation of DLs as in [9], one can show that satisfiability checking of $\mathcal{ALCH}^3$ concepts w.r.t. a $\mathcal{ALCH}$ TBox can be reduced to satisfiability checking of a unary predicate w.r.t. a simple CoLP. Thus, satisfiability checking of unary predicates w.r.t. simple CoLPs is EXPTIME-complete.

---

[3] For a definition of $\mathcal{ALCH}$, we refer the reader to [3].

4   Illustration of the Algorithm

Before formally defining the algorithm for satisfiability checking of a unary predicate p w.r.t a simple CoLP P, we show an example run of the algorithm.

As in tableaux algorithms for Description Logics, the algorithm's basic data structure is a *labeled tree* where the labels are sets of positive and negative predicates. Indeed, the algorithm essentially tries to construct a tableau for the predicate and the program, thus representing an open answer set by a finite structure. Additionally – to take care of minimality – we keep track of the dependencies between atoms by means of a dependency graph.

Expansion rules expand the labeled tree in accordance with the simple CoLP, to construct a partial open answer set, based on the following principles.

1. The occurrence of a positive predicate p in a label has to be motivated by making the body of a rule with head predicate p true in the labeled tree. This principle is similar to the principle of *foundedness* in ordinary Answer Set Programming. We keep track of those dependencies in a positive atom dependency graph that must be acyclic (no atom can motivate itself).
2. The occurrence of a negative predicate *not p* has to be justified by showing that no body of a rule with head predicate p is true in the labeled tree. This ensures *satisfaction* of rules.

Applicability rules constrain the use of expansion rules:

– We can only expand nodes that have a *saturated* parent node, i.e., the parent node has to be fully expanded: it should contain either positive or negative information about all unary predicates in its label and about all binary predicates in its outgoing edges, and no expansion rules are applicable on that parent node.
– Similarly as in DL tableaux, we have a *blocking* rule, that takes care of stopping an expansion on a node x if the label of an ancestor y of x, $y < x$, subsumes the label of x. We thus avoid infinite expansions, and represent a possibly infinite open answer set by a finite structure.
– Additionally, the *caching rule* prohibits to expand a node if the label of a node somewhere else in the tree (thus, not necessarily an ancestor) subsumes the current label. They are not necessary to make the algorithm sound, complete, and terminating, but they make the completion tree smaller.

The algorithm succeeds (p is satisfiable w.r.t. to P), if a labeled tree T and a dependency graph G can be built such that the tree does not contain labels with contradiction (for example, *not p* and p in one label) and the dependency graph is not cyclic.

As an example we take the simple CoLP P and check satisfiability of $a \in \mathcal{P}^1(P)$:

$$r_1 : a(X) \leftarrow f(X, Y_1), b(Y_1), not\ f(X, Y_2), g(X, Y_2), b(Y_2)$$
$$r_2 : b(X) \leftarrow f(X, Y), not\ c(Y)$$
$$r_3 : c(X) \leftarrow not\ b(X)$$

with f and g free.

The initially labeled tree consists of a single node ε with label a:

The dependency graph G contains the corresponding atom $a(\varepsilon)$. Note that we draw the tree with each predicate in the label superscripted with an indication of its expansion status – a^u means a is *unexpanded*.

Recall that we want to construct a (partial) open answer set. As in ordinary ASP, an atom that is in an (open) answer set has to be *motivated* by a rule, i.e., there has to be a rule with head predicate a that has a true body. The *Expand Unary Positive* rule does exactly this (see (i) in Section 5.1): it selects a rule with head predicate a and expands the labeled tree according to the body of the rule. In the example, the only relevant rule is r_1. To make its body true, we extend the tree with 2 successors (corresponding to Y_1 and Y_2), such that both successors are labeled with b and its edges with f, and *not* f and g respectively:

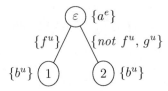

The predicate a is now expanded; the root has two children, 1 and 2, with the unexpanded b in their label as well as unexpanded binary predicates on the outgoing edges.

The root node is not saturated yet (see (vii) in Section 5.2), i.e., there are unary predicates in P of which we do not have a positive or negative occurrence in the label, neither have all choices been made for ε's outgoing edges and the binary predicates in P. Moreover, there are predicates in the successors of the root that are not expanded yet. Thus, we cannot start expanding the root's children.

Note that all binary positive predicates in the outgoing edges from ε are free such that they are trivially minimally motivated (we do not need to make any bodies true to motivate the presence of either f or g). Remains the negative predicate *not* f on the edge $(\varepsilon, 2)$. In order to justify the presence of *not* f, one has to negate the bodies of *all* binary rules that have f as head predicate. As there are no such rules, *not* f can be set to expanded and considered justified.

The root is still not saturated after the above operations (it does not contain all positive or negative versions of the unary predicates). In order to mend this, we *choose a unary predicate* (see (iii) in Section 5.1). The algorithm picks a predicate, b for example, and adds its negation to the content of ε. Note that the algorithm can pick either b or *not* b. Currently, we use the naive heuristics, though, that it is more likely that something is not in a node.

Additionally to its unexpanded superscript u, we keep track of all the rules with head b (in this case only r_2). Recall that we are trying to construct a partial open answer set with as a universe (part of) the tree we are constructing. In the example, the universe is currently $\varepsilon, 1$, and 2. Intuitively, the negative presence of b has to be justified by making sure all bodies, ground with this universe, of rules that have head predicate b are made false. Otherwise, there would be a rule that has a true body and thus forces us to introduce b instead of $not\ b$. In the example program, the body of r_2 is $f(X,Y), not\ c(Y)$ which becomes true for the current tree, if there is a successor of ε that connects with ε via f and where $not\ c$ holds. Thus, in order to make sure that this body does not become true, we have to enforce that for each successor of ε either it is not connected via f or its label contains c. The example is simplistic: if the body would be $f(X,Y_1), c(Y_1), g(X,Y_2), d(Y_2)$ one would need to show for each 2 successors y_1, y_2 of ε (the node with which X is unified) that f is not present on the outgoing edge to y_1 or that c is not in the label of y_1 or that g is not present on the outgoing edge to y_2 or that d is not in the label of y_2.

Note that in order to justify a negative unary predicate, we need knowledge of all possible successors of a node. We only obtain this knowledge after all positive unary predicates that are or will be appearing in this node have been expanded (recall that positive unary predicates might introduce new successors, as did a). The algorithm thus *tries to complete the node first* with either negative or positive predicates, and only starts expanding negative predicates if all positive ones have been expanded. In the current tree, we are thus still missing a choice for c. By default, we again choose $not\ c$ which has to be justified by r_3.

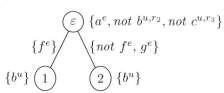

Now, all unary predicates are present in ε and all positive ones (a) are expanded, i.e., with the current label no more successors can be introduced, such that we can start expanding the negative predicates. We choose to justify $not\ c$ in the root ε by making the body of r_3 false, i.e., b has to be in ε. Clearly, this would cause a contradiction, such that we backtrack on the choice for c and include c in ε. Now, there is a positive predicate c that is not yet expanded, such that before expanding $not\ b^{u,r_2}$, we have to expand c as c might introduce new successors that influence the justification of $not\ b$. Clearly, c can be motivated, using r_2, as $not\ b$ is present in ε.

Thus, we can now justify $not\ b^{u,r_2}$, i.e., for each successor of ε we need that either f is not present in the outgoing edge or c has to be present in the label of that successor (see rule (ii) in Section 5.1). Thus, as $not\ f$ is already in the label of the edge from ε to 2 and f is in the label of the edge from ε to 1 , we only have to add the unexpanded c to 1. We have the following tree:

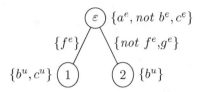

To finally saturate ε one can see that either g or *not* g is missing in the edge $(\varepsilon, 1)$. The *Choose a Binary Predicate* rule (see (vi) in Section 5.1) adds *not* g.

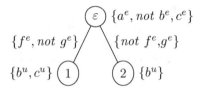

Now, one can see that to expand c^u in node 1 one needs *not* b in 1, by r_3. However, b^u is already present. The algorithm backtracks, i.e., removes *not* g^e again, removes c^e again, up until the wrong choice of the *Choose a unary predi-cate* rule, to introduce b^u instead of *not* b^{u, r_2}.[4] Expanding this b^u using rule r_2 introduces *not* c^u in node 1, and choosing *not* c in ε and 1 and *not* g in the edge $(\varepsilon, 1)$, leads to

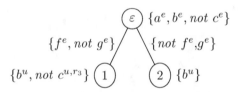

Now that ε is saturated, one can consider node 1. One sees that the label of 1 is a subset of the label of ε (not taking into account the expansion status), and, intuitively, one can use the expansions used on ε to further expand 1 simi-larly. This technique is called *blocking* and is similar to the blocking used in DL tableaux methods: node 1 is blocked by ε.

Due to the condition on simple CoLPs that $D(P)$ does not contain cycles with marked edges, one can construct an open answer set by rolling out at 1 the subtree that resides at ε. Similarly, one can see that node 2 is blocked by ε as well.

The constructed dependency graph G is

$$\{(a(\varepsilon), f(\varepsilon, 1)), (a(\varepsilon), b(1)), (a(\varepsilon), f(\varepsilon, 2)), (a(\varepsilon), b(2)), (b(\varepsilon), f(\varepsilon, 1))\}$$

i.e., the graph that keeps track of the positive dependencies. We thus have con-structed a labeled tree where no rules are applicable anymore, that does not

---

[4] Note that if we would have picked *not* c before *not* b above, we could have avoided this backtracking. It is future work to investigate how to optimize backtracking.

contain contradictions, and where the dependency graph is acyclic. The algorithm concludes that a is satisfiable w.r.t. P. This is indeed correct.

If we replace in P the rule r_1 with

$$r_1' : a(X) \leftarrow a(X), f(X, Y_1), b(Y_1),$$
$$not\ f(X, Y_2), g(X, Y_2), b(Y_2)$$

we end up with the same labeled tree; however, the dependency graph G' is $G \cup \{(a(\varepsilon), a(\varepsilon))\}$ and thus cyclic. One can check that a is indeed not satisfiable w.r.t. P, as no open answer set exists that contains some $a(x)$.

Note that we make extensive use of the *choose* rules and the concept of *saturation* to complete the labels of nodes such that they contain either the negative or positive versions of all unary predicates in the program (and similarly for binary predicates in the edges). Assume we would not do such a completion, i.e., we would drive the expansion of the nodes purely on what is entailed by the predicate p to satisfy and would not choose predicates that we apparently do not need to satisfy p.

Such a driven computation would not guarantee the global satisfaction of the program. Instead, it explores a partial solution pattern that would be able to satisfy a predicate. To make sure that this partial solution pattern can be extended to an open answer set, we have to complete it repeatedly. For example, if in the example program, we would add a rule $d(X) \leftarrow not\ d(X)$ the program would not have any open answer sets. However, an expansion that starts with a and does not make choices for the "non-relevant" predicate d would wrongly succeed. Thus, we need to make a choice for d or $not\ d$ in every node. Once we do, one would not be able to construct a labeled tree without contradictions.

In tableaux methods for DLs, one does not have this problem. The TBox (the program in our setting) is usually internalized and the satisfiability of the resulting concept is checked. There is no need to make extensive nondeterministic choices for each concept name in this concept expression.

5   Algorithm

In this section, we define a sound, complete, and terminating algorithm for satisfiability checking w.r.t. simple CoLPs.

For every non-free predicate q and a simple CoLP P, let P_q be the rules of P that have q as a head predicate. For a predicate p, $\pm p$ denotes p or $not\ p$, whereby multiple occurrences of $\pm p$ in the same context refer to the same symbol (either p or $not\ p$). The negation of $\pm p$ (in a given context) is $\mp p$, that is, $\mp p = not\ p$ if $\pm p = p$ and $\mp p = p$ if $\pm p = not\ p$.

The basic data structure for our algorithm is a *completion structure*.

Definition 2. *A completion structure for a simple CoLP P is a tuple $\langle T, G,$ CT, ST$\rangle$ where $T = (N_T, A_T)$ is a tree, $G = \langle V, A \rangle$ is a directed graph with nodes $V \subseteq \mathcal{B}_{P_{N_T}}$ and edges $A \subseteq V \times V$, and* CT $: N_T \cup A_T \to 2^{\mathcal{P}(P) \cup not(\mathcal{P}(P))}$ *and* ST $: \{(x, \pm q) \mid \pm q \in$ CT$(x), x \in A_T\} \cup \{(x, q) \mid q \in$ CT$(x), x \in N_T\} \cup$

$\{(x, not\ q, r) \mid not\ q \in \text{CT}(x), x \in N_T, r \subset P_q\}$ → $\{exp, unexp\}$ *are labeling functions.*

The tree T together with the labeling functions is used to represent/construct a tentative tree-shaped open answer set, where N_T represents the tentative universe. $G = \langle V, A \rangle$ is a directed graph which helps to keep track of dependencies between atoms in the constructed model, where V represents the tentative model (such a structure enables checking of the minimality requirement: no atom should depend on itself). The role of the labeling functions is as follows:

- The *content* function CT maps a node of the tree to a set of (possibly negated) unary predicates and an edge of the tree to a set of (possibly negated) binary predicates such that $\text{CT}(x) \subseteq \mathcal{P}^1(P) \cup not(\mathcal{P}^1(P))$ if $x \in N_T$, and $\text{CT}(x) \subseteq \mathcal{P}^2(P) \cup not(\mathcal{P}^2(P))$ if $x \in A_T$. The presence of a predicate symbol p (resp. negated predicate symbol *not p*) in the content of some node/edge x of T indicates that $p(x)$ is part (resp. not part) of the tentative model represented by T.
- The *status* function ST attaches to every (possibly negated) predicate which appears in the content of an edge x and every positive predicate in the content of a node x a status value which indicates whether the predicate has already been expanded in that node/edge. As indicated in Section 4, for negative predicates in nodes, we additionally keep track of the rule that justifies the negative occurrence.

The algorithm starts with defining an *initial completion structure* which basically captures the constraint that p, the predicate checked to be satisfiable is in the content of some node x, or in other words $p(x)$ is in the open answer set for some individual x.

Definition 3. *An* initial completion structure *given a unary predicate p and a simple CoLP P is a completion structure $\langle T, G, \text{CT}, \text{ST} \rangle$ with $T = (N_T, A_T)$, $N_T = \{\varepsilon\}$, $A_T = \emptyset$, $G = \langle V, A \rangle$, $V = \{p(\varepsilon)\}$, $A = \emptyset$, $\text{CT}(\varepsilon) = \{p\}$, and $\text{ST}(\varepsilon, p) = unexp$.*

Next, we show how to evolve by means of *expansion rules* an initial completion structure of p and P to an expanded clash-free structure that corresponds to a finite representation of an open answer set in case p is satisfiable w.r.t. P. *Applicability rules* state the necessary conditions to to apply these expansion rules. Note that when multiple expansion rules can be applied, one is chosen non-deterministically.

5.1   Expansion Rules

The expansion rules update the completion structure by making explicit what is needed for justifying the presence or absence of a certain atom in the partial model represented by the current completion. We first define a recurring operation in the expansion rules which describes the necessary updates in the completion structure whenever justifying a literal l in the current model imposes the presence of a new literal $\pm p(z)$ in the model. In such a case $\pm p$ is inserted in the content of z if it is not already there and marked as unexpanded, and in case $\pm p(z)$ is an atom, it should be a node in G. Moreover, if l is also an atom, a new edge from l to $\pm p(z)$ should be created to indicate the dependency of l on $\pm p(z)$ in the model. Formally:

- if $\pm p \notin \text{CT}(z)$, then $\text{CT}(z) = \text{CT}(z) \cup \{\pm p\}$ and $\text{ST}(z, \pm p) = unexp$,
- if $\pm p = p$ and $\pm p(z) \notin V$, then $V = V \cup \{\pm p(z)\}$,
- if $l \in \mathcal{B}_{P_{N_T}}$ and $\pm p = p$, then $A = A \cup \{(l, \pm p(z))\}$.

As a shorthand, we denote this sequence of operations as $update(l, \pm p, z)$; more general, $update(l, \beta, z)$ for a set of (possibly negated) predicates β, denotes $\forall \pm a \in \beta$, $update(l, \pm a, z)$.

In the following, for a completion structure $\langle T, G, \text{CT}, \text{ST} \rangle$, let $x \in N_T$ and $(x, y) \in A_T$ be the node, resp. edge, under consideration.

(i) Expand unary positive. For a unary positive (non-free) $p \in \text{CT}(x)$ such that $\text{ST}(x, p) = unexp$,

- nondeterministically choose a rule $r \in P_p$ of the form (1). The rule will motivate the presence of $p(x)$ in the tentative open answer set. To this end we continue by enforcing the body of this rule to be true in the constructed completion structure.
- for the β in the body of r, $update(p(x), \beta, x)$,
- nondeterministically pick up (or define when needed) k successors for x, $(y_m)_{1 \leq m \leq k}$, such that for every $1 \leq m \leq k$: $y_m \in succ_T(x)$ or y_m is a new successor of x and T is updated: $N_T = N_T \cup \{y_m\}$, $A_T = A_T \cup \{x, y_m\}$,
- for every successor y_m of x, $1 \leq m \leq k$: $update(p(x), \gamma_m, (x, y_m))$ and $update(p(x), \delta_m, y_m)$,
- set $\text{ST}(x, p) = exp$.

(ii) Expand unary negative. Justifying a negative unary predicate $not\ p \in \text{CT}(x)$ (the absence of $p(x)$ in the constructed model) means refuting the body of every ground rule which defines $p(x)$ (a body that is true in the constructed model would otherwise enforce the presence of $p(x)$, a contradiction with the fact that $not\ p \in \text{CT}(x)$). Formally, for a unary negative $not\ p \in \text{CT}(x)$ and a rule $r \in P_p$ of the form (1) and $\text{ST}(x, not\ p, r) = unexp$ do one of:

- choose some $\pm q \in \beta$, $update(not\ p(x), \mp q, x)$, and set $\text{ST}(x, not\ p, r) = exp$, or
- if for all $p \in \mathcal{P}^1(P)$, $p \in \text{CT}(x)$ or $not\ p \in \text{CT}(x)$, and for all $p \in \text{CT}(x)$, $\text{ST}(x, p) = exp$, then for all $y_{i_1}, \ldots, y_{i_k}$ such that $(1 \leq i_j \leq n)_{1 \leq j \leq k}$, where $succ_T(x) = \{y_1, \ldots y_n\}$, do one of the following:
  - for some m, $1 \leq m \leq k$, pick up a binary (possibly negated) predicate symbol $\pm f$ from γ_m and $update(not\ p(x), \mp f, (x, y_{i_m}))$, or
  - for some m, $1 \leq m \leq k$, pick up a unary negated predicate symbol $not\ q$ from δ_m and $update(not\ p(x), q, y_{i_m})$.

Set $\text{ST}(x, not\ p, r) - exp$.

One can see that once the body of a ground version of a unary rule $r \in P_p$, for which the head term X is substituted with the current node x, is locally refuted, the bodies of all ground versions of this rule, for which X is substituted with x, are locally refuted, too. For the other refutation case, all possible groundings of a rule have to be considered and this is not possible until all successors of x are known. This is the case when all positive predicates in the content of the current node have been expanded and no positive predicate will be further inserted in $\text{CT}(x)$. If this condition is met, an iteration over all possible groundings of the rule r is triggered. For every possible grounding, one of the body literals from the non-local part of the rule (γs or δs) has to refuted.

(iii) Choose a unary predicate. If for all $q \in \text{CT}(x)$, $\text{ST}(x, q) = exp$, and for all $(x, y) \in A_T$, and for all $\pm f \in \text{CT}(x, y)$, $\text{ST}((x, y), \pm f) = exp$, and there is a $p \in \mathcal{P}^1(()P)$ such that $p \notin \text{CT}(x)$, and $not\ p \notin \text{CT}(x)$, then either add p to $\text{CT}(x)$ with $\text{ST}(x, p) = unexp$, or add $not\ p$ to $\text{CT}(x)$ with $\text{ST}(x, not\ p, r) = unexp$, for every rule $r \in P_p$.

In other words, if there is a node x for which all positive predicates in its content and all predicates in the contents of its outgoing edges have been expanded, but there are still unary predicates p which do not appear in $\text{CT}(x)$, one has to pick such a p and inject either p or $not\ p$ in $\text{CT}(x)$. This is needed for consistency: it does not suffice to find a justification for the predicate to satisfy, but one also has to show that this justification is part of an actual open answer set, which is done by effectively constructing it (cf. end of section 4). We do not impose that all negative predicate symbols are expanded as that would constrain all the ensuing literals to be locally refuted.

Similarly to rules (i), (ii), and (iii) one can define the expansion rules for binary predicates: (iv) *Expand binary positive*, (v) *Expand binary negative*, and (vi) *Choose binary*.

5.2   Applicability Rules

The applicability rules restrict the use of the expansion rules.

(vii) Saturation. A node $x \in N_T$ is *saturated*, if for all $p \in \mathcal{P}^1(P)$, $p \in \text{CT}(x)$ or $not\ p \in \text{CT}(x)$, and no $\pm q \in \text{CT}(x)$ can be expanded with rules (i-iii), and for all $(x, y) \in A_T$ and $f \in \mathcal{P}^2(()P)$, $f \in \text{CT}(x, y)$ or $not\ f \in \text{CT}(x, y)$, and no $\pm f \in \text{CT}(x, y)$ can be expanded with (iv-vi). No expansions should be performed on a node from T until its predecessor is saturated.

(viii) Blocking. A node $x \in N_T$ is *blocked*, if its predecessor is saturated and there is an ancestor y of x, $y < x$, s.t. $\text{CT}(x) \subseteq \text{CT}(y)$.

No expansions can be performed on a blocked node. Intuitively, if there is an ancestor y of x whose content includes the content of x one can reuse the justification for y when dealing with x.

(ix) Caching. A node $x \in N_T$ is *cached*, if its predecessor is saturated and there is a non-cached node $y \in N_T$ such that $y \not< x$, $x \not< y$, and $\text{CT}(x) \subseteq \text{CT}(y)$.

No expansions can be performed on a cached node. Intuitively, x is not further expanded, as one can reuse the (cached) justification for y when dealing with x.

5.3   Termination, Soundness, and Completeness

A completion structure is *contradictory* if either (i) for some $x \in N_T$ and $a \in \mathcal{P}^1(P)$, $\{a, not\ a\} \subseteq \mathrm{CT}(x)$ or (ii) for some $(x, y) \in A_T$ and $f \in \mathcal{P}^2(P)$, $\{f, not\ f\} \subseteq \mathrm{CT}(x, y)$. An *expanded completion structure* for a simple CoLP P and $p \in \mathcal{P}^1(P)$, is a completion structure that results from applying the expansion rules to the initial completion structure for p and P, taking into account the applicability rules, s.t. no expansion rules can be further applied. An expanded completion structure $CS = \langle T, G, \mathrm{CT}, \mathrm{ST} \rangle$ is *clash-free* if: (1) CS is not contradictory, (2) G does not contain cycles.

One can show that an initial completion structure for a unary predicate p and a simple CoLP P can always be expanded to an expanded completion structure (*termination*), such that, if p is satisfiable w.r.t. P, there is a clash-free expanded completion structure (*completeness*), and, finally, that, if there is a clash-free expanded completion structure, p is satisfiable w.r.t. P (*soundness*).

Theorem 1. *Let P be simple CoLP and $p \in \mathcal{P}^1(P)$. Then, (1) one can construct a finite expanded completion structure by a finite number of applications of the expansion rules to the initial completion structure for p w.r.t. P, taking into account the applicability rules, and (2) there exists a clash-free expanded completion structure for p w.r.t. P iff p is satisfiable w.r.t. P.*

The OASP-R system implements the above algorithm in BProlog [2]. The source code for the program together with some example input programs is available at http://www.kr.tuwien.ac.at/staff/heymans/priv/oasp-r/.

The implementation is a straightforward translation of the algorithm into BProlog, using BProlog's backtracking mechanism to take care of the nondeterministic choices in our algorithm. We chose a Prolog engine for its fast prototype capabilities and BProlog in particular for it being one of the fastest performing Prolog engines currently available.[5]

6   Complexity Results

Let $CS = \langle T, G, \mathrm{CT}, \mathrm{ST} \rangle$ be a completion structure and let CS' be the completion structure from CS by removing from N_T all blocked and cached nodes y. There are at most $k \times l$ such nodes, where k is bound by $|\mathcal{P}^1(P)|$ and the number of non-empty γ_m (resp. γ) of rules of the form (1) (resp. (2)) and l is the number of nodes in CS'. If CS' has more than 2^n nodes, then there must be two nodes $x \neq y$ such that $\mathrm{CT}(x) = \mathrm{CT}(y)$; if $x < y$ or $y < x$, either x or y is blocked, which contradicts the construction of CS'. If $x \not< y$ and $y \not< x$, x or y is cached, again a contradiction. Thus, CS' contains at most 2^n nodes, so $l \leq 2^n$. Since CS' resulted from CS by removing at most $k \times l$ nodes, the number of nodes in CS is at most $(k + 1)2^n$, and the algorithm has to visit a number of nodes that is exponential in the size of P. At each visit, executing an expansion rule or checking an applicability rule can be done in exponential

---

time. The graph G has as well a number of nodes that is exponential in the size of P. Since checking for cycles in a directed graph can be done in linear time, we obtain the following result: the algorithm runs in nondeterministic exponential time, a nondeterministic variant of the worst-case complexity characterization. Note that such an increase in complexity is expected. For example, although satisfiability checking in $\mathcal{SHIQ}$ is EXPTIME-complete, practical algorithms run in double nondeterministic exponential time [17].

6.1   Experimental Evaluation

We investigated the performance of our BProlog implementation on some example programs: a set of rules describing *family* relations and a set of rules describing a *game* environment.[6]

The *family* program contains 64 rules and 88 predicates; the *game* program contains 265 rules and 544 predicates. A run of 1000 satisfiability checks results in an average of 0.131 seconds for the *family* program and 15.919 seconds for the *game* program, where each satisfiability check resulted in a positive answer. Time spent goes significantly up when more rules/predicates are present. This is not surprising as the number of nondeterministic choices increases with the rules/predicates present. In case predicates are not satisfiable, the location of the rules that cause the inconsistency is vital. If the inconsistency arises within the rules high up in the program, satisfiability checking stays under 0.2 seconds for both example programs; if the inconsistency arises within rules low in the program, our reasoner does not return within 300 seconds. This difference in behavior depending on the location of the inconsistency is due to the BProlog backtracking mechanism and the order in which it solves goals.

Note that adding more rules can actually lead to better results in OASP-R. For example, using the rules from *game+* which extends *game* by adding rules in the beginning, one gets an average of 13.686 seconds per satisfiability check, i.e., 2 seconds better than without those extra rules.

7   Outlook

We intend to investigate several optimizations of the algorithm originating from both the DL tableaux as well as ASP reasoning algorithms. For example, dependency-directed backtracking will allow to backtrack on the choices that caused an inconsistency instead of backtracking on the last choice the BProlog engine made. Similar to DL tableaux, we will investigate whether we can internalize a program to a form that reduces the amount of nondeterminism in the algorithm. A Java implementation will allow us to more flexibly implement optimization strategies.

---

[6] Experiments were done on a QuadCore Intel(R) Xeon(R) CPU E5450 at 3GHz under Linux (openSUSE 11.0 (X86-64)). All example programs can be found at `http://www.kr.tuwien.ac.at/staff/heymans/priv/oasp-r/` and originated from ontologies that accompanied the RacerPro DL reasoner [1].

Acknowledgement

This work is partially supported by the Austrian Science Fund (FWF) under the projects P20305 and P20840, and by the European Commission under the project OntoRule (IST-2009-231875). We would like to thank Uwe Keller for his valuable comments.

References

1. RacerPro 1.9.0. Racer Systems GmbH & Co. KG,
   http://www.racer-systems.com/index.phtml
2. BProlog 7.1. Afany software, http://www.probp.com/
3. Baader, F., Calvanese, D., McGuinness, D.L., Nardi, D., Patel-Schneider, P.F. (eds.): The Description Logic Handbook: Theory, Implementation, and Applications. Cambridge University Press, Cambridge (2003)
4. Eiter, T., Ianni, G., Lukasiewicz, T., Schindlauer, R., Tompits, H.: Combining answer set programming with description logics for the semantic web. Artificial Intelligence 172(12-13), 1495–1539 (2008)
5. Feier, C., Heymans, S.: A sound and complete algorithm for simple conceptual logic programs. Technical Report INFSYS Research Report 184-08-10, KBS Group, Technical University Vienna, Austria (October 2008),
   http://www.kr.tuwien.ac.at/staff/heymans/priv/projects/fwf-doasp/alpsws2008-tr.pdf
6. Gelfond, M., Lifschitz, V.: The Stable Model Semantics for Logic Programming. In: Proc. of ICLP 1988, Cambridge, Massachusetts, pp. 1070–1080 (1988)
7. Grosof, B.N., Horrocks, I., Volz, R., Decker, S.: Description logic programs: combining logic programs with description logic. In: Proc. of the World Wide Web Conference (WWW), pp. 48–57. ACM, New York (2003)
8. Heymans, S., de Bruijn, J., Predoiu, L., Feier, C., Van Nieuwenborgh, D.: Guarded hybrid knowledge bases. Theory and Practice of Logic Programming 8(3), 411–429 (2008)
9. Heymans, S., Van Nieuwenborgh, D., Vermeir, D.: Conceptual logic programs. Annals of Mathematics and Artificial Intelligence (Special Issue on Answer Set Programming) 47(1–2), 103–137 (2006)
10. Heymans, S., Van Nieuwenborgh, D., Vermeir, D.: Open answer set programming for the semantic web. Journal of Applied Logic 5(1), 144–169 (2007)
11. Heymans, S., Van Nieuwenborgh, D., Vermeir, D.: Open answer set programming with guarded programs. ACM Transactions on Computational Logic (TOCL) 9(4) (October 2008)
12. Krötzsch, M., Rudolph, S., Hitzler, P.: Description logic rules. In: Proc. 18th European Conf. on Artificial Intelligence (ECAI 2008), pp. 80–84. IOS Press, Amsterdam (2008)
13. Lukasiewicz, T.: A novel combination of answer set programming with description logics for the semantic web. In: Franconi, E., Kifer, M., May, W. (eds.) ESWC 2007. LNCS, vol. 4519, pp. 384–398. Springer, Heidelberg (2007)

14. Motik, B., Sattler, U., Studer, R.: Query answering for OWL DL with rules. Journal of Web Semantics 3(1), 41–60 (2005)
15. Rosati, R.: DL+log: Tight integration of description logics and disjunctive datalog. In: Proc. KR, pp. 68–78 (2006)
16. Rosati, R.: On the decidability and complexity of integrating ontologies and rules. Journal of Web Semantics 3(1), 61–73 (2005)
17. Tobies, S.: Complexity Results and Practical Algorithms for Logics in Knowledge Representation. PhD thesis (2001)

Search for More Declarativity
Backward Reasoning for Rule Languages Reconsidered

Simon Brodt, François Bry, and Norbert Eisinger

Institute for Informatics, University of Munich,
Oettingenstraße 67, D-80538 München, Germany
http://www.pms.ifi.lmu.de/

Abstract. Good tree search algorithms are a key requirement for inference engines of rule languages. As Prolog exemplifies, inference engines based on traditional uninformed search methods with their well-known deficiencies are prone to compromise declarativity, the primary concern of rule languages. The paper presents a new family of uninformed search algorithms that combine the advantages of the traditional ones while avoiding their shortcomings. Moreover, the paper introduces a formal framework based on partial orderings, which allows precise and elegant analysis of such algorithms.

1 Introduction

The foremost advantage of rule languages is their *declarativity*. It allows problem-solving by specifying a problem's "what" without bothering about its "how". This separation of concerns makes it easy for rule authors to add or modify rules, thus supporting rapid prototyping, stepwise refinement, adaptation and evolution in application areas with unknown solution algorithms and/or frequently changing prerequisites.

Such unburdening of rule authors from control issues depends on a well-designed inference engine. Assuming that the underlying logical system features reasonable soundness and completeness properties, which it usually does, the most tricky design decision is to combine it with a search method that preserves all or most of these properties while still ensuring an adequate degree of efficiency.

The exact criteria for such design decisions are subject to several fundamental assumptions about the reasoning process, such as tuple-oriented vs. set-oriented or forward vs. backward reasoning. But we need not place special emphasis on those assumptions for the purpose of this paper. Although our motivation examples will use backward reasoning with definite rules, our concern is not the evaluation of this particular kind of rules, but a complete and space-efficient search method for rule engines in general. Such a search method is not only applicable to backward reasoning with and without memoization [13,15], but also to forward reasoning approaches using some goal guidance [3,4,6].

Given the wealth of research results on search [1,10,11,12,16, among many others], soberingly few actually come into consideration as candidates for rule inference engines. Their bulk has been on *informed* search methods, on incorporating

A. Polleres and T. Swift (Eds.): RR 2009, LNCS 5837, pp. 71–86, 2009.
© Springer-Verlag Berlin Heidelberg 2009

domain-specific knowledge into the search. But this is at odds with the very idea of rule-based systems: rules may represent domain-specific knowledge, but the inference engine evaluating them needs to be applicable to arbitrary rule sets and is therefore, for better or worse, generic and domain-independent. The same holds all the more for rules used in reasoning on the Web, where domain-specific knowledge is hardly available.

This narrows down the choice to *uninformed* search methods, of which there are barely a handful: breadth-first and depth-first search [7], iterative-deepening, [8], iterative broadening [5]. All of them have weak points: storage requirements for breadth-first search can become prohibitive already for medium-size problems, depth-first search is incomplete in search spaces with infinite branches, the iterative variants re-evaluate parts of the search space over and over again.

Under these circumstances a sensible compromise seems to be the one chosen for Prolog: to use depth-first search and to give rule authors some control to avoid infinite dead ends, for example by ordering the rules. However, this compromise wreaks havoc on declarativity.

Assume a term representation for natural numbers where `zero` represents 0 and `succ(X,Y)` can provide the predecessor X to any Y representing a nonzero natural number. Consider the straightforward rules defining for this representation the predicates `nat`, nat_2 and `less`, together with four queries:

```
nat(zero)   ←
nat(Y)      ← succ(X,Y) ∧ nat(X)
nat₂(X,Y)   ← nat(X) ∧ nat(Y)
less(X,Y)   ← "reasonably defined"
```

```
1 ← nat(X)
2 ← nat₂(X,Y)
3 ← less(zero,X) ∧ nat₂(X,Y)
4 ← nat₂(X,Y) ∧ less(zero,X)
```

Problem 1: Incomplete Enumeration. Query 1 results in an enumeration of $\mathbb{N}$, which is fine. One would expect query 2 to result in an enumeration of $\mathbb{N} \times \mathbb{N}$, but it only enumerates $\{0\} \times \mathbb{N}$. The reason is that depth-first backtracking search never reaches branches to the right of the first infinite one. Note that reorderings of rules or literals would not affect the problem.

Problem 2: Non-Commutativity of Logical Connectives. Assume single-answer mode[1] for queries 3 and 4. Then both queries ask about the existence of an $X > 0$ with $(X,Y) \in \mathbb{N} \times \mathbb{N}$ for some Y. The two queries are logically equivalent, but query 3 results in an affirmative answer and query 4 in a nonterminating evaluation giving no answer at all.

Such blatant infringements on declarativity are sometimes wrongly attributed to SLD-resolution, although it is perfectly sound and complete with any literal selection function [9]. The only cause of the problems is depth-first backtracking search. With a complete search method the problems would not arise.

Consequently, one way to avoid them is to replace depth-first search by iterative-deepening [14]. Unfortunately, this approach introduces a new problem.

```
even(zero)  ←
even(Y)     ← succ(X,Y) ∧ odd(X)
odd(Y)      ← succ(X,Y) ∧ even(X)
```

```
5 ← constant(X) ∧ even(X)
```

---

[1] Single-answer mode in Prolog can be achieved by a cut at the end of each query.

Problem 3: Inefficiency on Functional Rule Sets. Let `constant(X)` bind X to the term representation of some fixed, large number $n \in \mathbb{N}$. The rules define relations that are functions. Evaluation of query 5 ought to require $O(n)$ steps, and so it does with depth-first backtracking search. Iterative-deepening, on the other hand, needs $O(n^2)$ steps.

Search should not slow down the evaluation of functional rules, which do not need any search in the first place. Some functional rules escape being slowed down thanks to the compiler's tail recursion optimisation. But this sidestepping the problem fails in "quasi tail recursive" cases like the above, which do not match typical tail recursion patterns but nevertheless induce almost linear search trees.

Desiderata for Search Methods. A search method for rule inference engines usually has to be *uninformed*, as discussed earlier. It ought to meet the following requirements, which are essentially a collection of all advantageous properties from traditional methods.

- *Completeness* (or *exhaustiveness/fairness*) on both finite and infinite search trees. It visits every node in the search space after finitely many steps. Recall that we want to apply it also for finding all solutions to a query, and if there are infinitely many, the method must be capable of a fair enumeration.

  Depth-first search and iterative broadening violate this requirement on infinite trees. Depth-bounded backtrack search and credit search [1] violate it even on finite trees.
- *Polynomial space complexity* $O(d^c)$ where c is a constant and d the maximum depth currently reached during the search (or of the entire tree, if it is finite). Breadth-first search has exponential worst-case space complexity $O(2^d)$.
- *Linear time complexity* $O(n)$ where n is the current number of nodes that have been visited at least once (or of the entire tree, if it is finite).

  Note that any non-repetitive method, which visits every node at most once, meets this requirement. Iterative-deepening does not, see problem 3 above.

Note that space and time complexity here depend on different variables. The desired space complexity $O(d^c)$ is *polynomial in depth* d. The desired time complexity $O(n)$ is *linear in size* n, and often $O(n) = O(b^d)$ for an upper bound b of the branching factor. Linear in size b^d is much larger than polynomial in depth d.

This paper introduces D&B-search, a new uninformed search method, which integrates depth-first and breadth-first search. It meets these desiderata, the basic algorithm even with space complexity linear in depth. D&B-search can be parameterised to turn it into a family of algorithms with breadth-first and depth-first search as its extremal cases. The parameter also allows to control the amount of storage provided for completeness.

The paper is organised as follows. Section 2 presents D&B-search. A formal framework for the analysis of search methods follows in Section 3. Then Section 4 analyses D&B-search with this framework showing that it meets the desiderata above. Finally, Section 5 reports about the current state of development and plans for improvements.

2   D&B-Search and Its Family of Algorithms

Let us abbreviate depth-first and breadth-first search by D-search and B-search, respectively. The idea of D&B-search is to alternate D-search with B-search, controlling their rotation by a sequence $f_0, f_1, f_2, \ldots$ of *depth bounds*. These are defined by a function $\mathbb{N} \to \mathbb{N}$, $i \mapsto f_i$ with $i < f_i < f_{i+1}$ for $i \in \mathbb{N}$.

D-search starts, but may expand nodes at depth f_{i+1} or beyond only if all nodes at depth $\leq i$ have been expanded. If they haven't, B-search takes over. It may expand nodes at depth $i + 1$ only if some node at depth f_{i+1} has been expanded before. If none has, D-search takes over again. And so on.

In this way no node is ever re-expanded, D&B-search is non-repetitive. Its principle bears some resemblance to the principle of A^*-search [10,11], which combines a heuristic estimate for fast advances into promising parts of the search space with a path-cost function ensuring a minimum degree of B-search behaviour and thus completeness. Likewise, D&B-search, which is uninformed and has no heuristics for "promising", combines fast D-search advancement with a minimum degree of B-search behaviour to ensure completeness. The following diagrams illustrate how D-search and B-search interact for $f_i = 2^i$.

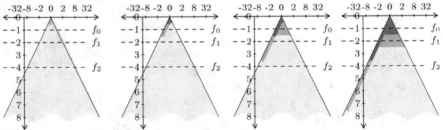

D-search advances exponentially faster than B-search. Hence the total number of nodes to be stored at any time (those on the branch traversed by D-search together with those at the deepest level reached by B-search) depends polynomially (for $f_i = 2^i$ even linearly) on the maximal depth reached up to that time. More details on space complexity will follow on page 76.

From an algorithm-oriented point of view it is better to focus not on the depth-bounds f_i, but on the nodes that serve as synchronisation points between D-search and B-search. Let us call a node "earlier" than a given one, if (unrestricted) D-search would expand it before expanding the given one.

For each depth-bound f_i its *pivot-node* s_i is the earliest (i. e., left-most) node at depth f_i. It is undefined if there are no nodes at depth f_i.

All other nodes are partitioned into finite sets. The *pre-pivot-set* S_0 is the set of nodes earlier than the pivot-node s_0. For each other pivot-node s_{i+1} let D_i be the set of nodes earlier than s_{i+1} and B_i the set of nodes at depth i. All nodes in these two sets must be expanded before expanding the pivot-node s_{i+1}, but some have already been expanded before earlier pivot-nodes. So the *inter-pivot-set*, i. e., the set of nodes expanded in-between s_i and s_{i+1}, is $S_{i+1} = (D_i \cup B_i) \backslash X_i$ where $X_0 = S_0 \cup \{s_0\}$ and $X_{i+1} = X_i \cup S_{i+1} \cup \{s_{i+1}\}$. Finally, the *post-pivot-set* R is empty if the tree is infinite. Otherwise there is a maximal $i_{\max}$ for which $s_{i_{\max}}$ is defined, and R is the set of all remaining nodes of the tree except $X_{i_{\max}}$.

Using these notions, reconsider the behaviour of D&B-search[2] for $f_i = 2^i$:

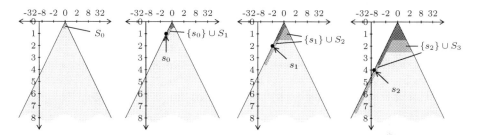

The general pattern is best seen for the last transition (the others are some-what special). The third diagram shows the snapshot where all nodes in $X_1 \cup S_2 = S_0 \cup \{s_0\} \cup S_1 \cup \{s_1\} \cup S_2$ have been expanded (indicated by shading) and D-search is ready to expand the pivot-node s_2.

Next, D-search expands s_2, at which point the set of expanded nodes is X_2, and then continues until its next step would be to expand the pivot-node s_3. During this continuation it expands all nodes in $D_2 \backslash X_2 \subseteq S_3$. At this point control passes to B-search for expanding the remaining nodes in $B_2 \backslash X_2 \subseteq S_3$, the as-yet unexpanded nodes at depth 2. When done, all nodes in S_3 have been expanded and control passes back to D-search, which is now ready to expand the pivot-node s_3. This is the snapshot in the last diagram, the darkest shade indicating $X_2 \backslash \{s_2\}$, the medium shade indicating S_3 with $D_2 \backslash X_2$ on the left-most branch and $B_2 \backslash X_2$ at depth 2.

Let us now turn to the initial stages. D&B-search starts with D-search ex-panding all nodes in the pre-pivot-set S_0 (which contains only the root node for $f_0 = 1$, but would contain more for $f_0 > 1$). D-search is ready to continue with pivot-node s_0. The first diagram shows the snapshot at this point.

Then D-search expands s_0, then all nodes in $D_0 \backslash X_0 \subseteq S_1$ (of which there aren't any for $f_1 - f_0 = 1$). Its next step would be to expand s_1. Now control passes to B-search for expanding the remaining nodes in $B_0 \backslash X_0 \subseteq S_1$ (of which at depth 0 there aren't any). D-search is ready to continue with pivot-node s_1. This is the snapshot in the second diagram.

Altogether, D&B-search expands the nodes of the search tree in the order $S_0, s_0, \ldots, S_i, s_i, \ldots, R$. For finite trees this has interesting consequences:

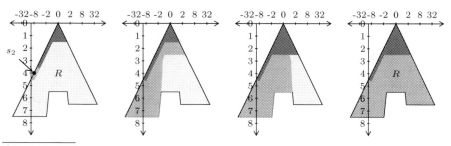

---

[2] The pivot-nodes explain the *behaviour* of D&B-search, but not its implementation, where they are not directly available. They play an important role indirectly, though.

In a finite tree there are $i_{\max}$ pivot-nodes and $i_{\max} - 1$ inter-pivot-sets as well as sets B_i, part of whose nodes is all B-search ever expands. But $i_{\max}$ is small ($O(\log d)$ for $f_i = 2^i$) compared to the maximum depth d. So B-search stops quite soon. The overall behaviour is dominated by D-search in the post-pivot-set R.

In an infinite tree D-search cannot leave the left-most infinite branch. Everything "to the right" of this branch, the largest part of the search tree as it increases in size much faster than in depth, is therefore handled by B-search.

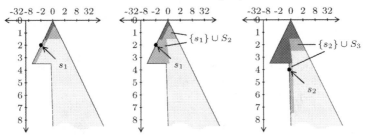

Although the merits of B-search for infinite trees are debatable, it is at least complete. So D&B-search has a kind of built-in adaptivity. Depending on the search tree it behaves essentially like the uninformed search method that best suits the tree, taking "best" with a pinch of salt.

This adaptivity effect would also apply to D&I-search (Section 5), a suggested combination of D-search with iterative-deepening search instead of B-search.

The D&B Family. Furthermore we can parameterise the function f_i with $c \in \mathbb{N} \cup \{\infty\}$ to define a family of algorithms.

Assume[3] that the tree's branching factor is bounded by $b \in \mathbb{N}$. An obvious idea is to let $f_{c,i} := \lfloor b^{\frac{i}{c}} \rfloor$ with $\frac{i}{\infty} := 0$, $\frac{i}{0} := \infty$ and $b^\infty := \infty$. However, these functions do not satisfy $i < f_{c,i} < f_{c,i+1}$ for $c \neq 1$. Therefore[4] let $f_{c,i} := \lfloor b^{\frac{i}{c}} \rfloor + i$. For this family of functions and algorithms we get:

- For $1 \leq c \leq \infty$ the algorithm is complete (for $c = 0$ it is not).
- For $1 \leq c < \infty$ its space complexity is $O(d^c)$, which is polynomial in depth.
- For $c = 0$ it corresponds to D-search because $f_{0,0} = \infty$.
  The pre-pivot-set S_0 contains all nodes of the whole tree.
- For $c = \infty$ it corresponds to B-search because $f_{\infty,i} = i + 1$, the slowest function with $i < f_i$. All sets $D_i \backslash X_i$ are empty, thus $S_{i+1} = B_i \backslash \{s_i\}$.

The parameter c is a means to express how much storage one is willing to invest into completeness. Between the two extremes "none" ($c = 0$, D-search) and "unlimited" ($c = \infty$, B-search) we now have available an almost arbitrary gradation of algorithms in-between, each of them with space complexity polynomial in depth and time complexity linear in size (since the algorithm is non-repetitive).

Moreover, the parameter c can easily be turned into a parameter of a single implementation for the whole family. It is even possible to adapt the parameter dynamically, i.e., during the traversal of the search tree.

---

[3] This assumption can be dropped for the implementation [2, Sec. 5.2].

[4] Alternatively, the requirement could be weakened to non-strict monotonicity. While possible in principle, this would make the formal analysis more complex.

3   A Framework for Analysing Tree Traversal Algorithms

Most of the definitions and theorems below refer to their counterparts in a technical report [2, http://www.pms.ifi.lmu.de/publikationen#PMS-FB-2009-7], which works out the formal framework in full detail.

An uninformed search algorithm cannot anticipate which parts of the search space contain or don't contain solutions. In order to be able to find *all* solutions it has to visit all nodes in the search tree, just like a traversal algorithm. Therefore our framework formalises *traversal* algorithms, gaining the advantage that it does not need to consider whether or not a node is a solution.

Let b be an upper bound for the out degree of trees under consideration. Let $\Sigma = \{0, \dots, b-1\}$ and Σ^* the set of all words over Σ.

Definition 1 (Traversability [2, Def. 2.1.1])
A set $\Omega \subseteq \Sigma^*$ is called *traversable*, iff $u \in \Omega$ holds for all $uv \in \Omega$. The set of all traversable $\Omega \subseteq \Sigma^*$ is denoted by $Trav_\Sigma$.

Definition 2 (Tree [2, Def. 2.1.2])
Let $E := \{(w, wi) \mid w \in \Sigma^*, i \in \Sigma\}$. Then (Σ^*, E) is a complete infinite tree with out degree b. Any tree with maximum out degree b can be obtained by choosing a traversable $\Omega \subseteq \Sigma^*$ and restricting the edge set E to Ω. The resulting tree $(\Omega, E|_\Omega) = (\Omega, \{(w, w') \in E \mid w, w' \in \Omega\})$ is simply written Ω from now on.

Notation 3.
$$\Omega_k := \{w \in \Omega \mid |w| = k\} = \Omega \cap \Sigma^k$$
$$\Omega_N := \{w \in \Omega \mid |w| \in N\} = \bigcup_{i \in N} \Omega_i \text{ for } N \subseteq \mathbb{N}$$
$$\Omega_{\geq k} := \Omega_{\{\geq k\}} \quad \text{where} \quad \{\geq k\} := \{i \in \mathbb{N} \mid i \geq k\}$$

Notation 4. In the following we often talk about an α with $\alpha \preccurlyeq \omega$. Such an ordinal number α may be considered just the set $\mathbb{N}$ in the infinite case ($\alpha = \omega$) and some set of the form $\{0, 1, \dots, n\}$ or $\emptyset$ in the finite case ($\alpha \prec \omega$), each together with the common (well-)ordering on natural numbers.[5]

The next three definitions model tree traversals at different levels of abstraction, each building on the former. A *traversal-sequence* assigns to each number corresponding to a "time point" the node of the tree visited at that time point. A *traversal-run* enriches a traversal-sequence by associating with each time point a subset of the nodes of the tree. This subset represents the nodes kept in memory at this point for later processing (one can reasonably assume this to be almost the only use of memory by a tree traversing algorithm). A *traversal-algorithm* can then be specified by assigning a traversal-run to each tree $\Omega \in Trav_\Sigma$.

Definition 5 (Traversal-sequence [2, Def. 2.2.3])
Let $a : \alpha \to \Omega$ be a finite or infinite sequence of nodes in Ω. Then a is called *traversal-sequence* iff for each w occurring in a also its parent occurs in a and the first occurrence of its parent is located before the first occurrence of w.

---

[5] The notation was chosen for two reasons: (1) it calls attention to the "succession" of the numbers and (2) it covers finite and infinite cases in a uniform way.

Definition 6 (Traversal-run[6] [2, Def. 2.2.4])
A *traversal-run* is a sequence $A : \alpha \rightarrow \Omega \times \mathcal{P}(\Omega)$ of pairs (node, node set) with:
1. The first node set contains exactly the root of the tree Ω.
2. For each pair the node is a member of the corresponding node set.
For all successive pairs $(\text{node}_n, \text{set}_n)$ and $(\text{node}_{n+1}, \text{set}_{n+1})$
3. the children of node_n are included in set_{n+1} (modelling expansion of node_n).
4. from set_n to set_{n+1} an arbitrary number of nodes may be dropped.
5. for any node w in set_{n+1}, either it is member of set_n or its parent is node_n.
A traversal-run *induces* the traversal-sequence obtained by omitting the sets.

Definition 7 (Traversal-algorithm[6] [2, Def. 2.2.4])
A *traversal-algorithm* is a family $(A_\Omega)_{\Omega \in Trav_\Sigma}$ of traversal-runs. In other words, the algorithm assigns to each tree $\Omega \in Trav_\Sigma$ a traversal-run over this tree.

As an example of a traversal-run, consider a typical queue-based implementation of B-search. For each time point n and pair $(\text{node}_n, \text{set}_n)$ the set_n consists of all nodes in the queue at this time point (they are the nodes needed for future expansion) and node_n is the first in the queue (the node to be expanded in the step from n to $n + 1$). This node is not a member of set_{n+1} because B-search removes the expanded node from the queue when inserting its children.

Modelling a stack-based implementation of iterative-deepening, each set_n consists of all nodes in the stack at this time point and node_n is the top of stack. This node is not a member of set_{n+1} unless it is the root ε, which needs to be kept in the stack as the bottom element for later re-expansion when starting another iteration.

Definition 8 (Completeness [2, Def. 2.2.9])
  - A traversal-sequence $a : \alpha \rightarrow \Omega$ is called *complete*, iff it is surjective.
  - A traversal-run is *complete*, iff its induced traversal-sequence is.
  - A traversal-algorithm is *complete*, iff its runs A_Ω are for all $\Omega \in Trav_\Sigma$.

Definition 9 (Weak completeness [2, Def. 2.2.14])
A traversal-sequence $a : \alpha \rightarrow \Omega$ is called *weakly complete*, iff $|a[\alpha]| = |\Omega|$.
The definition of weak completeness for traversal-runs and traversal-algorithms is analogous to the definition of their completeness.

Obviously completeness implies weak completeness, but conversely only in the finite case. In the infinite case weak completeness intuitively means that traversal does not artificially stop when there are still unexpanded nodes in the search space. Weak completeness follows from some simple criteria [2, Lem. 2.2.16 & 3.3.3], which are easier to test than the condition defining (full) completeness.

When analysing tree traversals, it is in most cases sufficient to know whether some node is visited earlier than another one, comparing only their first visits. Time points of revisits, though represented in traversal-sequences, are usually irrelevant. This observation leads to the final abstraction level in our framework.

---

[6] The definition does not depend on any computability requirements. They are not needed for the framework and would not simplify anything either.

Definition 10 (Representing ordering [2, Def. 2.2.12])
A partial ordering $\lhd$ on Ω is called *representing ordering* of a traversal-sequence
a, iff all nodes occurring in a are

1. ordered by their first occurrence.
2. smaller than any node that does not occur in a, but is $\lhd$-comparable to the
   root ε of Ω.[7]

If the representing ordering of a is unique, it is referred to as $\lhd_a$.

Definition 11 (Traversal-ordering[8] [2, Def. 2.3.2])
A partial ordering $\lhd$ on traversable $\Omega \subseteq \Sigma^*$ is called *traversal-ordering*, iff

1. it is compatible with the tree structure of Ω, i.e., for all $uvw \in \Omega$
   $$u \not\leq uvw \quad \text{and} \quad u \leq uvw \Rightarrow u \leq uv \leq uvw \quad \text{hold.}$$
2. all nodes $\lhd$-comparable to the root ε are totally ordered by $\lhd$.[7]
3. no node not $\lhd$-comparable to ε is smaller than any node $\lhd$-comparable to ε.[7]

At least one of the representing orderings of any traversal-sequence is a traversal-
ordering. Those that are not, can be disregarded. [2, Rem. 2.3.3]

Notation 12. For a partial ordering $(\Omega, \lhd)$, sets $M, N \subseteq \Omega$ and $w \in \Omega$ define
$$M \lhd N \quad :\Leftrightarrow \quad \forall m \in M \, \forall n \in N : m \lhd n$$
$$\lhd(w) \quad := \quad \{w' \in \Omega \mid w' \lhd w\}$$

Definition 13 (Completeness of an ordering [2, Def. 2.3.1])
A partial ordering $\lhd$ on $\Omega \subseteq \Sigma^*$ is called *complete*, iff $(\Omega, \lhd) \cong \alpha$ for some $\alpha \preccurlyeq \omega$.

Theorem 14 (Characterisation of completeness [2, Thm. 2.3.5])
A total ordering $(\Omega, \lhd)$ is complete iff $\exists f : \mathbb{N} \to \mathbb{N}$ with $\Omega_k \lhd \Omega_{\geq f(k)}$.

Theorem 15 (Characterisation of completeness [2, Thm. 2.3.8])
A partial ordering $(\Omega, \lhd)$ is complete, iff it is isomorphic to a finite sum of complete ordinal numbers, where only the last[9] summand may be infinite (i.e., $= \omega$).
Equivalently, $(\Omega, \lhd)$ is complete iff $(\Omega, \lhd)$ is isomorphic to a countably infinite
sum of finite ordinal numbers.

The main result now is that these criteria essentially characterise also the completeness of traversal-sequences and thus of traversal-algorithms.

Theorem 16 (Equivalence of completeness definitions [2, Thm. 2.3.13])

1. A traversal-sequence is complete in the sense of Definition 8 iff all its (traversable) representing orderings are complete according to Definition 13.
2. A weakly complete traversal-sequence is complete in the sense of Definition 8
   iff its representing ordering is complete according to Definition 13.

---

[7] This requirement is mainly due to technical reasons. It excludes irrelevant but potentially troublesome cases that could otherwise be formally construed.
[8] A total traversal-ordering is a topological ordering of Ω.
[9] Recall that addition is generally not commutative in ordinal number arithmetic.

4   Analysis of Tree Traversal Algorithms

In this section we first analyse some well-known algorithms in order to illustrate that the proofs of their (in)completeness are significantly more concise when based on our framework than the proofs found in the literature. The main part of the section is then devoted to the analysis of D&B-search, which would be hardly possible without the framework.

4.1   Known Algorithms

This subsection demonstrates the expressive and analytic power of our framework and the degree to which it makes (in)completeness proofs more concise. It shows by the example of D-search and of B/A^*-search how Theorem 14 and 15 may be used to prove (in-)completeness.

In both cases we define some total traversal-ordering characterising the desired algorithm. Then we show the (in-)completeness of the traversal-ordering. The algorithm itself can be obtained by means of the induced algorithm of the ordering, which mainly traverses the nodes in order, i.e., starting with the minimum of the ordering and then moving to the next greater node each step.[10]

D-Search. In the representation introduced by Definition 2 each node w of a tree Ω can be considered a word over the alphabet Σ. Seen this way D-search traverses the nodes of Ω in lexicographical order. So we define $\lhd_{depth} := \lhd_{lex}$.

Obviously D-search is incomplete on most infinite trees. This can be shown easily even without the framework by some counterexample. But if you want to explain why D-search is incomplete, things become more complicated. Probably one would say something like "D-Search will never return from the first infinite branch". This statement is based on the reader's common understanding of an algorithm's behaviour. How to make it more precise beyond intuition, however, is not obvious. Our framework allows to formulate the statement precisely.

Let Ω contain an infinite number of nodes and thus an infinite branch (lemma by König). Let $t \in \Sigma^\omega$ be the lexicographically first infinite branch in Ω. The set of prefixes $T \subseteq \Omega$ of t is just the set of nodes on t. Define $S := \{w \in \Omega \mid T \rhd_{depth} w\}$ and $U := \{w \in \Omega \mid T \lhd_{depth} w\}$. Intuitively, S consists of all nodes "to the left" of the first infinite branch T and U of all nodes "to the right" of T. If $U \neq \emptyset$ then $\lhd_{depth}$ is incomplete on Ω:

Possibility 1: $\exists w \in \Omega \forall n \in \mathbb{N} \exists w' \in \Omega_n : w' \lhd_{depth} w$ since $\exists w \in \Omega : T \lhd_{depth} w$ and $T \cap \Omega_n \neq \emptyset$ for all $n \in \mathbb{N}$. Consequently $\exists k \in \mathbb{N} \forall n \in \mathbb{N} : \Omega_k \npreceq_{depth} \Omega_n$ holds (set $k = |w|$). Incompleteness follows by Theorem 14.

Possibility 2: $S \lhd_{depth} T \lhd_{depth} U$ holds. The smallest β that fulfils this condition[11] is $\beta = \underset{\cong k \prec \omega}{|S|} + \underset{\cong \omega}{|T|} + \underset{\cong \alpha \succ 0}{|U|} \succ \omega$. Incomplete[12] by Theorem 15.

---

[10] "minimum" and "next greater node" are well-defined, see [2, Def. 2.3.9, Proof].

[11] An ordinal number β *fulfils* the condition $S \lhd_{depth} T \lhd_{depth} U$, if there is an isomorphic well-ordering $(\Omega, \lhd)$ that fulfils the condition.

[12] $\lhd_{lex}$ is generally not a well-ordering. Particularly $(\Omega, \lhd_{depth}) \not\cong \beta$ in general. But even if $(\Omega, \lhd_{depth})$ were well-ordered it would still be incomplete as shown above.

B-Search and A^*-Search. The informed A^*-search uses an optimistic cost estimation function[13] $F(w) = G(w) + H(w)$ to prioritise more promising nodes. With $G(w) = |w|$ and $H(w) = 1$ we obtain B-search as special-case.

A^*-search prefers nodes with smaller estimated costs and takes the lexicographically smaller one first if the estimated costs are equal. Consequently $w \vartriangleleft_{A^*} w' :\Leftrightarrow F(w) < F(w')$ or $F(w) = F(w') \wedge w \vartriangleleft_{lex} w'$ defines the order in which A^*-search traverses the nodes.

To prove the completeness of A^*-search we apply Theorem 14 using the function $k \mapsto max_<(F[\Omega_k]) + 1$. We must show that $\Omega_k \vartriangleleft_{A^*} \Omega_{(max_<(F[\Omega_k]) + 1)}$, meaning $w \vartriangleleft_{A^*} w'$ for $w \in \Omega_k$ and $w' \in \Omega_{(max_<(F[\Omega_k]) + 1)}$. This is true because $F(w') \geq G(w') \geq |w'| = max_<(F[\Omega_k]) + 1 > max_<(F[\Omega_k]) \geq F(w)$.

In the special case of B-search completeness can be proved even faster using Theorem 15. One only has to convince oneself that the following is true:

$$\vartriangleleft_{breadth} \cong |\Omega_0| + |\Omega_1| + |\Omega_2| + \ldots \cong \sum_{i=0}^{+\infty} |\Omega_i| \preccurlyeq \omega$$

Compared to the proofs in [10,11] the argumentation above is extremely short and precise. Due to its formal character it doesn't even need a deeper understanding of the concrete procedure of A^*-search. At this point we benefit from the abstract level of our analytic framework.

4.2 D&B-Search

The analysis of D&B-search is based on its traversal-ordering $\vartriangleleft_{d\&b}$ too. First we give a constructive definition of $\vartriangleleft_{d\&b}$ which corresponds directly to the description of D&B-search in Section 2. Second an alternative axiomatic definition of $\vartriangleleft_{d\&b}$ is presented. We show the equivalence of the two definitions and then alternate between them when proving completeness and space complexity.

Definition 17 (pivot-nodes & pre/inter/post-pivot-sets [2, Def. 4.1.1])

$$i_{max} := \begin{cases} -1 & \text{if } \Omega = \emptyset \\ max(\{i \mid \Omega_{f_i} \neq \emptyset\}) & \text{if } |\Omega| < \infty \\ \infty & \text{if } |\Omega| = \infty \end{cases}$$

$s_i := min_{lex}(\Omega_{f_i})$ for $0 \leq i \leq i_{max}$

$D_i := \vartriangleleft_{lex}(s_{i+1})$

$B_i := \Omega_i$

$S_0 := \vartriangleleft_{lex}(s_0)$

$S_{i+1} := (D_i \cup B_i) \setminus X_i$

$R := \Omega \setminus X_{i_{max}}$

$X_0 := S_0 \cup \{s_0\}$

$X_{i+1} := X_i \cup S_{i+1} \cup \{s_{i+1}\}$

$$X_{i_{max}} := \bigcup_{j=0}^{i_{max}} X_j$$

The pivot-nodes and pre/inter/post-pivot-sets should look familiar from page 74. By means of these nodes and sets the next definition constructs $\vartriangleleft_{d\&b}$.

---

[13] G denotes the cost incurred so far on the path to w, and H denotes the optimistically estimated cost remaining for the path from w to a goal.

Definition 18 (D&B-ordering, constructive [2, Def. 4.3.1]**)**

1. $S_0 \lhd_{d\&b} s_0 \lhd_{d\&b} S_1 \lhd_{d\&b} s_1 \lhd_{d\&b} \cdots \lhd_{d\&b} S_{i_{max}} \lhd_{d\&b} s_{i_{max}} \lhd_{d\&b} R$
2. $\forall w, w' \in S_0 : \qquad\qquad w \lhd_{d\&b} w' \Leftrightarrow w \lhd_{lex} w'$
3. $\forall w, w' \in S_{i+1} \cap \Omega_i : \quad w \lhd_{d\&b} w' \Leftrightarrow w \lhd_{lex} w'$
4. $\forall w, w' \in S_{i+1} \backslash \Omega_i : \quad w \lhd_{d\&b} w' \Leftrightarrow w \lhd_{lex} w'$
5. $\forall w, w' \in R \cap \Omega_{i_{max}} : \quad w \lhd_{d\&b} w' \Leftrightarrow w \lhd_{lex} w'$
6. $\forall w, w' \in R \backslash \Omega_{i_{max}} : \quad w \lhd_{d\&b} w' \Leftrightarrow w \lhd_{lex} w'$

Condition 1 already appeared on page 75. It defines the order between the pivot-nodes and sets and can be read as *D&B-search expands all members of the pre-pivot-set S_0 before expanding the pivot-node s_0, and it expands s_0 before expanding all members of the inter-pivot-set S_1, and so on.*

The rest affects the inner order of the sets. All equivalences can be read as *D&B-search expands w before w' iff depth-first search would*. Condition 2 matches exactly the description of D&B-search on page 75. Conditions 3 to 6 are a little bit less restrictive than the informal description. There D-search always had to finish work on some inter- or post-pivot-set before B-search could start, the nodes in the subset $S_{i+1} \backslash \Omega_i$ had to be expanded before the nodes in the subset $S_{i+1} \cap \Omega_i$. Here the two may interleave their work on such a set, the order between the two subsets is not restricted by the conditions above.

Recall the original view of D&B-search as introduced at the very beginning of Section 2. It was the view of alternating D-search and B-search, controlling their rotation by a sequence $f_0, f_1, f_2, \ldots$ of depth bounds. For the axiomatic definition of $\lhd_{d\&b}$ we reuse this view.

Definition 19 (D&B-ordering, axiomatic [2, Sec. 4.3 (D&B)]**)**

(Ax1) $\Omega_k \lhd_{d\&b} \Omega_{f_{k+1}}$

(Ax2) $\forall w, w' \in \Omega_k : w \lhd_{d\&b} w' \Leftrightarrow w \lhd_{lex} w'$ (D&B)

(Ax3) $\forall w \in \Omega_k : \underbrace{\lhd_{lex}(w) \lhd_{d\&b} w}_{(Ax3a)} \vee \underbrace{\exists w' \in \Omega_{f_k} : w' \lhd_{d\&b} w}_{(Ax3b)}$

(Ax1) signifies that none of the nodes in $\Omega_{f_{k+1}}$ is expanded before all nodes of Ω_k have been expanded. Therefore it limits the depth of the depth-first-traversal.

(Ax3b) disjunction concerns the breadth-first-traversal. It means that a node may only be expanded if some node at sufficient depth has been expanded (by depth-first-traversal) before. This implies a depth limit for breadth-first-traversal because (Ax3) requires (Ax3b) or (Ax3a) to hold for each node.

(Ax2) and (Ax3a) are less interesting. (Ax3a) says that a node may be expanded if all lexicographically smaller nodes have been expanded before. This is the specification of D-search. If at some time (Ax3a) becomes false because of (Ax1), i.e., (Ax3b) is true, (Ax2) enforces exactly B-search because it requires every level to be traversed in lexicographical order.

We have to show that the two definitions of $\lhd_{d\&b}$ are equivalent. Though our framework is a great help when formulating the arguments, the proof is still more extensive than can be presented within the space limitations of this paper. Without the framework it would be quite impossible. The results are as follows.

Theorem 20 (D&B-ordering, constructive $\Rightarrow$ axiomatic [2, Thm. 4.3.8]**)**
If $\lhd$ satisfies Definition 17 and 18 then $\lhd$ is a model of (D&B).

Theorem 21 (D&B-ordering, axiomatic $\Rightarrow$ constructive [2, Thm. 4.3.9]**)**
If $\lhd$ is a model of (D&B) then it satisfies Definition 17 and 18.

Thus, the two definitions are equivalent. But what does this help? In particular, why do we need the axiomatic definition? We will see one reason[14] in the next theorem. Moreover the theorem emphasises again the power of our framework for completeness proofs as the proof uses one of its main results.

Theorem 22 (D&B-ordering, completeness [2, Thm. 4.3.10]**).** If $\lhd$ satisfies Definition 17 and 18 or is a model of (D&B), then $\lhd$ is complete.

Proof. Follows immediately from (Ax1) and Theorems 14 and 20.   □

Finally, we are interested in the space complexity of D&B-search. In this context we find the constructive definition to be very helpful. Let us start with the general result for any function f with $i < f_i < f_{i+1}$. This result is independent from the family defined in Section 2.

Theorem 23 (D&B-search, general space complexity [2, Cor. 4.3.15]**)**
The space complexity $M(d)$ at depth $d \in \mathbb{N}$ of the algorithm induced by $\lhd_{d\&b}$ is

$$M(d) \leq \begin{cases} b \cdot (d+1) & \text{if } d < f_0 \\ b \cdot \left(d + 1 + b^i\right) & \text{else} \end{cases} \qquad \text{where } i := \underset{j}{argmax} \{f_j \mid f_j \leq d\}.$$

But of course we are most interested in the space complexity of the family defined in Section 2. We obtain their complexity from Theorem 23 by specialising f_i to the corresponding functions:

Theorem 24 (D&B family $c=0$, linear space complexity [2, Thm. 4.3.17]**)**
Let $c = 0$, $f_i = f_{0,i} = \lfloor b^{\frac{i}{0}} \rfloor + i = \infty$ and $\lhd_0$ be the corresponding ordering. The space complexity of the induced algorithm is $M_0(d) \leq b \cdot (d+1)$.

Theorem 25 (D&B family, polynomial space complexity [2, Thm. 4.3.16]**)**
Let $1 \leq c < \infty$, $f_i = f_{c,i} = \lfloor b^{\frac{i}{c}} \rfloor + i$ and $\lhd_c$ be the corresponding ordering. The space complexity of the induced algorithm is $M_c(d) \leq b \cdot (d+1+d^c)$.

5   Conclusion

In this paper we have presented D&B-search, a new uninformed search method based on integrating depth-first and breadth-first search into one. We have shown that the ratio of depth-first to breadth-first search can be balanced by a parameter, thus defining a family of search methods with depth-first and breadth-first search as its borderline cases.

---

[14] The second reason is that the axiomatic definitions provide the invariants for our implementation [2, Sec. 5.1].

We have also introduced a formal framework for analysing informed or uninformed search methods, which is based on partial orderings and uniformly covers finite and infinite search trees. We have illustrated its analytic power by giving very concise, yet formally precise proofs for well-known (in)completeness results on depth-first search, breadth-first search, and A^*-search.

Finally, we have analysed D&B-search using the formal framework. In the borderline cases the results are the known ones for depth-first and breadth-first search. In all non-borderline cases D&B-search is complete (exhaustive), it is non-repetitive and thus its time complexity is linear in size, and its space complexity is polynomial in depth. The polynomial is of degree c for the very parameter defining the D&B family, which therefore allows to control the amount of storage to be spent for the sake of completeness.

It should be noted that D&B-search is intrinsically better than running depth-first and breadth-first search in parallel, be it by round robin scheduling or more advanced time-sharing techniques or physically parallel on different processors. With all of these approaches the space complexity is exponential in depth for the process running breadth-first search. In contrast to that, D&B-search has space complexity polynomial in depth. It is this property that made necessary the somewhat involved form of integrating depth-first with breadth-first search.

In this paper we have not addressed implementation issues. The technical report [2] on which the paper is based also presents two implementation approaches to the level of detail of pseudo code showing that the required data structures are essentially a doubly-linked list of doubly-linked lists. Coding this pseudo code in a real programming language is rather straightforward.

We are about to start work on a prototype implementation of one of these approaches and plan to use it for empirical comparisons with other uninformed search methods. We intend to focus especially on logic programming applications using backward reasoning approaches without and with memoization [15,13].

At the conceptual level, we plan to follow-up the observation that the form of integrating depth-first with breadth-first search results in a kind of built-in adaptivity as explained in Section 2. The predominant behaviour of D&B-search corresponds to depth-first search if the search tree is finite and to breadth-first search if the search tree is infinite. This effect can be maintained if depth-first search is integrated with other complete search methods in the same way.

Its space complexity being exponential in depth, breadth-first search, although theoretically complete on infinite trees, cannot advance to very deep levels in practice. Iterative-deepening usually does better and is also complete. However, as pointed out in Section 1, iterative-deepening deteriorates time complexity in those cases in which depth-first search *is* complete. It would therefore be alluring if there was a possibility to use depth-first search whenever it is complete and iterative-deepening only when needed to ensure completeness. Alas, these conditions are not decidable as they stand.

But we can come very close to such a combination by transferring the principle of integration used for D&B-search to a combination of depth-first search and iterative-deepening. The result, called D&I-search, behaves predominantly like depth-first search if the search tree is finite and like iterative-deepening if the search tree is infinite.

Technically, this can be achieved by the same depth bounds f_i as with D&B-search. D&I-search even has the same representing ordering as D&B-search, so its completeness is just a corollary. In order to make sure that iterative-deepening does not expand any nodes that have already been expanded by depth-first search, iterative-deepening's algorithm needs to be slightly modified and the underlying data structure becomes slightly more complicated. This optimisation even results in the converse effect: with D&I-search, iterative-deepening can to some extent also prune the search space of depth-first search.

We plan to investigate D&I-search and also, using the same principle, other promising combinations of search methods.

Acknowledgements. We are grateful to Tim Furche, who read a draft of this paper and gave many valuable hints for its improvement. We thank all of our colleagues in the group for stimulating discussions about the work presented here.

References

1. Barták, R.: Incomplete depth-first search techniques: A short survey. In: Proceedings of 6th Workshop on Constraint Programming for Decision and Control (CPDC 2004), pp. 7–14 (2004)
2. Brodt, S.: Tree-search, partial orderings, and a new family of uninformed algorithms. Research report PMS-FB-2009-7, Institute for Informatics, University of Munich, Oettingenstraße 67, D-80538, München, Germany (2009), http://www.pms.ifi.lmu.de/publikationen#PMS-FB-2009-7
3. Bry, F.: Query evaluation in recursive databases: Bottom-up and top-down reconciled. Data and Knowledge Engineering 5(4), 289–312 (1990)
4. Debray, S., Ramakrishnan, R.: Abstract interpretation of logic programs using magic transformations. Journal of Logic Programming 18, 149–176 (1994)
5. Ginsberg, M.L., Harvey, W.D.: Iterative broadening. In: Proc. Eighth National Conference on Artificial Intelligence (AAAI 1990), pp. 216–220 (1990)
6. Kerisit, J.-M.: A relational approach to logic programming: the extended Alexander method. Theoretical Computer Science 69, 55–68 (1989)
7. Knuth, D.E.: The Art of Computer Programming, 3rd edn., vol. 1. Addison-Wesley Publishing Co., Reading (1997)
8. Korf, R.E.: Depth-first iterative-deepening: An optimal admissible tree search. Artificial Intelligence 27(1), 97–109 (1985)
9. Lloyd, J.W.: Foundations of Logic Programming, 2nd edn. Springer, Heidelberg (1987)
10. Nilsson, N.J.: Principles of Artificial Intelligence. Springer, Heidelberg (1982)
11. Pearl, J.: Heuristics: Intelligent Search Strategies for Computer Problem Solving. Addison-Wesley Publishing Co., Reading (1984)
12. Ruml, W.: Heuristic search in bounded-depth trees: Best-leaf-first search. Technical report, Harvard University (2002)

13. Shen, Y.-D., Yuan, L.-Y., You, J.-H.: SLT resolution for the well-founded seman-
    tics. Journal of Automated Reasoning 28, 53–97 (2002)
14. Stickel, M.E.: A Prolog technology theorem prover: Implementation by an extended
    Prolog compiler. Journal of Automated Reasoning 4(4), 353–380 (1988)
15. Tamaki, H., Sato, T.: OLDT resolution with tablulation. In: Wada, E. (ed.) Logic
    Programming 1986. LNCS, vol. 264, pp. 84–98. Springer, Heidelberg (1987)
16. Winston, P.H.: Artificial Intelligence, 3rd edn. Addison-Wesley Publishing Co.,
    Reading (1992)

Distributed Resolution
for Expressive Ontology Networks

Anne Schlicht and Heiner Stuckenschmidt

Knowledge Representation and Knowledge Management Research Group
Computer Science Institute
University of Mannheim
{anne,heiner}@informatik.uni-mannheim.de

Abstract. The Semantic Web is commonly perceived as a web of partially interlinked machine readable data. This data is inherently distributed and resembles
the structure of the web in terms of resources being provided by different parties at different physical locations. A number of infrastructures for storing and
querying distributed semantic web data, primarily encoded in RDF have been
developed but almost all the work on description logic reasoning as a basis for
implementing inference in the Web Ontology Language OWL still assumes a centralized approach where the complete terminology has to be present on a single
system and all inference steps are carried out on this system.

We propose a distributed reasoning method that preserves soundness and completeness of reasoning under the original OWL import semantics. The method is
based on resolution methods for $\mathcal{ALCHIQ}$ ontologies that we modify to work
in a distributed setting. Results show a promising runtime decrease compared to
centralized reasoning and indicate that benefits from parallel computation trade
off the overhead caused by communication between the local reasoners.

1   Introduction

Almost all the work on description logic reasoning as a basis for implementing inference in the Web Ontology Language OWL still assumes a centralized approach where
the complete terminology has to be present on a single system and all inference steps
are carried out on this system. This approach has a number of severe drawbacks. First of
all, the complete, possibly very large data sets have to be transferred to the central reasoning system creating a lot of network traffic. Furthermore, transferring the complete
models to a single reasoner also makes this a major bottleneck in the system. This can
go as far as reaching the limit of processable data of the reasoning system. A number of
approaches for distributed reasoning about interlinked ontologies have been proposed
that do not require the models to be sent to a central reasoner [5,7,10]. All these approaches rely on strong restrictions on the types of links between ontologies or the way
concepts defined in another ontology may be used and refined and thus introduce special kinds of links between data sets stored in different locations. In particular, none
of these approaches supports the standard definition of logical import from the OWL
specification, limiting their usefulness on real data sets. We illustrate these problems
using a small example.

A. Polleres and T. Swift (Eds.): RR 2009, LNCS 5837, pp. 87–101, 2009.

Example 1. We assume the two small ontologies depicted below are connected by an owl:imports statement in ontology B.

Ontology A	Ontology B
$A{:}Car \sqsubseteq A{:}Vehicle$	$B{:}HybridCar \sqsubseteq A{:}Vehicle$
$A{:}Car \sqsubseteq \exists_{\leq 1} A{:}hasEngine$	$B{:}HybridCar \sqsubseteq \exists_{\geq 2} A{:}hasEngine$

As we can easily see, $A{:}Car \sqsubseteq \neg B{:}HybridCar$, and hence adding the assertion "$A{:}Car \sqcap B{:}HybridCar(a)$" would yield an inconsistency.

Our goal is to have a distributed reasoning method that performs local reasoning on the two ontologies and that is still able to detect the inconsistency. Looking at the previous proposals for distributed reasoning mentioned above, we notice that none of them meets these requirements. The framework of ϵ-connections[7] does not apply in this scenario as it does not allow the specification of subsumption relationships between interlinked ontologies. Using the framework of conservative extensions[10] does not provide any advantages in terms of local reasoning. In particular, the overall model is neither a conservative extension of ontology A nor of ontology B as in both cases, the additional information in the other parts can be used to derive new information concerning the signature of ontology A or ontology B, respectively. Encoding the ontology network in distributed description logics, finally, the domains of A and B are disjoint by definition and the assertion "$A{:}Car \sqcap B{:}HybridCar(a)$" is not expressible.

Note that a set of ontologies linked by mapping axioms can also be represented in terms of OWL imports. In this case, the mapping axioms would be part of any of the two ontologies and and the other ontology would be imported by the one containing the mapping axioms.

Our aim is to develop a method for reasoning about description logic ontologies that overcomes the disadvantages of existing methods. We have designed and implemented a distributed reasoning method that 1) preserves soundness and completeness of reasoning under the original OWL import semantics 2) avoids restrictions on the use of definitions from remote models in local definitions or on the way knowledge is distributed a priori 3) decreases runtime by parallel computation, trading off the overhead caused by communication.

In previous work [15] we proposed a distributed resolution method for $\mathcal{ALC}$. The extension to $\mathcal{ALCHIQ}$ is complex because the ordered resolution calculus we used for $\mathcal{ALC}$ cannot handle the equality literals introduced by number restrictions. A resolution calculus that decides $\mathcal{ALCHIQ}$ is much more sophisticated and requires a more involved strategy for exchanging axioms between reasoning peers. The paper is structured as follows: In the next section we address the distribution principles and distributed resolution in general. Section 3 reviews the idea proposed in [15] and presents the details of our distributed reasoning method. In Section 4 we investigate the properties of the method with respect to number of derivations, communication effort and degree of parallelization and show that these parameters are promising.

2   Distributing Logical Resolution

2.1   Distribution Principles

There are various options for distributing the process of logical reasoning. Many of these options have been investigated in the field of automated theorem proving for first-order logics [4,3]. In the following we discuss these options and their pros and cons with respect to the requirements and goals defined in the introduction. In particular, we have to make two choices:

1. We have to choose a reasoning method that is sound and complete for description logics and permits distribution.
2. We have to choose a distribution principle that supports local reasoning and minimizes reasoning and communication costs.

Concerning the reasoning method, analytic tableaux are the dominant method for implementing sound and complete inference systems for description logics [8]. It has been shown, however, that sound and complete resolution methods for expressive description logics can be defined [16,9]. We exclude other existing methods such as a reduction of DL reasoning to logic programming from our investigation because these approaches are not sound and complete for the languages we are interested in. Because tableaux-based as well as resolution-based methods meet our requirements with respect to language coverage and completeness, the decisive factor is their suitability for distributed reasoning.

The survey [3] discusses different strategies for parallelizing logical inference. In particular, the authors distinguish between parallelism at the term-, clause and the search level where paralellism at the search level is further distinguished into multi-search and distributed search approaches. Parallelism at the term- and clause level is not suitable for our purposes as it speeds up basic reasoning functions such as matching or unification using a shared memory. The idea of multi-search approaches is to try different heuristics or starting points in parallel and require the complete logical model to be available to all reasoners. The distributed search paradigm naturally fits the distributed storage of parts of the model and therefore represents a paradigm that fits the goals of our research as it allows to assign the part of the search space relevant for a specific model to a local reasoner instance that interacts with other local reasoners if necessary. The choice of the distributed search paradigm has consequences for the choice of the reasoning method. In particular, it has been shown that distributed search can be used in combination with ordering-based methods [6,2] to support parallel execution of logical reasoning. We build on top of these results by proposing distributed reasoning methods based on the principles of resolution. Our proposal extends beyond the state of the art in distributed theorem proving as it addresses specific decidable subsets of first-order logics that have not yet been investigated in the context of distributed theorem proving. Furthermore, existing strategies for assigning inference steps to reasoners such as the ancestor-graph criterion [2] cannot avoid redundancy. We propose a method based on ordered resolution that takes advantage of the special structure of clauses in the description logic $\mathcal{ALCHIQ}$ for efficiently deciding satisfiability in a distributed setting.

2.2 Resolution Theorem Proving

Before describing our distributed resolution method for ontologies, we first briefly review standard resolution reasoning and present the basic idea for distributed resolution.
Resolution is a very popular reasoning method for first order logic (FOL) provers. As
description logics are a strict subset of first order logic, resolution can be applied to description logic ontologies as well [17]. For this purpose the DL ontology is transformed
into a set of first order clauses as defined in Section 3.2. This translation can be done
on a per axiom basis independently of other parts of the model. It can be shown that the
ontology is satisfiable if and only if the set of clauses is satisfiable. The set of clauses is
satisfiable iff exhaustive application of the rule standard resolution with factoring does
not derive an empty clause.

Definition 1 (Standard Resolution). *For clauses C and D and literals A and $\neg B$,
standard resolution with factoring is defined by the rule*

$$\text{Standard Resolution with Factoring} \quad \frac{C \vee A_1 \vee \cdots \vee A_n \quad D \vee \neg B}{C\sigma \vee D\sigma}$$

where the substitution σ is the most general unifier of $A_1, \ldots, A_n$ and B.

2.3 Distributed Resolution

The implementation of a resolution algorithm is described in [18]. For an input set of
clauses, it systematically applies resolution rules to appropriate pairs of clauses and
adds the derived new clauses to the clause set. If an empty clause is derived or no new
and non-redundant clause can be derived, the algorithm terminates. An essential part of
a resolution prover and the most time consuming component [18] are reduction rules
that delete clauses that are not necessary for the decision process. Without reduction,
the number of clauses generally increases infinitely and it may be impossible to saturate
even a small set of clauses.

   As described in [15], a standard resolution algorithm can be modified to support
distributed reasoning. In particular, the inferences can be distributed across different
reasoners by separating the set of input clauses and running provers on separate parts
of the set:

- Every reasoner separately saturates the clause set assigned to it.
- Newly derived clauses are propagated to other reasoners if necessary.

Instead of adding every clause that is derived to the local set of clauses, some new
clauses are propagated to other reasoners and deleted locally.

Definition 2 (Allocation). *An* allocation *for a set $\mathcal{C}$ of clauses and a set of ontology
modules M is a relation $a \in (\mathcal{C} \times M)$ such that*

$$\forall c \in \mathcal{C} \colon \exists m \in M \colon a(c, m)$$

The set of modules a clause c is allocated to by the allocation a is defined by

$$a(c) := \{m \in M \mid a(c, m)\}$$

If the allocation relation is functional we may omit the parenthesis and write $a(c) = m$.

In addition to the propagation of clauses we have to add a second modification to the algorithm to turn it into a distributed resolution algorithm. In contrast to the centralized case, a reasoner that has saturated the local clause set may have to continue reasoning once a new clause is received from another reasoner. The whole system of connected reasoners stops if the empty clause is derived by one of the reasoners or all are saturated.

After this intuitive description of a distributed resolution algorithm, we define distributed resolution formally:

Definition 3 (Distributed Resolution Calculus). *A* distributed resolution calculus $R(a)$ *is a resolution calculus that depends on an allocation relation* $a \colon C \to M$ *such that each rule* r *of* $R(a)$ *is restricted to premises* $P \subset C$ *with*

$$\exists m \in M \colon \forall c \in P \colon a(c, m)$$

We call this restriction allocation restriction.

Hence, the rules of a distributed resolution calculus are restricted to premises allocated to the same module. A distributed calculus can be obtained from any resolution calculus by defining an allocation relation and adding the allocation restriction to each rule of the calculus.

Obviously, termination of the underlying calculus is preserved by distribution if it does not depend on reduction rules. In the worst case, each inference of the original calculus is performed once in every module of the distributed calculus. The results presented in Section 4 also indicate that local reduction (i.e. deleting clauses that are redundant with respect to the reasoner they are processed by) is sufficient in practice.

Preserving completeness without allocating each clause to every reasoner is more difficult, we have to make sure the allocation restriction never excludes inferences that are possible in the original calculus. For standard resolution, a given clause C has to be propagated to any reasoner whose clause set contains a clause with a literal that matches (i.e. is unifiable and of opposite polarity) any of the literals in C. This would lead to a substantial communication overhead and potentially redundant inference steps.

To avoid redundancy, we aim at allocating every clause to only a single reasoner. A functional allocation guarantees that the same resolution step is never carried out twice, because equivalent clauses are always assigned to the same unique reasoner which takes care of avoiding local redundancy.

3 Distributed Resolution for Description Logic

As we have seen above, the ability to define a sound and complete distributed reasoning method relies on two requirements: (1) the existence of a sound and complete resolution calculus and (2) the ability to find a corresponding allocation that satisfies the allocation restriction. In this section, we show that for the case of ontologies defined in $\mathcal{ALCHIQ}$ both of these requirements can be satisfied leading to a sound and complete distributed resolution method. We do not address reduction rules in this section because reduction is not necessary to guarantee the theoretical properties of the proposed calculus. However, for efficient reasoning reduction is essential and hence the practical effects distribution has on reduction are discussed in the experimental section.

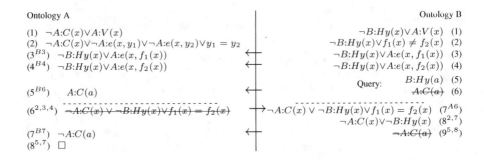

Ontology A

(1) $\neg A{:}C(x)\vee A{:}V(x)$
(2) $\neg A{:}C(x)\vee\neg A{:}e(x,y_1)\vee\neg A{:}e(x,y_2)\vee y_1=y_2$
(3^{B3}) $\neg B{:}Hy(x)\vee A{:}e(x,f_1(x))$
(4^{B4}) $\neg B{:}Hy(x)\vee A{:}e(x,f_2(x))$

(5^{B6}) $A{:}C(a)$

$(6^{2,3,4})$ $\neg A{:}C(x)\vee\neg B{:}Hy(x)\vee f_1(x)=f_2(x)$

(7^{B7}) $\neg A{:}C(a)$
$(8^{5,7})$ $\square$

Ontology B

$\neg B{:}Hy(x)\vee A{:}V(x)$ (1)
$\neg B{:}Hy(x)\vee f_1(x)\neq f_2(x)$ (2)
$\neg B{:}Hy(x)\vee A{:}e(x,f_1(x))$ (3)
$\neg B{:}Hy(x)\vee A{:}e(x,f_2(x))$ (4)

Query: $B{:}Hy(a)$ (5)
$A{:}C(a)$ (6)

$\neg A{:}C(x)\vee\neg B{:}Hy(x)\vee f_1(x)=f_2(x)$ (7^{A6})
$\neg A{:}C(x)\vee\neg B{:}Hy(x)$ $(8^{2,7})$
$\neg A{:}C(a)$ $(9^{5,8})$

Fig. 1. Distributed refutation example. The designer of ontology B from Example 1 wants to check satisfiability of the concept "$B{:}HybridCar\sqcap A{:}Car$" and adds the the appropriate query. Since the concept is unsatisfiable, an empty clause is derived. Predicates are abbreviated to simplify presentation, derived clauses are noted below the dashed line, arrows denote propagation of a clause. Literals that are not resolvable literals are grayed out (assuming predicates from B precede predicates from A and $A{:}V > A{:}C$). Clauses that are striked out are locally deleted on propagation.

3.1   Distribution Principle

The idea for our distributed reasoning approach is to take advantage of the restrictions description logic imposes on first order logic. In particular, we identify a property that holds for many efficient resolution calculi and use it for defining a distribution principle. The important property of a resolution calculus is, that each clause contains only one resolvable literal. I.e. for every possible inference the resolvable literal (or a subterm of it) is unified with a literal of another premise, the other literals are (possible with substituted variables) passed to the conclusion. Formally, the resolvable literal and its uniqueness are defined as follows:

Definition 4 (Resolvable Literal). *A literal* lit *of a clause* $C\vee$ lit *is a* resolvable literal *of* $C\vee$ lit *with respect to a calculus* R *and logical language* L *iff there is a clause* $D\vee$ lit$'\in L$, *such that* R *can be applied to the premises* $C\vee$ lit *and* $D\vee$ lit$'$ *deriving the clause* $(C\vee D\vee$ lit$'')\sigma$ *with appropriate substitution* σ *and literal* lit$''$[1].

In standard resolution all literals of a clause are resolvable literals, but more advanced calculi restrict the applicability of resolution rules such that there is only one resolvable literal in each clause. In particular, for the ordered resolution calculus defined in [11] for $\mathcal{ALC}$ description logic, each clause contains an unique resolvable literal [15]. Based on the uniqueness of the resolvable literal and an allocation of symbols to reasoners we can define an allocation function that allocates every clause to one module of the networked ontology. Note that for ontologies linked by import statements, the namespaces define the allocation of symbols. For distributed reasoning on a single ontology, the ontology is first partitioned into linked modules as described in [14].

---

[1] For ordered resolution lit'' is false and may be omitted.

Definition 5 (Allocation a(c))

$$a(c) := \{alloc(topSymbol(lit)) \mid lit \text{ is resolvable literal of } c\}$$

where topSymbol of a possibly negated predicate literal $(\neg)P(t_1, ..., t_n)$ *is* P.

If all clauses have unique resolvable literals, then the allocation a is functional, too. For determining where (and if) a derived clauses is propagated, we first pick the unique resolvable literal of the clause, then the top symbol of this literal and finally the reasoner this symbol is allocated to. If a symbol s is allocated to a module m we say that module m is *responsible* for s. Figure 1 illustrates distributed resolution on the ontologies from Example 1. In [15] we showed that ordered resolution with this allocation function is a complete distributed method for deciding $\mathcal{ALC}$ satisfiability because the inferences are the same for distributed and centralized resolution. However, for supporting more expressive ontologies, a more complex calculus is required and the allocation function defined above is not sufficient to guarantee completeness.

3.2 Preliminaries

Before presenting the calculus our distributed method is based on, we define the description logic we use and the translation of description logic axioms to first order clauses.

The Description Logic $\mathcal{SHIQ}$. A $\mathcal{SHIQ}$ ontology is a set $\mathcal{O}$ of axioms α of the following syntax in BNF:

$$
\begin{aligned}
\alpha ::= &C \sqsubseteq C \mid C \equiv C \mid C(x) \mid R(x,x) & n ::= &number \\
&\mid Trans(R) \mid R \sqsubseteq R & A ::= &concept\_name \\
C ::= &\top \mid \bot \mid A \mid \neg C \mid C \sqcap C \mid \exists R.C \mid \forall R.C & R ::= &role\_name \mid Inv(role\_name) \\
&\mid \exists_{\leq n} R.C \mid \exists_{\geq n} R.C & x ::= &individual\_name
\end{aligned}
$$

The *signature* of an ontology $Sig(\mathcal{O})$ is the disjoint union of concept names, role names and individual names.

Normalization. The resolution calculus we apply requires first order clauses as input, hence the first order formulas obtained from an ontology are translated to clauses. To guarantee termination of the applied resolution calculus, the ontology has to be normalized prior to clausification. This ensures that only certain types of axioms and corresponding clauses occur in the reasoning procedure. For simplicity, we assume the ontology contains only subsumption axioms $A \sqsubseteq C$ where A is not a complex concept and no equivalence axioms. Complex subsumptions $C \sqsubseteq D$ are equivalent to $\top \sqsubseteq \neg C \sqcup D$ and equivalences $C \equiv D$ can be replaced by two subsumptions $C \sqsubseteq D$ and $D \sqsubseteq C$. The definitorial form normalization we use replaces complex concepts C in the right hand side of an axiom by a new concept name A and adds the axiom $A \sqsubseteq C$ to the ontology. Thus, it splits up nested axioms into simple ones by introducing new concepts.

Definition 6 (Definitorial Form). *For simple subsumptions* $A \sqsubseteq D$ *with atomic concept* A *the Definitorial Form is defined by*

$$Def(A \sqsubseteq D) := \begin{cases} \{A \sqsubseteq D\} & \text{if all subterms of } D \text{ are literal concepts} \\ \{Q \sqsubseteq D|_p\} \cup Def(A \sqsubseteq D[Q]_p) & \text{if } D|_p \text{ is not a literal concept} \end{cases}$$

where $D|_p$ denotes a certain[2] subterm of D and $D[Q]_p$ is the term obtained by replacing this subterm with Q.

Clausification. After normalization, the ontology contains only simple axioms that can be translated to first order clauses as follows:

$A \sqsubseteq B$	$\neg A(x) \vee B(x)$	$A \sqsubseteq \exists_{\leq n} r.B$	$\neg A(x) \vee \neg r(x, y_i) \vee y_i = y_j \vee \neg B(y_i)$
$A \sqsubseteq B \sqcap C$	$\neg A(x) \vee B(x)$		$i = 1..n+1 \quad j = 1..i-1$
	$\neg A(x) \vee C(x)$	$A \sqsubseteq \exists_{\geq n} r.B$	$\neg A(x) \vee r(x, f_i(x)) \quad i = 1..n$
$A \sqsubseteq B \sqcup C$	$\neg A(x) \vee B(x) \vee C(x)$		$\neg A(x) \vee f_i(x) \neq f_j(x) \quad j = 1..i-1$
$A \sqsubseteq \exists r.B$	$\neg A(x) \vee r(x, f(x))$		$\neg A(x) \vee B(f_i(x))$
	$\neg A(x) \vee B(f(x))$	$r \sqsubseteq s$	$\neg r(x, y) \vee s(x, y)$
$A \sqsubseteq \forall r.B$	$\neg A(x) \vee \neg r(x, y) \vee B(y)$	$r \equiv Inv(s)$	$\neg r(x, y) \vee s(y, x)$
			$\neg s(x, y) \vee r(y, x)$

The clauses resulting from the ontologies of Example 1 are depicted in Figure 1.

3.3 Distributed Resolution for $\mathcal{ALCHIQ}$

When an ontology contains qualified($\mathcal{Q}$) or unqualified($\mathcal{N}$) number restrictions or functional properties ($\mathcal{F}$), the translation to clauses contains equalities. To deal with these equalities, a much more sophisticated calculus than ordered resolution is required which in turn requires a more involved allocation of clauses to ontology modules. Before we present the necessary adaptions and extensions to the distributed resolution method, we briefly describe the calculus our method is based on. The DL expressivity that can be covered with this calculus is $\mathcal{ALCHIQ}^-$ which is $\mathcal{SHIQ}$ without transitive properties with the additional restriction that number restrictions are only allowed on roles that do not have subroles. Extension of the method for supporting $\mathcal{SHOIQ(D)}$ is discussed in the next subsection.

Resolution Calculus for $\mathcal{ALCHIQ}^-$. A complete calculus that terminates on clauses obtained from ontologies that contain number restrictions is *basic superposition* [1,9], an extension of ordered resolution. Like ordered resolution, basic superposition uses two parameters, a *selection function* and *ordering of literals* that restrict applicability of the resolution rules.

As usual for theories containing equalities, we assume a translation of predicates to general function symbols such that all literals are equalities (e.g. the literal $P(x)$ translates to $P(x) \approx \top$), we may still write $P(x)$ for readability purpose and call these literals *predicate literals*. Clauses are split into skeleton clause C and substitution σ representing all substitutions introduced by previous unifications. The clause $C\sigma$ is

---

[2] The exact definition of $|_p$ (*position*) is not relevant in this paper, please refer to [11] for detail.

denoted as closure $C \cdot \sigma$ or alternatively a closure is denoted by enclosing non-variable subterms of $C\sigma$ that correspond to variables in C in brackets (e.g. $P([f(y)])$ for $P(x) \cdot \{x \mapsto f(y)\}$). For distributing basic superposition, the rules we have to take care of are positive and negative superposition, the other rules contain only one premise and hence distribution of the input clauses into separate sets does not restrict application of these rules.

Definition 7 (Superposition)

$$\text{Positive superposition} \quad \frac{(C \vee s \approx t) \cdot \rho \quad D \vee (w \approx v) \cdot \rho}{(C \vee D \vee w[t]_p \approx v) \cdot \theta}$$

where

1. σ is the most general unifier of $s\rho$ and $w\rho|_p$ and $\theta = \rho\sigma$
2. $t\theta \not\preceq s\theta$ and $v\theta \not\preceq w\theta$
3. in $(C \vee s \approx t) \cdot \theta$ nothing is selected and $(s \approx t) \cdot \theta$ is strictly maximal
4. in $D \vee (w \approx v) \cdot \theta$ nothing is selected and $(w \approx v) \cdot \theta$ is strictly maximal
5. $w|_p$ is not a variable.
6. $s\theta \approx t\theta \not\preceq w\theta \approx v\theta$

$$\text{Negative superposition} \quad \frac{(C \vee s \approx t) \cdot \rho \quad D \vee (w \not\approx v) \cdot \rho}{(C \vee D \vee w[t]_p \not\approx v) \cdot \theta}$$

where

1. σ is the most general unifier of $s\rho$ and $w\rho|_p$ and $\theta = \rho\sigma$
2. $t\theta \not\preceq s\theta$ and $v\theta \not\preceq w\theta$
3. in $(C \vee s \approx t) \cdot \theta$ nothing is selected and $(s \approx t) \cdot \theta$ is strictly maximal
4. $(w \not\approx v) \cdot \theta$ is selected or maximal and no other literal is selected in $D \vee (w \not\approx v) \cdot \theta$
5. $w|_p$ is not a variable.

In addition to the superposition rules, two rules with only one premise are necessary to deal with equalities (see [9] for details). Ordered resolution is a special case of positive superposition, where $w|_p = w$, i.e. p is the root position. A sequence of ordered resolution inferences can be combined into a *ordered hyperresolution* inference by deleting intermediate conclusions. We assume that ordered hyperresolution and not ordered resolution is applied to clauses containing multiple resolvable literals (see derivation of clause A6 in Figure 1).

Definition 8 (Resolution Calculus R_Q [9])
R_Q *is the calculus with 1) rules positive and negative superposition, reflexivity resolution and equality factoring, 2) selection of every negative binary literal, 3) the term ordering $\succ_Q$ is a lexicographic path ordering (LPO, [12]) based on a total precedence $>$ of function, constant and predicate symbols with $f > c > P > \top$ for every function f constant c and predicate P.*

Literals containing different variables are $\succ_Q$-incomparable because otherwise the ordering would depend on the substitution. Literals that contain a function symbol are

Table 1. The 8 types of $\mathcal{ALCHIQ}$ closures [9]

1	$\neg R(x, y) \vee Inv(R)(x, y)$
2	$\neg R(x, y) \vee S(x, y)$
3	$\mathbf{P^f}(x) \vee R(x, \langle f(x) \rangle)$
4	$\mathbf{P^f}(x) \vee R([f(x)], x)$
5	$\mathbf{P_1}(x) \vee \mathbf{P_2}(\langle \mathbf{f}(x) \rangle) \vee \bigvee \langle f_i(x) \rangle \approx/\napprox \langle f_j(x) \rangle$
6	$\mathbf{P_1}(x) \vee \mathbf{P_2}([g(x)]) \vee \mathbf{P_3}(\langle \mathbf{f}[g(x)] \rangle) \vee \bigvee \langle t_i \rangle \approx/\napprox \langle t_j \rangle$
7	$\mathbf{P_1}(x) \vee \bigvee_{i=1}^{n} \neg R(x, y_i) \bigvee_{i=1}^{n} \mathbf{P_2}(y_i) \vee \bigvee_{i=1\ j=i+1}^{n\ \ n} y_i \approx y_j$
8	$\mathbf{R}(\langle \mathbf{a} \rangle, \langle \mathbf{b} \rangle) \vee \mathbf{P}(\langle \mathbf{t} \rangle) \vee \bigvee \langle t_i \rangle \approx/\napprox \langle t_j \rangle$

$\mathbf{P}(t)$, where t is a term, denotes a possibly empty disjunction of the form $(\neg)P_1(t) \vee \cdots \vee (\neg)P_n(t)$. $\mathbf{P}(\mathbf{f}(x))$ denotes a disjunction of the form $\mathbf{P_1}(f_1(x)) \vee \cdots \vee \mathbf{P_m}(f_m(x))$. Note that this definition allows each $\mathbf{P_i}(f_i(x))$ to contain positive and negative literals. $\langle t \rangle$ denotes that term t may but need not be marked (i.e. has been introduced by a previous unification), $\approx/\napprox$ denotes a positive or negative equality predicate. For clauses of type 6 t_i and t_j are either of the form $f([g(x)])$ or of the form x and the clause contains at least one term $f(g(x))$.

ordered first to avoid substituting the arguments of functions with function terms. Limited nesting depth of literal terms is necessary to guarantee termination of the calculus, it makes sure only the types of clauses depicted in Table 1 occur when basic superposition is applied to clauses obtained from an $\mathcal{ALCHIQ}$ ontology (i.e. the set of $\mathcal{ALCHIQ}$ closures is closed under basic superposition).

Because the set of clause types is finite and the set of symbols is finite for every given ontology, the number of clauses that can be derived is finite, too and hence basic superposition terminates for $\mathcal{ALCHIQ}$ input[9].

Allocation for $\mathcal{ALCHIQ}$. The first consideration for distributing basic superposition is the number of resolvable literals. A close look to Definition 7 reveals that the resolvable literals $s \approx t \cdot \rho$ and $w \approx/\napprox v \cdot \rho$ are necessarily either selected or maximal. Furthermore, the $\mathcal{ALCHIQ}$ closures of types 3-6 and 8 are totally ordered and types 1 and 2 contain exactly one selected literal. Only closures of type 7 may contain multiple resolvable literals, but since all resolvable literals have the same top symbol, the allocation from Definition 5 is still functional. Before allocating $\mathcal{ALCHIQ}$ closures, we have to extend the definition of *topSymbol* to equality literals:

Definition 9 (Top Symbol)
The top Symbol of an equality literal $f(t_1) \approx/\napprox g(t_2)$ is f if $f(t_1) \succ g(t_2)$.

Note that arguments of equalities that are resolvable literal of an $\mathcal{ALCHIQ}$ clause are always comparable. With this extended definition, we could use the allocation function a for distributed resolution on $\mathcal{ALCHIQ}$. Unfortunately, this calculus would not be complete, because some inferences of basic superposition are prevented in the distributed setting. We illustrate this problem on an example inference:

$$\text{Positive superposition} \quad \frac{C \vee f(x) \approx g(x) \qquad D \vee P(f(y))}{C \vee D \vee P(g(x))}$$

Here, $f(x)$ is unified with $f(y)$ but f is not the top symbol of the resolvable literal $P(f(y))$. Hence, if f and P are allocated to different ontology modules, the rule is not applied and the clause $C \vee D \vee P(g(x))$ is missing in the reasoning process. Due to superposition of equalities into predicate literals, we have to extend the allocation to guarantee completeness of the decision procedure.

Definition 10 (Allocation for $\mathcal{ALCHIQ}$). *The clause allocation $a_+(c)$ for the distributed calculus $R_Q(a_+)$ is defined by $a_+(c) := a(c) \cup a_f(c)$ with*

$$a_f(c) := \{alloc(funSymbol(lit)) \mid lit \text{ is resolvable literal of } c\}$$

where

- *$funSymbol(lit) := f$ for every literal $lit = (\neg)P(f(t))$ or $lit = (\neg)P(f(x), x)$ or $lit = (\neg)P(x, f(x))$ with unary or binary predicate symbol P. For other literals $funSymbol(lit)$ is null.*
- *$alloc \colon Sig(\mathcal{O}) \to M$ is an allocation of the signature symbols of the input ontology $\mathcal{O}$, including concepts introduced by the definitorial form transformation. $alloc(null) := \emptyset$*

Note that resolvable literals of closures of type 7 never contain function symbols, hence for all $\mathcal{ALCHIQ}$ closures c the allocation $a_f(c)$ is a function and the set $a_+(c)$ consists of at most two modules. The allocation a_+ solves the problem of the example depicted above. But, it remains to be proved that no other pair of premises that could be resolved in a basic superposition inference is allocated to different ontology modules and hence completeness of the calculus for $\mathcal{ALCHIQ}$ is preserved by distribution.

Theorem 1 (Completeness of Distributed Resolution for $\mathcal{ALCHIQ}$). *The distributed resolution calculus $R_Q(a_+)$ decides $\mathcal{ALCHIQ}$ satisfiability.*

Since R_Q decides $\mathcal{ALCHIQ}$, it remains to be shown that every inference in the original calculus is performed in the distributed calculus, too. Let us first consider superposition into root position (i.e. $w|_p = w$). In this case, basic superposition is equivalent to ordered resolution, both premises are allocated to the same module because the two resolvable literals have the same top symbol. Superposition at other positions is only possible for function equalities into predicate literals i.e. $s \approx t$ is an equality literal $f(x) \approx / \not\approx g(x)$ and $w \approx / \not\approx v$ is a predicate literal $(\neg)P(f(x))$ or $(\neg)R(x, f(x))$. Variable equations are never selected or maximal and hence no resolvable literals. If $s \approx t$ is a predicate literal or $w \approx / \not\approx v$ is an equality, superposition is only possible at root position, otherwise unification is impossible or w would be a variable which is not allowed according to Definition 7.

Hence, for every application of a rule in R_Q, the allocation a_+ ensures all premises meet in one module. A clause is allocated to at most two modules, local saturation of the local clause sets is enough to guarantee completeness of the method. Note that only clauses are duplicated, duplication of inferences can be avoided by restricting basic superposition such that only the module responsible for the top symbol of $s \approx t$ in Definition 7 performs the inference.

3.4 Extension to $\mathcal{SHOIQ(D)}$

The expressivity that can be handled by basic superposition while guaranteeing termination is $\mathcal{ALCHIQ}^-$ which is $\mathcal{ALCQ}$ plus role hierarchies and inverse roles, with the restriction that number restrictions are only allowed on roles that do not have subroles. For extending the expressivity to $\mathcal{ALCHIQ}$ the *decomposition rule* has to be added to the calculus[9]. Decomposition is an reduction rule that is applied to newly derived clauses eagerly. Since the decomposition rule has only one premise, it can be added to our approach without restricting the possible inferences.

Transitivity axioms contained in a $\mathcal{SHIQ}$ ontology can be eliminated by a well known transformation, reducing the expressivity to $\mathcal{ALCHIQ}$. Hence, with some pre-processing we can decide satisfiability of a $\mathcal{SHIQ}$ ontology. The transformation is polynomial in the size of the input, but the adaption to the distributed setting is not trivial. In contrast to the transformation of description logic axioms to first order clauses mentioned so far, the translation of transitivity depends on the whole ontology and not only on the transitivity axioms. Hence, the linked ontologies cannot be transformed independently. Nominals are concepts with a single instance, e.g. the concept $\{Erdös\}$ with instance *Erdös* is used in the concept description $\exists coauthorOf\{Erdös\}$. Nominals are replaced by common concepts for many applications: Each nominal $\{nom\}$ is replaced by a new concept Nom and the axiom $Nom(nom)$ is added to the ontology. The restriction that Nom may not contain another instance is not expressible in description logic without nominals. However, it can be expressed by the first order clause $\neg Nom(x) \lor x = nom$.

Datatypes ($\mathcal{D}$) can be eliminated without changing the semantics by moving the datatypes into the abstract domain. In practice, sorts (datatype and abstract) are handled different from the other predicates to speed up reasoning. Built-in datatype predicates can be added to support e.g. the *greater* relation between integers.

4   Experiments

Our distributed resolution implementation is based on the first order prover SPASS[3] [19]. A number of different resolution strategies including ordered resolution and basic superposition are supported, precedence and selection are specified in the input file. We implemented definitorial form normalization and clausification in a separate tool. Clauses are stored in separate files for each ontology and include precedence and selection in every input file. Apart from compliance to the requirements of R_Q (Definition 8) the precedence was random. The applied reduction rules include forward and backward subsumption reduction[4]. For turning SPASS into a distributed reasoner (i.e. adding the "Distributed" option) we added support for sending and receiving clauses. A set of received clauses is treated like a set derived from a given clause, i.e. it is forward and backward reduced with respect to the local worked off clause list before adding the non redundant received clauses to the usable list. All reasoners are connected at startup, clause communication is performed in separate processes to avoid the local reasoning being blocked on sending a clause. The priority of the reasoning and communication

---

[3] http://www.spass-prover.org
[4] The complete configuration for Spass is: Distributed=1 Auto=0 Splits=0 Ordering=1 Sorts=0 Select=3 FullRed=1 IORe=1 IOFc=1 IEmS=0 ISoR=0 IOHy=0 RFSub=1 RBSub=1 RInput=0 RSSi=0 RObv=1 RCon=1 RTaut=1 RUnC=1 RSST=0 RBMRR=1 RFMRR=1.

Table 2. Results of tests on the chem ontology from SWEET project

Query	# Parts	Runtime/ms	# Derivations	# Propagations	Busy Factor
Satisfiability	1	230	608	-	100%
	13	146	610	432	20%
Subsumption	1	132	154	-	100%
(positive)	13	36	578	252	60%

processes is adjusted such that while not saturated locally, a clause is only send to another reasoner if this destination reasoner messages that it is idle. New clauses are only received when the local clause set is completely saturated. Startup and shutdown of the system is initialized by a central control process. In a fully decentralized P2P system this job is performed by the peer that receives a query. The control process starts the separate machines on their respective input clauses files. Apart from passing clauses between each other, the reasoners send status messages whenever they are locally saturated, when they continue reasoning on newly received clauses and when they derive an empty clause. When one reasoner finds a proof or all reasoners are saturated for an interval longer than the maximal time necessary for clause propagation the query is answered and the reasoners are shut down.

Dataset. Our implementation is tested on the Semantic Web for Earth and Environmental Terminology (SWEET [13]), a set of linked ontologies published by the NASA Jet Propulsion Laboratory. We used the chemical ontology chem[5] and the ontologies that are directly or indirectly imported by chem. In total, our dataset consists of 13 ontologies liked by 34 import statements. The ontology network describes 480 classes and 99 individuals, translation to first order logic yields 930 clauses. We replaced datatype properties by object properties and nominals by common concepts because the current version of our system does not support them. The expressivity of the obtained test ontology network is $\mathcal{SHIN}$.

Results. For comparing the runtimes in the centralized and distributed setting, we ran a satisfiability check and tested all 456 positive subsumption queries (i.e. axioms $A_1 \sqsubseteq A_2$ derivable from the ontology network). The runtime of negative queries is in general similar to the runtime of a satisfiability check [15]. We used standard 1.6GHz desktop machines with 3GB RAM, the denoted runtimes do not include the time needed for establishing the TCP connections. For simplicity, we connect all reasoners prior to the distributed reasoning process, it would be much more efficient to connect the peers only on demand. The connection time is not relevant for our investigations because it can be expected to increase only linear with the number of ontologies if the network is sparsely connected. The source code and original and preprocessed dataset is available online[6]. The most important result is that distributed resolution is considerably faster than conventional resolution. The runtime for checking satisfiability of the knowledge base is decreased by one third. Answering a positive subsumption query took only about a quarter of the runtime when computed in the distributed setting.

---

[5] http://sweet.jpl.nasa.gov/2.0/chem.owl

[6] http://ki.informatik.uni-mannheim.de/dire.html

The number of derived clauses shows the effect of distribution on application of reduction rules. For the satisfiability query, almost all redundant clauses are detected and deleted also in the distributed setting. However, positive queries cause much more derivations than necessary. Runtime is not affected by the redundant derivations because they are performed by reasoners that would have been idle otherwise. The number of propagation is important when the network connection is slow e.g. due to physical distance between the reasoning peers.

The most important factor for scalability of our approach is the amount of computation that is actually performed in parallel. In the worst case, only one reasoning peer is active at the same time while the others are idle and waiting for new input. For technical reasons, the timer for computing the busy factor starts when all reasoners are connected. The busy factor depicted in table 2 is the average (weighted by runtime) percentage of the runtime each reasoner is active. The first couple of milliseconds all reasoners are busy, but after local saturation only those that received new clauses continue reasoning. For some positive queries the busy factor reached 100% because the empty clause is derived in one reasoner before one of the others is locally saturated.

5   Conclusions

In this paper, we have shown that the principle of distributed resolution as a basis for reasoning about interlinked ontologies that has been proposed in previous work [15] can be extended to expressive ontology languages. This result is non-trivial, because ordered resolution, that has been used as a basis for previous work cannot be applied to expressive ontology languages due to the existence of equality induced by number restrictions. Our work extends previous results both on a theoretical and a practical level. On the theoretical level, we have developed a distributed version of the basic superposition calculus presented by [9] and have shown that the distributed version of the calculus decides satisfiability for $\mathcal{ALCHIQ}$. On the practical level, we have extended the implementation of our distributed reasoning engine with this new calculus and have tested it on a set of expressive real world ontologies. By conducting experiments, we have shown that the distributed version of the algorithm significantly outperforms centralized reasoning. Further, we have investigated the ability of the method to support parallelization with promising results. In summary, we have shown that the principle of distributed resolution can be applied to expressive ontologies and that distributed resolution is a real alternative to tableaux-based methods when it comes to distributed reasoning in the presence of OWL ontologies. In the future, we will investigate optimizations of the methods, primarily in terms of advanced redundancy checking. Further, we plan to exploit the advantages of resolution being a bottom-up reasoning method by investigating the use of our method for supporting incremental reasoning.

References

1. Bachmair, L., Ganzinger, H., Lynch, C., Snyder, W.: Basic paramodulation. Inf. Comput. 121(2), 172–192 (1995)
2. Bonacina, M.P.: The clause-diffusion theorem prover peers-mcd (system description). In: McCune, W. (ed.) CADE 1997. LNCS, vol. 1249, pp. 53–56. Springer, Heidelberg (1997)

3. Bonacina, M.P.: A taxonomy of parallel strategies for deduction. Annals of Mathematics and Artificial Intelligence 29(1–4), 223–257 (2001) (Published in February 2001)
4. Bonacina, M.P., Hsiang, J.: Parallelization of deduction strategies: An analytical study. J. Autom. Reasoning 13(1), 1–33 (1994)
5. Borgida, A., Serafini, L.: Distributed description logics: Assimilating information from peer sources. Journal of Data Semantics 1, 153–184 (2003)
6. Conry, S.E., MacIntosh, D.J., Meyer, R.A.: Dares: A distributed automated reasoning system. In: Proc. AAAI 1990, pp. 78–85 (1990)
7. Grau, B.C., Parsia, B., Sirin, E.: Combining owl ontologies using e-connections. Journal of Web Semantics 4(1) (2005)
8. Donini, F., Lenzerini, M., Nardi, D., Schaerf, A.: Reasoning in description logics. In: Brewka, G. (ed.) Principles of Knowledge Representation and Reasoning. Studies in Logic, Language and Information, pp. 193–238. CLSI Publications, Stanford (1996)
9. Hustadt, U., Motik, B., Sattler, U.: Reasoning in Description Logics by a Reduction to Disjunctive Datalog. Journal of Automated Reasoning 39(3), 351–384 (2007)
10. Lutz, C., Walther, D., Wolter, F.: Conservative extensions in expressive description logics. In: Twentieth International Joint Conference on Artificial Intelligence IJCAI 2007 (2007)
11. Motik, B.: Reasoning in Description Logics using Resolution and Deductive Databases. PhD thesis, Universität Karlsruhe (TH), Karlsruhe, Germany (January 2006)
12. Nieuwenhuis, R., Rubio, A.: Theorem proving with ordering and equality constrained clauses. Journal of Symbolic Computation 19, 321–351 (1995)
13. Raskin, R.G., Pan, M.J.: Knowledge representation in the semantic web for Earth and environmental terminology SWEET. Computers & Geosciences 31(9), 1119–1125 (2005)
14. Schlicht, A., Stuckenschmidt, H.: A flexible partitioning tool for large ontologies. In: International Conference on Web Intelligence and Intelligent Agent Technology, WI/IAT (2008)
15. Schlicht, A., Stuckenschmidt, H.: Peer-to-peer reasoning for interlinked ontologies. International Journal of Semantic Computing, Special Issue on Web Scale Reasoning (2010) (to be published)
16. Tammet, T.: Resolution methods for Decision Problems and Finite Model Building. PhD thesis, Chalmers University of Technology and University of Göteborg (1992)
17. Tsarkov, D., Riazanov, A., Bechhofer, S., Horrocks, I.: Using vampire to reason with OWL. In: McIlraith, S.A., Plexousakis, D., van Harmelen, F. (eds.) ISWC 2004. LNCS, vol. 3298, pp. 471–485. Springer, Heidelberg (2004)
18. Weidenbach, C.: Combining superposition, sorts and splitting. In: Robinson, A., Voronkov, A. (eds.) Handbook of Automated Reasoning, ch. 27, vol. II. Elsevier, Amsterdam (2001)
19. Weidenbach, C., Brahm, U., Hillenbrand, T., Keen, E., Theobald, C., Topic, D.: SPASS Version 2.0. In: Voronkov, A. (ed.) CADE 2002. LNCS (LNAI), vol. 2392, p. 275. Springer, Heidelberg (2002)

Scalable Web Reasoning Using Logic Programming Techniques*

Gergely Lukácsy[1] and Péter Szeredi[2]

[1] Digital Enterprise Research Institute, Galway, Ireland
[2] Budapest University of Technology and Economics, Budapest, Hungary
gergely.lukacsy@deri.org, szeredi@cs.bme.hu

Abstract. One of the key issues for the uptake of the Semantic Web idea is the availability of reasoning techniques that are usable on a large scale and that offer rich modelling capabilities by providing comprehensive coverage of the OWL language. In this paper we present a scalable extension of our ABox reasoning framework called DLog.

DLog performs query-driven execution whereby the terminological part of the description logic knowledge base is converted into a Logic Program and the assertional facts are accessed dynamically from a database. The problem of instance retrieval is reduced to a series of instance checks over a set of individuals containing all solutions for the query. Such a superset is calculated by using static-code analysis on the generated program.

We identify two kinds of parallelism within DLog execution: (1) the instances in the superset can be independently checked in parallel and (2) a specific instance check can be executed in parallel by specialising well-established techniques from Logic Programming. Moreover, for efficiency reasons, we propose to use a specialised abstract machine rather than relying on the more generic WAM execution model. We describe the architecture of a distributed framework in which the above mentioned techniques are integrated. We compare our results to existing approaches.

Keywords: Scalability, Parallelism, OWL, DL, Logic Programming.

1 Introduction

In this paper we describe extensions of DLog, a $\mathcal{SHIQ}$ Description Logic (DL) ABox reasoner [1], using the unique name assumption. We are interested in scenarios that have large numbers of individuals and a relatively small terminology and where query answering is the most important reasoning task. This is in line with the aims of the upcoming OWL 2 QL profile. In this setup, DLog already proved to be very efficient thanks to its query-oriented top down execution model that ensures that only those parts of the ABox are accessed that are relevant to

* This work has been funded in part by Science Foundation Ireland under Grant No. SFI/08/CE/I1380 (Lion-2) and by the Irish Research Council for Science, Engineering and Technology (IRCSET). Earlier development work on the DLog system was supported by the Hungarian NKFP programme under Grant No. 2/052/2004.

A. Polleres and T. Swift (Eds.): RR 2009, LNCS 5837, pp. 102–117, 2009.

the given query. However, the DLog execution is sequential which turns out to be a bottleneck when working with really large datasets, as DLog is simply not able to feed the underlying database with queries fast enough.

The contributions of the present paper include a new abstract machine designed to efficiently execute logic programs generated by the DLog system as well as a new parallel architecture of DLog.

We acknowledge that having a scalable DL reasoner does not solve all the problems of the Semantic Web. Specifically, we still need solutions for handling the heterogeneity of web ontologies, to provide support for ontology evolution and time management and to create a solid foundation of trust. We also acknowledge that data complexity results on the $\mathcal{SHIQ}$ language suggest [2] that sound and complete query answering is unlikely to be efficient on really large amounts of data. However, we believe that by providing the basis of a scalable reasoning framework on an expressive but still decidable OWL fragment we make a step towards turning the Semantic Web idea into reality.

The paper is structured as follows. In Section 2 we introduce the DLog system in a nutshell and summarise those features that are relevant for the rest of the paper. Section 3 discusses the design of the abstract machine for the execution of simple logic programs generated from DL knowledge bases. Section 4 presents the workflow and the architecture of the parallel DLog system. Section 5 gives a brief overview of related work, while in Section 6 we conclude with the discussion of future work and the summary of our results. Throughout the paper we assume basic level knowledge on Description Logics [3] and Prolog [4].

2   Overview of the DLog Approach

In this section we give a brief overview of the DLog reasoning process using an example (see [1] for formal details). The main idea is to transform the DL knowledge base to a Prolog program and use normal Prolog execution on it to answer instance retrieval queries. Let us consider the following knowledge base.

```
1  ∃hasFriend.Alcoholic  ⊑  ¬Alcoholic
2  ∃hasParent.¬Alcoholic  ⊑  ¬Alcoholic

3  hasParent(joe, bill). hasParent(joe, eva). hasFriend(bill, eva).
```

This TBox states that if someone has a friend who is alcoholic then she is not alcoholic (line 1). Furthermore, if someone has a non-alcoholic parent then she is not alcoholic either (line 2). The ABox contains several role assertions, but nothing about someone being alcoholic or non-alcoholic. Thus, in the database world, looking for non-alcoholic people would yield no results. In DL however, we can conclude that joe is non-alcoholic as one of his parents is bound to be non-alcoholic (as at least one of two people who are friends has to be non-alcoholic).

The common property of such problems is that solving them requires *case analysis* and therefore the trivial Prolog translation usually does not work. There are many other examples showing how incomplete knowledge is handled during DL reasoning, some of them do not even require a TBox [5].

The first step in the sound and complete DLog reasoning process is to convert a $\mathcal{SHIQ}$ TBox to a set of first order clauses containing no function symbols, called *DL clauses* [6]. This allows us to break the reasoning into two parts: an ABox independent TBox transformation followed by the actual data reasoning.

The second step deals with the transformation of DL clauses to a Prolog program. This is based on the Prolog Technology Theorem Proving (PTTP) approach, which provides a generic first-order theorem prover on top of Prolog [7]. PTTP uses *contrapositives* to compensate for the simple literal selection rule of Prolog; *ancestor resolution* for implementing the factoring resolution rule; and *iterative deepening* to ensure termination. For efficiency reasons in DLog we specialised this approach for the case of DL clauses. Specifically, for the simple function-free Prolog code generated from DL clauses, normal Prolog execution extended with *loop elimination* can be used instead of iterative deepening.

DL clauses are transformed to a *DL predicate* format by generating certain contrapositives and grouping these into predicates according to the functor of the clause head. Negations are eliminated by introducing new predicate names. DL predicates can be executed by an interpreter or, alternatively, DL predicates can be compiled into directly executable Prolog code, by adding an argument for storing the list of ancestors and including loop elimination and ancestor resolution in the DL predicates themselves. As an example, the DL predicate format of the above alcoholic problem is shown below:

```
1 alcoholic(A) :- hasParent(B, A), alcoholic(B).          ⟶  contrapositive

2 not_alcoholic(A) :- hasParent(A, B), not_alcoholic(B).  ⟶  original clause
3 not_alcoholic(A) :- hasFriend(A, B), alcoholic(B).       ⟶  original clause
4 not_alcoholic(A) :- hasFriend(B, A), alcoholic(B).       ⟶  contrapositive
```

We now present the Prolog code generated for the DL predicate `not_alcoholic`, as shown in lines 2-4 above (we use the abbreviation `not_alc` for compactness).

```
1 not_alc(A, L0) :- member(B, L0), B == not_alc(A), !, fail.
2 not_alc(A, L0) :- member(alcoholic(A), L0), !.
3 not_alc(A, L0) :- L1=[not_alc(A)|L0], hasParent(A, B), not_alc(B, L1).
4 not_alc(A, L0) :- L1=[not_alc(A)|L0], hasFriend(A, B), alcoholic(B, L1).
5 not_alc(A, L0) :- L1=[not_alc(A)|L0], hasFriend(B, A), alcoholic(B, L1).
```

Lines 1 and 2 implement loop elimination and ancestor resolution, respectively. Lines 3-5 are derived from the clauses of `not_alcoholic`, by extending the head and appropriate body calls with an additional argument, storing the ancestor list (variables L0 and L1). Similar code is generated for predicate `alcoholic`. Although we proved that the translation exemplified above is complete it is not efficient. In DLog we use a series of optimisations that result in more efficient (and more complex) Prolog translation; these are described in detail in [1].

Here we only mention two optimisations, decomposition and superset. The goal of *decomposition* is to split a body into independent components and make sure that the truth value of each component is only calculated once. Decomposition is achieved by a recursive process that uncovers the dependencies between the goals of the body. This optimisation results in clause bodies where independent components are separated using conditional Prolog structures. For example, the axiom that "someone is happy if she has a child having both a clever and a pretty child" results in the following translation (excluding the management of ancestor lists).

```
1  Happy(A) :-
2      hasChild(A, B),
3      (   hasChild(B, C),
4          Clever(C) -> true              ⟶  first component
5      ),
6      (   hasChild(B, D),
7          Pretty(D) -> true              ⟶  second component
8      ).
```

The idea of *superset* is to determine for each predicate P a set of instances S for which $I(P) \subseteq S$ holds, where $I(P)$ denotes the set of solutions of P. If the size of S is not significantly greater than the size of $I(P)$, then we can use S to efficiently reduce the initial instance retrieval problem to a finite number of *deterministic* instance checks. Technically this generic schema can be implemented by creating a "choice" predicate for each concept Concept that invokes the deterministic variant of Concept for each individual in the superset.

```
1  choice_Concept(A, AL) :-
2          (   nonvar(A) -> deterministic_Concept(A, AL)
3          ;   member_of_superset_Concept(A),
4              deterministic_Concept(A, AL)
5          ).

6  deterministic_Concept(A, AL) :- ..., !.     ⟶  A is a specific instance
7  ...
```

Note that the deterministic translation of a DL concept has Prolog cuts (!) at the end of each of its clauses. This results in pruning the rest of the search space after a successful execution of any of the clauses.

A superset of predicate P is calculated by applying static program analysis as described in detail in [1]. For example, the superset of predicate not_Alcoholic includes all individuals having a parent or a friend, or being a friend of someone.

The optimisations applied in the DLog system guarantee that ancestor goals are always ground. This opens up the possibility to use hash tables rather than lists for managing ancestor resolution and loop elimination. For this purpose we have implemented in C a *backtrackable hash table* [8]. This, besides the obvious efficiency advantage, also makes DLog programs structure free.

3   The DLog Abstract Machine

The Warren Abstract Machine (WAM) [9] has become a de facto target platform
for Prolog compilers. Most sequential and parallel implementations of Prolog rely
directly on WAM, or on a variant of it. For efficiency reasons we suggest to use
a much simpler abstract machine, called the DLog Abstract Machine (DAM), to
execute Prolog programs generated using the DLog approach. In the following
we sketch the main design principles of DAM.

Compared to generic Prolog, DLog programs for instance checking are considerably simpler as: (1) predicates can only be unary or binary; (2) there are no
compound data structures – unification is trivial; (3) predicate invocations are
ground; (4) concept predicates are deterministic – no need for deep backtracking
into concept predicates; (5) no need for the heap and the trail stack; (6) no need
for cell tagging, as all constants are numeric.

3.1   Architecture of DAM

The DAM is a three-stack machine. It has a *control stack* containing *frames* of
fixed size. A frame is created when entering a predicate and is used to store the
local environment of the predicate and return address information. A predicate
can be viewed as a function which receives its arguments implicitly in a frame
and returns a Boolean value. The second stack, the *choice point stack*, is used to
support deep backtracking in cases related to role predicates and also to ensure
efficient communication with the database/triple store[1]. The third stack is used
as a backtrackable hash table according to the principles discussed in [8].

Four pieces of information are stored in a frame of the DAM control stack:

1. the return address of the predicate (virtual register R);
2. the actual instance being checked, represented by a URI (virtual register A);
3. the ancestor list, represented as an index in the backtrackable hash table
   (virtual register H);
4. a pointer to the corresponding choice stack frame (virtual register P).

The fields of the current frame serve as (virtual) registers of the DAM. As the
frames are of fixed size, accessing a field of e.g. the frame preceding or following
the current one incurs no overhead.

The following information is stored in a frame of the choice point stack:

1. a counter used in implementing number restrictions (virtual register C);
2. a handle used for interfacing with the triple store;
3. a buffer for instances returned by the triple store.

Further to the virtual registers on the stacks, DAM has the following (global)
registers: V – the Boolean return value of a procedure invocation; PC – the
program counter variable; T – the current frame of the control stack.

---

[1] Triple stores are specialised databases for storing semantic web metadata. In the
following we use the phrases "triple store" and "database" as synonyms.

DAM operates only with three control structures: conjunction, disjunction and loops (used for counting instances in a qualified number restriction, including existential restrictions as a special case). In discussing the DAM we assume that each predicate contains exactly one of the three control structures; this can be achieved by introducing auxiliary predicates. As an example we show below how the Happy example from Section 2 can be transformed to satisfy this assumption.

```
1  Happy(A) :-
2      aux_1(A).                ⟶  a conjunction with a single member
3  ...                         ⟶  possible other clauses of Happy

4  aux_1(A) :-                  ⟶  existential restriction
5      hasChild(A, B),
6      aux_2(A).

7  aux_2(A) :-                  ⟶  a conjunction with two members
8      aux_3(A),
9      aux_4(A).

10 aux_3(A) :- ⟶ existential restr.    aux_4(A) :- ⟶ existential restr.
11     hasChild(A, B),                     hasChild(A, B),
12     Clever(B), !.                       Pretty(B), !.
```

3.2   The Instruction Set

The instruction set of the DAM is fairly limited. Each instruction consists of an operation code with typically zero, one, or two operands. For example, the instruction call pred invokes predicate pred, while exit_on_failure (with no arguments) exits the given predicate if register V contains the value FAILURE. The set of instructions of the DAM is summarised in Table 1.

In Figure 1 we give the operational semantics of the DAM instructions using pseudo-code with C syntax. Here we use capitals to refer to DAM registers, lower case names for parameters and local variables, while the keywords previous and next refer to the frames preceding and following the current one, respectively. A register reference can be used on its own (e.g. A), referring to the appropriate field of the current control frame; or it can be used together with the keyword previous or next, to refer to the appropriate field of a neighbouring frame. The instruction exit_with is invoked within other instructions: this is considered as a macro to be expanded, i.e. the invocation should be replaced by the definition.

To simplify the presentation we do not deal with memory management issues, assuming that the stacks have enough memory allocated to perform the computation. Thus creating or removing a stack frame is simply done by incrementing or decrementing register T (which points to the current control frame).

We assume that the DAM-triple store interface works as follows. Once a query has been posed to the triple store it returns a handle which is stored in the actual choice-stack frame. Using this handle the DAM can ask for the first batch of solutions, which is then stored in the buffer part of the relevant choice-stack

Table 1. The instruction set of the DAM

Instruction	Arguments	Description
put_ancestor	N	extend the ancestor list in the local frame by the term with name N and argument A
check_ancestor	N	succeeds if the ancestor list contains a term with name N and argument A
fail_on_loop	N	fails if a loop occurred, i.e. the term with name N and argument A is present on the ancestor list
call	P	invokes procedure P in a new control frame
execute	P	invokes procedure P in the existing control frame
exit_with	S	returns from a procedure with status S, continues execution according to register R
exit_on_failure	–	returns from procedure if V = FAILURE
exit_on_success	–	returns from procedure if V = SUCCESS
jump	L	jumps to label L
has_n_successors	R, n	checks if instance A has at least n R successors; creates a choice point; loads the first choice to A
count_and_exit	–	decreases counter C if the previous instruction was successful; returns with success if C is 0
next_choice	–	loads the next solution from the choice stack to A
abox_query	Q	returns success if A is a solution of query Q

frame. The buffer involves a header specifying the buffer length and the number of solutions not yet processed. This setup makes it possible to return query solutions one by one to the DAM code which issued the given query. When a solution is requested and the buffer is empty, a request is sent to the triple store (using the query handle) to provide the next batch of solutions.

We now describe the auxiliary procedures used in Figure 1. Procedures add_to_hash and hash_search perform the extension and search of the backtrackable hash table [8], representing the ancestor list. The procedure cardinality_check(i, r, n) returns true if the triple store contains at least n r-successors for the instance i. The procedure create_choice(i, r) issues a query to the triple store to find all r-successors of the individual i, and returns the first solution found. The procedure has_choice returns true if the current choice frame has more solutions, while next_choice returns the next solution.

Finally, the procedure abox_query(i, q) checks if instance i belongs to query predicate q according to the triple store. Query predicates are defined in terms of ABox predicates using conjunction and disjunction only; they can be thought of as complex database queries. In its most simple case a query predicate corresponds to a simple ABox concept predicate.

3.3   Transforming into DAM

Now we turn our attention to discuss how certain parts of the DLog programs are transformed into DAM code.

```
 1 put_ancestor n:          ⟶ inserts term n(A) into the hash table
 2     H = add_to_hash(A, n, H);

 3 check_ancestor n:        ⟶ checks if term n(A) is in the hash table
 4     if (hash_search(A, n, H)) exit_with SUCCESS;

 5 fail_on_loop n:          ⟶ checks if term n(A) is in the hash table
 6     if (hash_search(A, n, H)) exit_with FAILURE;

 7 call p:
 8     T++; A = previous->A; H = previous->H; R = PC + 1;
 9     PC = &p;              ⟶ invokes procedure in new frame

10 execute p:
11     PC = &p;             ⟶ invokes procedure in the current frame

12 exit_with s:
13     T--; V = s; PC = next->R;  ⟶ drops frame; jumps to return address

14 exit_on_failure:
15     if (V == FAILURE) exit_with V;

16 exit_on_success:
17     if (V == SUCCESS) exit_with V;

18 jump l:                  ⟶ jumps to label l
19     PC = l;

20 has_n_successors r n:    ⟶ loads successors of A to the choice stack
21     if (!cardinality_check(A, r, n)) exit_with FAILURE;
22     A = create_choice(A, r);

23 count_and_exit:          ⟶ counts and exists if counter reaches zero
24     if (V == SUCCESS) P->C--;
25     if (P->C == 0) exit_with SUCCESS

26 next_choice:
27     if (!has_choice()) exit_with FAILURE;
28     A = next_choice();   ⟶ sets the next solution instance to A

29 abox_query q:
30     V = abox_query(A, q);  ⟶ executes a (complex) database query
```

Fig. 1. Operational semantics of the DAM instructions

Role axioms are handled partly by the DLog framework (the transitivity axioms are eliminated) and partly by the underlying triple store. Namely, we assume that the database "understands" the notion of role hierarchy and is able to answer queries such as hasSpouse(bill, Y) (Y is the spouse of bill) based on hierarchical relation between hasSpouse and hasWife, for example.

Conjunctions of concept predicates are transformed into a series of call and exit_on_failure instructions. That is, a conjunction consisting of goals...

```
1 g₁(X), ..., gₖ₋₁(X), gₖ(X)
```

... is directly transformed into the following DAM form:

```
1 call g₁
2 exit_on_failure
3 ...
4 call gₖ₋₁
5 exit_on_failure
6 execute gₖ
```

Note that the last goal of the conjunction is invoked using the execute instruction rather than the call instruction. This is an optimisation which allows us to use the current frame to invoke g_k rather than creating a new one. For recursive predicates this is known as the *tail-recursion optimisation* which basically transforms a recursive call into an iteration.

Analogously to conjunctions, disjunctions of concept predicates are transformed into a series of call and exit_on_success instructions, with execute for the last goal. Note that both transformation schemata use the fact that we work with deterministic predicates, e.g. a disjunction immediately succeeds if one of its members completes successfully. Alternatively, one could use an even more compact schema, where each pair of call and exit_on_success instructions is replaced by a single one, similar to the try instruction of the WAM [9]. However, this would require that two separate return addresses – one for success and one for failure – were stored on the control stack for each predicate.

Finally, a qualified number restriction $(\geq nRC)$ is transformed into a *loop*, where we first check if the instance at hand has at least n R-successors, then we enumerate the successors until we find at-least n successors belonging to concept C. Specifically, $(\geq nRC)$ is transformed into the following DAM program.

```
1    has_n_successors R n    ⟶  fails if A has not enough successors, sets A, C
2 label1:
3    call C                  ⟶  returns with success or failure
4    count_and_exit          ⟶  if success : C--, returns success if C = 0
5    next_choice             ⟶  set A to next successor, return fail if no more
6    jump label1
```

Note that this technique handles the only case where deep backtracking is needed in the Prolog code. As an optimisation for the above schema we can use techniques that allow us to reuse a frame rather than repeatedly build it with call C in each iteration.

As an example for the DLog to DAM translation, let us show below parts of the DAM code for the predicate `not_alc` from Section 2.

```
 1  predicate(not_alc):            ⟶  A contains the instance to check
 2      fail_on_loop not_alc       ⟶  return fail if within not_alc(A)
 3      check_ancestor alcoholic   ⟶  return success if within alcoholic(A)
 4      call aux_1
 5      exit_on_success
 6      call aux_3
 7      exit_on_success
 8      execute aux_4

 9  predicate(aux_1):
10      put_ancestor not_alc          ⟶  uses A, sets H
11      has_n_successors has_parent 1 ⟶  return fail if A has no parent at all
12  label(1):
13      call not_alc,
14      count_and_exit                ⟶  returns if we found a not_alc parent
15      next_choice                   ⟶  return fail if no parent belongs to not_alc
16      jump 1
```

4   The Parallel DLog Architecture

We now identify several parallelisation possibilities in executing DLog programs. First, we briefly summarise the main ideas behind how to turn the DAM into a parallel execution engine – this is a very high level discussion and its purpose is to give an insight to the possibilities. Next, we introduce in detail a new parallel architecture for the DLog system.

4.1   Kinds of Parallelism Available in DLog

The DLog Abstract Machine, discussed in the previous section, can be viewed as a simplification of the WAM for the case of special Prolog programs, produced by the DLog transformation. Analogously, we can simplify well-studied parallelisation techniques for DLog programs. Combining these two ideas produces the *Parallel DLog Abstract Machine* (PADAM).

Logic programming offers an excellent ground for parallel execution. On one hand, the operational semantics of Logic Programming includes *non-determinism* which leads to a very natural parallelisation. On the other hand, logic programs use single assignments for variables, which makes it possible to avoid the problems of certain types of flow dependencies present in more traditional languages.

There are two main kinds of parallelism applicable for Logic Programs: AND- and OR- parallelism. The former consists of the simultaneous computation of several goals in a body, while the latter relies on executing the clauses of a predicate in parallel. Because of the decomposition DLog applies at compile time we already have the independent components of a clause body allowing us to apply independent AND-parallelism (IAP) techniques directly.

An interesting feature of DLog programs is that each clause contains a *cut* operation at the very end: thus once a clause succeeds we can terminate the computation of all the other clauses being executed in parallel. This means that exploiting OR-parallelism within a single instance check involves *speculative work* [10], i.e. wasted computer efforts, which may be lost because of being on a branch of computation pruned by a cut operation.

As OR-parallelism is easier to exploit, our initial efforts go in this direction. Because of the speculative nature of fine grained OR-parallelism, we decided to first address the issue of coarse grained parallelism, by executing multiple instance check problems in parallel.

4.2   An Initial Coarse-Grained Model of Parallelism

We now focus on exploiting coarse-grained parallelism, which does not require the modification of the sequential DAM engine, and still seems to promise good scalability. The proposed architecture of the parallel DLog system is presented in Figure 2. We explain the workflow of the system, which starts with the receipt of the input and completes with producing the answer to the instance retrieval query. We refer to certain parts of the architecture by using the numbers and letters in Figure 2.

The content of the ABox is stored either in a triple store or in a relational database (arrow 1). As a general consideration, we assume that the ABox is extensionally reduced, i.e. beside roles, it contains only atomic concepts and their negations. When using a database we need to create appropriate tables for the concept and role assertions, taking care that tables should also be created for negated concepts (to properly model incomplete knowledge). If we use the DLog system over an existing database we need to create links between the concept and role names in the TBox and the database tables [11].

The content of the TBox is first transformed into DL clauses (arrow 2). These clauses are then further transformed into DAM byte code (arrow A) using the specialised PTTP techniques and optimisations outlined in Section 2. Specifically, the DAM code contains the component clause bodies and the superset expressions (i.e. an expression whose *evaluation* gives the superset) for each predicate. The DAM programs are stored in a repository (arrow B) allowing us to re-use them without performing the transformation steps again.

The conjunctive query, i.e. a conjunction of possibly negated atomic concepts and unnegated (positive) binary roles, first undergoes an optimisation phase (arrow 3) inspired by database join optimisations and [12]. Here, using heuristics, the conjunctions can be re-ordered and grouped in order to avoid cross product computations and ensure efficient execution.

As the last step of the ABox independent transformations, the generated DAM program is simplified/partially evaluated with respect to the given query (arrow C). Practically this means that we leave out those parts of the DAM program

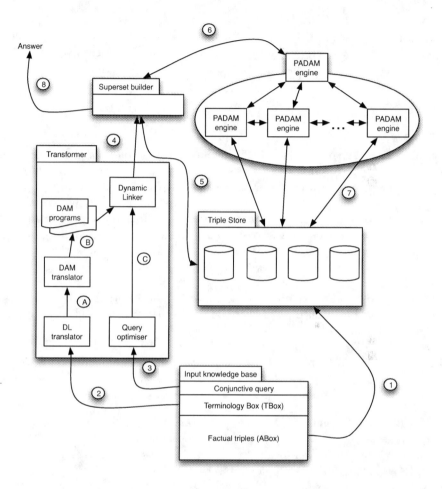

Fig. 2. The architecture of the Parallel DLog system

that do not play any role when executing the specific query, thus reducing the size of the byte code that needs to be transferred between workers in a later stage of the execution. We also "pack" the query itself into the DAM code together with a newly constructed superset expression belonging to it.

In the next step, the Superset Builder receives the simplified byte code (arrow 4) and calculates the superset for the given query. One of the key ideas here is that (i) the superset expression is evaluated in parallel and (ii) the instances in the superset are checked in parallel. Note that this scheme can be used for supersets which do not fit into memory: the Superset Builder basically acts as a mediator in the producer-consumer setup between the database and the PADAM execution engines (arrows 5 and 6). The parallel DAM engines also use the database when checking particular individuals (arrow 7).

During the execution the Superset Builder receives notifications from the PADAM engines about whether particular individuals are solutions or not. This information is forwarded to the output of the reasoning process (arrow 8) where another consumer can pick it up and use it for various purposes.

5   Related Work

Because of the separation of the TBox and ABox reasoning and the usage of a database to store ABox facts, the suggested parallel DLog architecture can be considered as a Deductive Database system. Here we heavily rely on the scalability capabilities of the underlying relational databases/triple stores. We argue that this is the right way to do as there are decades long expertise and proved implementations for databases and triple stores that scale up to several billions of records [13]. Note that this setup is actually suggested by the database literature as well. For example, [14] discusses the problem of handling massive amounts of data in a relational database with support for recursive queries. The suggested solution exploits optimisation techniques both from the relational database as well as from the deductive database theory – much like the parallel DLog architecture.

From the DL point of view, several techniques have emerged for dealing with large scale ABox-reasoning. We can divide these into two main groups, based on whether they support some full-featured DL subset or whether they pose restrictions on the terminology axioms in order to retain even better scalability.

Full reasoners. To make traditional tableau-based reasoning more efficient on large data sets, several techniques have been developed in recent years [12]. These are used by the state-of-the-art DL reasoners, such as RacerPro or Pellet.

In [15], a resolution-based inference algorithm is described, which is not as sensitive to the increase of the ABox size as the tableau-based methods. However, this approach still requires the input of the *whole content* of the ABox before attempting to answer any queries. The KAON2 system implements this method and provides reasoning services over the description logic language $\mathcal{SHIQ}$ by transforming the knowledge base into a disjunctive datalog program.

Article [16] introduces the notion of distributed ordered resolution which is a parallelised variant of ordered resolution usable for reasoning over already distributed $\mathcal{ALC}$ knowledge bases.

DLog belongs to this first group: it provides $\mathcal{SHIQ}$ support while retains as much scalability as possible. The original DLog system was evaluated in depth in [1]. There we compared the performance of DLog with that of the RacerPro, Pellet and KAON2 systems on publicly available benchmark ontologies. The test results showed that DLog is significantly faster than traditional tableau-based reasoners and it also outperforms KAON2 in most of the test cases.

Restricted reasoners. Extreme cases involve serious restrictions on the knowledge base to ensure efficient execution with large amounts of instances. For example,

[17] suggests a solution called the *instance store*, where the ABox is stored externally, and is accessed in a very efficient way. The drawback is that the ABox may contain only axioms of form $C(a)$, i.e. we cannot make role assertions.

The YARS2 framework [13] aims to provide efficient support for the upcoming OWL2 QL profile that is based on a DL-Lite variant. DL-Lite [18] is a family of Description Logics that allows the separation of TBox and ABox during reasoning and provides polynomial-time data complexity for query answering. On the other hand, DL-Lite poses restrictions on the terminology axioms, e.g. cardinality constraints are not allowed.

Article [19] introduces the term Description Logic Programming (DLP). This idea uses a direct transformation of $\mathcal{ALC}$ description logic concepts into definite Horn-clauses, and poses some restrictions on the form of the knowledge base, which disallow axioms requiring disjunctive reasoning. As an extension, [20] introduces a fragment of the $\mathcal{SHIQ}$ language that can be transformed into Horn-clauses where queries can be answered with polynomial data complexity.

Work in [21] presents the DLDB2 system based on the DLP idea that delegates certain reasoning tasks to an external TBox reasoner. Similarly to our approach, DLDB2 takes advantage of the scalability of the underlying relational database. However it poses serious restrictions on the supported language (e.g. universal restrictions are not allowed).

6   Conclusion and Future Work

In this paper we have presented the design of the scalable extension of the description logic $\mathcal{SHIQ}$ reasoning system DLog. Unlike the traditional tableau-based approach, we answer conjunctive queries by transforming the knowledge base into a logic program that is executed in a distributed fashion. This technique allows us to use top-down query execution and to store the content of the ABox externally in a scalable/distributed database.

Following an overview of the DLog system we presented the design considerations of an abstract machine aimed to execute DLog programs. Based on this we then proposed a parallel architecture that introduces parallelism at several stages of the execution process. The key ideas were to calculate the superset of a query in parallel and to check the individuals in the superset using instances of our proposed abstract machine communicating in a peer-to-peer fashion.

Future work involves designing the details of the communication between DLog and the triple store, the implementation and the performance evaluation of our initial parallel DLog model. Building on these results the model should be further refined, including the exploration of parallelism within a single instance check reasoning task (cf. Section 4.1).

As an overall conclusion, we argue that resolution-based techniques are very promising in practical applications, with relatively small TBox, but large ABox. Specifically, we believe that translating to Logic Programs and using the parallel DLog architecture provides a viable framework for scalable DL reasoning.

References

1. Lukácsy, G., Szeredi, P.: Efficient description logic reasoning in Prolog: The DLog system. Theory and Practice of Logic Programming 09(03), 343–414 (2009)
2. Hustadt, U., Motik, B., Sattler, U.: Data Complexity of Reasoning in Very Expressive Description Logics. In: Kaelbling, L.P., Saffiotti, A. (eds.) Proc. of the 19th Int. Joint Conference on Artificial Intelligence (IJCAI 2005), Edinburgh, UK, pp. 466–471. Morgan Kaufmann Publishers, San Francisco (2005)
3. Baader, F., Calvanese, D., McGuinness, D., Nardi, D., Patel-Schneider, P.F. (eds.): The Description Logic Handbook: Theory, Implementation and Applications. Cambridge University Press, Cambridge (2004)
4. Nilsson, U., Maluszynski, J. (eds.): Logic, Programming and Prolog. John Wiley and Sons Ltd., Chichester (1990)
5. Nagy, Z., Lukácsy, G., Szeredi, P.: Translating description logic queries to Prolog. In: Van Hentenryck, P. (ed.) PADL 2006. LNCS, vol. 3819, pp. 168–182. Springer, Heidelberg (2005)
6. Zombori, Z.: Efficient two-phase data reasoning for description logics. In: Bramer, M. (ed.) IFIP AI. IFIP, vol. 276, pp. 393–402. Springer, Heidelberg (2008)
7. Stickel, M.E.: A Prolog technology theorem prover: A new exposition and implementation in Prolog. Theoretical Computer Science 104(1), 109–128 (1992)
8. Kádár, B.: Architectural extensions of the dlog description logic reasoning system MSc Thesis, http://sintagma.szit.bme.hu/lukacsy/docs/kadarMSc.pdf
9. Warren, D.H.D.: An abstract Prolog instruction set. Technical Note 309, SRI International, Menlo Park, CA (October 1983)
10. Gupta, G., Pontelli, E., Ali, K.A., Carlsson, M., Hermenegildo, M.V.: Parallel execution of Prolog programs: A survey. ACM Trans. Program. Lang. Syst. 23(4), 472–602 (2001)
11. Kádár, B., Lukácsy, G., Szeredi, P.: Large scale semantic web reasoning. In: Proceedings of the 3rd International Workshop on Applications of Logic Programming to the Web, Semantic Web and Semantic Web Services (ALPSWS 2008), Udine, Italy, December 2008, pp. 57–70 (2008)
12. Haarslev, V., Möller, R.: On the scalability of description logic instance retrieval. Journal of Automated Reasoning 41(2), 99–142 (2008)
13. Harth, A., Umbrich, J., Hogan, A., Decker, S.: YARS2: A federated repository for querying graph structured data from the web. In: Aberer, K., Choi, K.-S., Noy, N., Allemang, D., Lee, K.-I., Nixon, L.J.B., Golbeck, J., Mika, P., Maynard, D., Mizoguchi, R., Schreiber, G., Cudré-Mauroux, P. (eds.) ASWC 2007 and ISWC 2007. LNCS, vol. 4825, pp. 211–224. Springer, Heidelberg (2007)
14. Terracina, G., Leone, N., Lio, V., Panetta, C.: Experimenting with recursive queries in database and logic programming systems. Theory Practice of Logic Programming 8(2), 129–165 (2008)
15. Hustadt, U., Motik, B., Sattler, U.: Reasoning for Description Logics around SHIQ in a resolution framework. Technical report, FZI, Karlsruhe (2004)
16. Schlicht, A., Stuckenschmidt, H.: Towards distributed ontology reasoning for the web. In: International Conference on Web Intelligence and Intelligent Agent Technology (WI-IAT 2008), December 2008, vol. 1, pp. 536–539 (2008)
17. Horrocks, I., Li, L., Turi, D., Bechhofer, S.: The Instance Store: DL reasoning with large numbers of individuals. In: Proceedings of DL 2004, British Columbia, Canada (2004)

18. Calvanese, D., Giacomo, G., Lembo, D., Lenzerini, M., Rosati, R.: Tractable reasoning and efficient query answering in description logics: The dl-lite family. J. Autom. Reason. 39(3), 385–429 (2007)
19. Grosof, B.N., Horrocks, I., Volz, R., Decker, S.: Description logic programs: Combining logic programs with description logic. In: Proc. of the Twelth International World Wide Web Conference (WWW 2003), pp. 48–57. ACM, New York (2003)
20. Hustadt, U., Motik, B., Sattler, U.: Data complexity of reasoning in very expressive description logics. In: Proceedings of the Nineteenth International Joint Conference on Artificial Intelligence (IJCAI 2005), International Joint Conferences on Artificial Intelligence, pp. 466–471 (2005)
21. Pan, Z., Zhang, X., Heflin, J.: DLDB2: A scalable multi-perspective semantic web repository. In: ACM International Conference on Web Intelligence, pp. 489–495. IEEE, Los Alamitos (2008)

On the Ostensibly Silent 'W' in OWL 2 RL*

Aidan Hogan and Stefan Decker

Digital Enterprise Research Institute,
National University of Ireland, Galway
firstname.lastname@deri.org

Abstract. In this paper, we discuss the draft OWL 2 RL profile from the perspective of applying the constituent rules over Web data. In particular, borrowing from previous work, we discuss (i) optimisations based on a separation of terminological data from assertional data and (ii) the application of authoritative analysis to constrain third party interference with popular ontology terms. We also provide discussion relating to the applicability of new OWL 2 constructs for two popular Semantic Web ontologies – namely FOAF and SIOC – and provide some evaluation of the proposed use-cases based on reasoning over a representative Web dataset of approx. 12 million statements.

1   Introduction

As more and more data becomes available on the Web, the Semantic Web movement aims to provide technologies which enable greater data-integration and query answering capabilities than the keyword/document centric models prevalent today. The core of these technologies is the Resource Description Framework (RDF) for publishing data in a machine-readable format, wherein there now exist millions of RDF data sources on the Web contributing billions of statements [9]. The Semantic Web technology stack also includes means to supplement instance (assertional) data being published in RDF with ontologies described in RDF Schema (RDFS) [3] and the Web Ontology Language (OWL) [23] (terminological data) providing machines a more sapient understanding of the information – in particular enabling deductive reasoning to be performed.

Reasoning over aggregated Web data is useful, for example, (i) to infer implicit knowledge and thus provide query-answering over a more complete dataset, (ii) to assert equality between equivalent resources resident in remote documents, (iii) to flag inconsistencies wherein conflicting data is provided by one or more parties; and (iv) to execute mappings, where they exist, between different data-models concerned with the same domain. However, reasoning over Web data is indeed an ambitious goal with many inherent difficulties, the most overt of which

---

* The work presented in this paper has been funded in part by Science Foundation Ireland under Grant No. SFI/08/CE/I1380 (Lion-2) and by an IRCSET Scholarship.

A. Polleres and T. Swift (Eds.): RR 2009, LNCS 5837, pp. 118–134, 2009.

are (i) the requirement for near-linear scale in execution and (ii) the requirement to be tolerant with respect to noisy and conflicting data (for a detailed treatment of noise in RDF Web data, we refer the interested reader to [15]).

With these requirements in mind, in previous work we introduced Scalable Authoritative OWL Reasoner (SAOR) [16]; we discussed the formulation and suitability of a set of rules inspired by pD* [24] – to cover a significant fragment of OWL Full reasoning – for forward-chaining materialisation over Web data. We gave particular focus to scalability and tolerance against noisy Web data showing that, by applying certain practical restrictions, materialisation over a diverse Web dataset – in the order of a billion statements – is feasible.

From the scalability perspective, we introduced a separation of terminological data from assertional data in our rule execution model, based on the premise that terminological data is the most frequently accessed segment of the knowledge base and represents only a small fraction of the overall data.

From the Web tolerance perspective, we presented many issues relating to the effects of noise – which is present in abundance on the Web – on reasoning. We particularly focused on the introduced problem of "ontology hijacking" wherein third-party sources redefine or subsume popular Web ontology terms. Our solution was to include consideration of the source or "context" of data, and provide "authoritative analysis" to curtail the privileges of third-parties.

Drawing on our experiences in reasoning over Web data, in this paper we discuss the new OWL 2 RL draft profile [18]. OWL 2 RL is a fragment of the new OWL 2 language for implementation within rule-based applications; hitherto, there existed only non-standard rule-implementable fragments of OWL reasoning, the mostly prominent thereof being Description Logic Programs (DLP) [11] and pD* [24]. OWL 2 RL extends upon both with a more complete list of rules, including support for a significant fragment of OWL 2 RDF-based semantics [21].

We subsequently present a number of Web use-cases for new OWL 2 terms in the context of two popular Web ontologies: Friend Of A Friend (FOAF) [4] and Semantically Interlinked Online Communities (SIOC) [1,2]; we evaluate our proposed use-cases based on reasoning over a 12m statement Web dataset.

Specifically, in this paper we: (i) discuss a separation of terminological data from assertional data in executing OWL 2 RL/RDF rules; (ii) discuss authoritative reasoning over OWL 2 RL/RDF rules; and (iii) present insights and evaluation on possible deployment of new OWL 2 constructs within two popular Web ontologies – viz.: SIOC and FOAF.

2   OWL 2 RL vs. SAOR

Before we continue, we recall pertinent high-level discussion relating to our previous work on SAOR, and draw parallels to OWL 2 RL (for a more extensive treatment of SAOR, we refer the interested reader to [16]; a full list of SAOR rules is available in [16, Table 2]). In doing so, we provide insights into possible obstacles and optimisations relating to applying OWL 2 RL for materialisation over an RDF dataset collected from the Web.[1] Please note that in Appendix A, we replicate the OWL 2 RL/RDF rules from [18] and denote certain characteristics which we will refer to in this section.

SAOR is designed to accept as input a Web knowledge-base in the form of a large body of statements collected by means of a Web crawl, and to output inferred statements by forward-chaining reasoning according to a tailored fragment of OWL; input and inferred statements can then be exploited by a consumer application, such as for query answering. We identified three main aspects around which our system and ruleset is designed and implemented: *computational feasibility* for scalability, *reduced output statements* such that consumer applications are not over-burdened, and finally *Web tolerance* for avoiding undesirable and potentially expensive "inflationary" inferences caused by noisy Web data.

In this section, we will introduce how our ruleset and implementation is designed to adhere to these requirements for Web reasoning, and contrast our approach with the OWL 2 RL ruleset; we begin by discussing general issues.

2.1   High-Level Issues

Firstly, SAOR ignores inconsistencies in the data; in OWL 2 RL, inconsistencies are flagged by means of a `false` inference which indicates that the original input graph is inconsistent – such rules could additionally be supported in SAOR. In both cases, the explosive nature of reasoning upon inconsistent data is avoided; i.e., inconsistencies do not lead to the inference of all possible triples.

Secondly, in SAOR we avoid inventing new anonymous individuals. Such invention breaks the upper bound on possible inferable statements from an input graph – $|T|^3$ where T is the union of the set of RDF terms in the input and the set of 'built-in' terms that appear in the rule consequents – and allows for the inference of infinite statements. For example, in [24], pD*sv was introduced which extended pD* with an additional rule based on `owl:someValuesFrom`:[2]

`?v someValuesFrom ?w ; onProperty ?p . ?u a ?v . ⇒ ?u ?p _:b . _:b a ?w .`

Here, `_:b` is a unique blank-node created for each set of variable bindings from the rule body. Now, given an input graph where a binding for `?w` is a subclass of the respective binding for `?v`, this rule will infer infinite statements; such rules are excluded from pD*, SAOR and OWL 2 RL due to such effects on termination.

In a related matter, in pD*, blank nodes are allowed in all positions in a form of partially-generalised triples; literals are not allowed in subject or predicate positions. Thus, and following RDFS entailment practices [13, Section 7.1], pD* includes rules which invent so called "surrogate blank nodes" to represent literals in subject and predicate positions where they would otherwise be disallowed.

---

[1] Although we focus on forward-chaining applications of OWL 2 RL, much of our discussion has a more general appeal.

[2] In this paper, we use prefixed names as prevalent in the literature and, following Turtle syntax, use 'a' as a shortcut for `rdf:type` and '(...)' for RDF lists.

Although these blank nodes are formed by a direct mapping from a finite set of literals, they still create new terms and thus in SAOR, we opted to allow literals and blank-nodes in all positions of a triple. This is analogous to the Rule Interchange Format (RIF) [6] and the OWL 2 RL notion of a generalised triple.

Thus far, OWL 2 RL maintains an upper bound of $|T|^3$ inferred generalised statements. However, rules `dt-type2`, `dt-eq` and `dt-diff` (Table 1) are based on an infinite set of literals independent of the input graph. Thus, materialisation according to these rules (which are clearly intended for backward-chaining) would lead to inference of infinite triples. One could curtail such inferences by omitting the rules or only applying the rules over literals which appear in the input graph: either would maintain the $|T|^3$ upper bound. Also, rule `dt-eq` could be used to infer equivalence between literals and their canonical versions, introducing at most $|CL|$ terms where CL is the set of literals in the input with lexically distinct canons: the upper bound would then be $(|CL| + |T|)^3$.

Continuing, in SAOR we also aim to omit inference of what we term "extended axiomatic" statements. These include: (i) the set of RDF(S) axiomatic triples [13, Section 4.1]; (ii) the set of additional OWL axiomatic triples listed for pD* [24, Table 6]; and (iii) inferences which apply to every RDF term in the graph. For the latter, we firstly omit rules which assert membership of `rdfs:Resource` for all terms, viz.: RDFS/pD* rules `rdfs4a/rdfs4b` [13, Section 7.3]. Secondly, we omit rules which mandate symmetric `owl:sameAs` inferences for all terms, viz: OWL 2 RL rule `eq-ref` (Table 3)[3] and pD* rules `rdfp5a/rdfp5b` [24, Table 6]. Such rules immediately add $|T|$ statements to the graph and could be considered inflationary; they are, perhaps, better suited to backward-chaining support (in an approach such as [17]) than materialisation.

Indeed, reasoning involving `owl:sameAs` relations is problematic on the Web: in [14] we found 85,803 equivalent individuals to be inferable from a Web dataset through the incongruous values `08445a31a78661b5c746feff39a9db6e4e2cc5cf` and `da39a3ee5e6b4b0d3255bfef95601890afd80709` for the prominent inverse-functional property `foaf:mbox_sha1sum` – the former value is the sha1-sum of an empty string and the latter is the sha1-sum of the 'mailto:' string, both of which are erroneously published by online FOAF exporters.[4] Thus, in SAOR, we cross-check the values of inverse-functional properties against a black-list of known noisy values. Also, we disallow `owl:sameAs` inferences from travelling to the predicate position of a triple or to the object position of an `rdf:type` triple: this is contrary to rule `eq-rep-p` in OWL 2 RL, and to the lack of a restriction on rule `eq-rep-o` where `rdf:type` predicates are allowed (Table 5).

Aside from noisy data, naïve materialisation over OWL 2 RL equality rules `eq-ref`, `eq-sym` (Table 3) and `eq-trans` (Table 6) – which axiomatise the reflexive, symmetric, and transitive nature of `owl:sameAs` resp. – leads to quadratic growth in inferences. Again, take for example the 85,803 equivalent individuals we had

---

[3] One important note: rule `eq-diff1` requires reflexive `owl:sameAs` statements to flag inconsistent reflexive `owl:differentFrom` statements; in the absence of the former, one should support the following rule: `?x :differentFrom ?x . ⇒ false`.

[4] See, for example http://blog.livedoor.jp/nkgw/foaf.rdf

previously found; naïvely, the OWL 2 RL rules would mandate $85,803^2 - 7.362b$ statements to represent the pair-wise equivalences. Also, assuming that each individual was mentioned in, on average, eight unique statements[5], the `eq-rep-*` rules would infer $7.362b * 8 = 59b$ statements, with massive repetition.

Although the above example again relies on noisy Web data, there do exist valid examples on the Web of large "equivalence chains" of individuals. Again in [14] we discovered a resource representing a "global user" on the `vox.com` blogging platform which exports FOAF data; this global user was identified by a blank node and was mentioned in the FOAF profiles of all users.[6] Thus, in our crawl we found 32,390 unique resources, in different documents, with the valid value `http://team.vox.com/` for inverse-functional property `foaf:weblog`. Again, such would lead to the inference of over a billion `owl:sameAs` statements and billions more statements in duplicative data.

Taking such considerations into account, in order to avoid an explosion of repetitive inferences in [14,16] we instead choose a single 'pivot' identifier for identifying equivalent individuals. Thus, we compress repetitive entries into one single entry; we also store equivalence relations from the pivot element to all other identifiers such that the fully expanded view can be realised by the consumer application using backward-chaining techniques.

Finally, there are two cardinality-related rules supported in SAOR for which no equivalent rule exists in OWL 2 RL; namely **rdfc2** (Table 9) and an exact-cardinality version of **cls-maxc2** (Table 6). Their omission relates to the constraint that OWL 2 RL documents must also be valid OWL 2 DL documents [22, Section 2.1] which enforces certain computational guarantees, e.g., for entailment checking. Thus, the OWL 2 RL ruleset omits exact-cardinality versions of rules for **cls-maxc*** and **cls-maxqc*** (Table 6) and support for minimum-cardinality; also missing are rules relating to disjoint-union expressions, which could be supported analogously to union-of and disjoint-class expressions (resp. **cls-duni1** and **cls-duni2** in Table 9). More puzzlingly, the ruleset omits support for self-restriction expressions which are supported by OWL 2 DL/EL; the omitted rules (**cls-hs*** in Table 9) are reciprocal of those for has-value expressions (**cls-hv*** in Table 4); motivation for the omission is missing from the draft documents.[7]

In terms of Web reasoning, one other notable consequence of enforcing OWL 2 DL restrictions in OWL 2 RL documents is the forbiddance of inverse-functional datatype-properties [19, Section 9.2.8]: the definition of such properties is common on the Web; examples include `foaf:mbox_sha1sum` and various FOAF chat ID properties whereby the former is commonly used as a primary means of identifying `foaf:Person` members without using URIs.

---

[5] Here, perhaps assuming uniqueness which also considers context.

[6] See `http://team.vox.com/profile/foaf.rdf` for the RDF description of the resource with `foaf:nick` "Team Vox" and, e.g.,
`http://danbri.vox.com/profile/foaf.rdf` as an example of a user profile, all of which reference the `Team Vox` user.

[7] We can only conjecture that this is perhaps related to a possible `owl:hasSelf` emulation of an `owl:ReflexiveProperty` expression which is not supported.

In summary, although the OWL 2 RL profile does not introduce new individuals, and although sound and complete when applied to a valid OWL 2 RL document, the profile is not immediately suited to application over Web data. Indeed, a Web reasoner should perhaps consider abandoning *completeness* guarantees for a more syntactically permissive, semantically inclusive and practicable (albeit, possibly *incomplete*) approach: e.g., allowing inverse-functional datatype-properties, including omitted rules as exemplified in Table 9 and curtailing quadratic equivalence inferencing on the Web.

2.2 Separating Terminological Data

The main optimisation of SAOR, and indeed the main divergence from traditional rules engines, is in considering a distinction between terminological data and assertional data. Herein, we refer to terminological data as the segment of the Web crawl which deals with class and property descriptions – using RDF(S) and OWL terms – that are supported by the given ruleset.

In [16], we showed that <1% of data in our large Web dataset represented terminological data; however, this small segment of data is the most frequently accessed for reasoning, with most rules including terminological expressions in their antecedents. For example, the FOAF ontology currently contains 559 triples, the majority of which we would consider to be terminological; however, there exists hundreds of millions[8] of statements on the Web which use the properties and classes defined by the former 559 triples. Based on such observations, we optimise access to the terminological data; we perform an initial scan of the dataset and extract terminological statements while building an in-memory hashtable representation of this information which we call our "TBox".

In creating an in-memory TBox, for which the terminological information required by each rule can be accessed in $\mathcal{O}(1)$ (in practical terms, considering our hashtable-based implementation), we significantly reduce the implementational complexity of all rules requiring terminological knowledge. Also, since we only index <1% of the data, the cost of building the hashtable is relatively low. In [16], we categorised our rules according to their terminological and assertional arity; i.e., the amount of patterns in the rule that could be answered by the TBox and the amount that could not. In particular, we identified eighteen rules which required zero or one assertional patterns and thus, could be serviced by statement-wise scan of the entire (possibly unsorted) dataset.

Take for example the following rule:

<u>?c owl:intersectionOf (?c₁ ... ?c_n)</u> . ?x a ?c . ⇒ ?x a ?c$_1$, ..., ?c$_n$.

Herein, the terminological patterns serviceable by the TBox are underlined. To execute this rule, the dataset can be scanned statement-by-statement, with triples satisfying the ?x a ?c . pattern joined with the TBox; inferred statements can be recursively joined with the TBox in the same fashion. Thus, we can execute such rules using two scans of the unsorted data; the first builds the TBox and the second executes the rules (again, cf. [16] for more detail).

---

[8] E.g., see http://vmlion25.deri.ie/; the figure could however be in the billions.

However, there exist a number of rules which contain more than one purely assertional pattern in the antecedent, and thus require execution of joins on the arbitrarily large ABox – and even worse – exhaustive application on all inferred ABox triples. Such rules are more expensive to compute and require indexing of a much larger portion of the data; in [16], we presented means to execute such rules using static sorted indices; however, such an approach encountered difficulties in achieving termination and is better suited to approximative reasoning. In any case, we showed that the majority of inferences for our Web dataset were covered by the set of rules with zero or one assertional pattern (<0.3% of inferences were found through rules with more than one assertional pattern[9]). Subsequently, using the rules with a low assertional arity, we demonstrating reasoning over 1.1b statements, crawled from 665k Web documents, in <16.5 hours.

Following from this, Appendix A lists OWL 2 RL rules in order of increasing complexity, starting with rules with no antecedent ($\mathcal{R}0$) and ending with rules with a variable number of assertional patterns ($\mathcal{R}6$-7). In practical terms, rules in $\mathcal{R}0$-3 present an opportunity for near-linear scale with respect to Web reasoning in a system such as SAOR (at least, given observable trends in Web data); rules in $\mathcal{R}4$-5 require assertional joins (with an upper-bound of five conjunctive patterns for rule **cls-maxqc4**), which are more expensive to compute at Web scale; rules in $\mathcal{R}6$-7 may present Web reasoners with the daunting task of computing arbitrarily-large conjunctive- assertional-patterns – Web reasoners would probably have to enforce maximally supported lengths for such expressions.

2.3   Authoritative Reasoning

In preliminary work on SAOR, we encountered a puzzling deluge of inferences from naïve reasoning over a Web dataset. For example, we found that reasoning on a single membership assertion of `owl:Thing` – apparently the "top-level" concept – caused 4,251 inferences when naïve reasoning was applied to the Web dataset.[10] Again for example, the document `http://www.eiao.net/rdf/1.0` defines 9 *properties* to be in the domain of `rdf:type` [15].[11]

The problem is more widespread than core RDF(S)/OWL terms; for one membership assertion of `foaf:Person`, naïve reasoning created 4,631 inferences (an additional 380 inferences on top of `owl:Thing`) as opposed to the six inferences mandated by the FOAF ontology. As another example, there are multiple documents which declare the class `foaf:Image` to be an `owl:ObjectProperty` [15].[12]

In [16], we termed such third party redefinition of ontology terms "ontology hijacking" and proposed a solution to counter such behaviour based upon analysis of "authoritative sources" for terminological data:

---

[9] This figure is, however, increasing with popular usage of transitive properties.

[10] E.g., the document
`http://lsdis.cs.uga.edu/~{}oldham/ontology/wsag/wsag.owl` accounts for 55 such inferences where `owl:Thing` is a member of 55 union class descriptions.

[11] Such usage of `owl:Thing` is prohibited by the structural syntax of OWL 2 RL [19].

[12] E.g., see
`http://wiki.sembase.at/index.php/Special:ExportRDF/Dieter_Fensel`

Definition 1 (Authoritative Source). *A Web document from source (context) c speaks authoritatively about an RDF term n iff:*

1. *n is a blank node; or*
2. *n is a URI and c coincides with, or is redirected to by, the namespace of n.*

We then defined our notion of an "authoritative rule application" whereby, here paraphrasing, each Web document satisfying a terminological pattern in the antecedent must speak authoritatively for at least one term appearing in both the assertional and terminological parts of the antecedent; e.g., take the rule:

$$\texttt{?p rdfs:domain ?c . ?x ?p ?y . } \Rightarrow \texttt{ ?x a ?c .}$$

Here, the term matched by `?p` must be authoritatively spoken for by the document serving the `rdfs:domain` triple. Therefore, taking the previous example where nine domains for `rdf:type` are non-authoritatively defined, the document `http://www.eiao.net/rdf/1.0` does not speak authoritatively for `rdf:type`, which is bound by `?p`: thus, no inference takes place.

Of course, we still allow extension of external ontologies, whereby memberships of local terms are translated into memberships of remote terms, but not vice-versa; e.g., for the above rule, authoritative reasoning will still permit a triple such as `ex:sibling rdfs:domain foaf:Person .` when served in a location authoritative for the `ex:` namespace, facilitating translation from subject-members of the local property `ex:sibling` to the remote class `foaf:Person`.

Along these lines, Tables 4-9 indicate authoritative variables for the OWL 2 RL rules in bold-face; when enforced, the document(s) serving the terminological statements must speak authoritatively for *at least one* binding of an authoritative variable for the rule to fire.

3   Web Use-Cases

The OWL 2 New Features and Rationale document [10] is intended to provide rationale and use-cases for novel OWL 2 features; in particular, the document defines 19 use-cases which motivate new features. However, the document focuses largely on domain-specific use-cases, with, e.g., nine use-cases tied to the Health Care and Life Sciences (HCLS) domain. In this section, we briefly look at how new OWL 2 features could be exploited on the Web by investigating possible pragmatic extensions of two prominent Web ontologies; namely Friend Of A Friend (FOAF) and Semantically Interlinked Online Communities (SIOC).

FOAF is a lightweight ontology providing classes and properties for describing personal information and resources; these terms are amongst the most commonly instantiated on the Web [9, Table 1&2], with many blogging platforms and social networks providing automatic exports of user profiles in FOAF. SIOC [2], similarly, is a lightweight ontology for describing and connecting resources relating to online social communities and the various platforms for information dissemination on the Web; SIOC reuses terms from other Web ontologies, including FOAF. SIOC terms have more recently seen a large growth in popularity as large-volume exporters have become available.[13]

Both ontologies are pragmatically lightweight to foster uptake amongst non-expert communities; we follow such precedent – e.g., we ignore the new disjoint-union qualified-cardinality and self-restriction constructs since both FOAF and SIOC have previously avoided complex class descriptions from the original OWL specification – and select the following novel OWL 2 constructs as possible targets for use in FOAF and/or SIOC: (i) `IrreflexiveProperty`/`AsymmetricProperty`; (ii) `propertyChainAxiom`; and (iii) `hasKey`.

In order to provide insights into the fecundity of our proposed extensions, we perform reasoning over a representative Web dataset. We retrieved this dataset from the Web in April 2009 by means of a Web crawl using MultiCrawler [12]; beginning with Tim Berners-Lee's FOAF file[14], we performed a seven-hop breadth first crawl for RDF/XML files; after each hop, we extracted all URIs from the crawled data as input for the next hop. Finally, we restricted the crawl according to pay-level-domains; we enforced a maximum of 5,000 crawled documents from each domain so as to ensure a diverse and representative dataset. The crawl consisted of access to 149,057 URIs, and acquired 54,836 (36.8%) valid RDF/XML documents containing 12,534,481 RDF statements; of these, 3,751,617 statements (29.9%) contain a URI in the FOAF namespace and 782,188 (6.2%) contain a URI in a SIOC namespace.

Property constraints. We now look at asymmetrical/irreflexive property constraints: we take precedent from the current `owl:disjointWith` assertions in FOAF and SIOC which analogously provide simple means of consistency checking. Firstly, please note that all asymmetric properties are irreflexive. Also, we only select properties whose irreflexivity was not already implicitly constrained by disjoint domain/range assertions; we exclude datatype properties, and, e.g., we exclude `workplaceHomepage` since the domain (`Person`) and range (`Document`) are disjoint and symmetric or reflexive relations thereof would already be flagged. We chose 6 FOAF properties and 17 SIOC properties as being implicitly assymetric/irreflexive and one SIOC property as being irreflexive alone: `sibling`.

Applying these constraints to our Web dataset, we found 319 `sioc:link`[15], 2 `foaf:holdsAcccount` and 1 `foaf:mbox` reflexive statements. We found no examples of symmetric statements for the asymmetric properties. Although the results are less than convincing, the asymmetric/irreflexive constraints would constitute a straightforward extension of the FOAF/SIOC ontologies; please note that for

---

[13] See http://vmlion25.deri.ie/ for a recent survey of the terms in a >1bn statement Web dataset.

[14] http://www.w3.org/People/Berners-Lee/card

[15] The textual description of the `link` property recommends irreflexive use [1]. Most such reflexive statements are produced by the SIOC WordPress exporter; e.g., see http://dowhatimean.net/?sioc_type=site

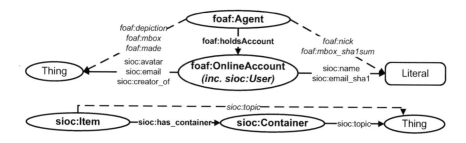

Fig. 1. FOAF/SIOC property chain translations

authoritative reasoning, these constraints should be provided in the FOAF/SIOC
ontologies for the FOAF/SIOC terms respectively.

Property chains. The `owl:propertyChainAxiom` allows for defining arbitrarily
long paths which, when present, succinctly infer membership of a single property.

The main use-case we envisage for this construct relates to translating SIOC
attributes attached to an instance of `sioc:User` into FOAF attributes attached
to `foaf:Agent`. The FOAF profile defines a class `foaf:OnlineAccount` intended to
represent the online presence of a `foaf:Agent` through the `foaf:holdsAccount` relation; SIOC defines `sioc:User`, a subclass of `foaf:OnlineAccount`, and provides a
more expressive vocabulary for defining and connecting the `sioc:User` with online
services. Thus, we can translate from the SIOC attributes for `sioc:User`/`foaf:On-`
`lineAccount` to equivalent FOAF properties with a domain of `foaf:Agent` or an
encompassing class thereof.

Another possible use-case is to formally realise the informal semantics of
`sioc:topic` which state that "...a Container will have an associated topic or set
of topics that can be propagated to the Items it contains" [1]: this propagation
can be implemented using OWL 2 property chains.

Figure 1 depicts the envisaged translations.; to take an example, the assertion
`foaf:nick :propertyChainAxiom (foaf:holdsAccount, sioc:name)` . made in the
FOAF or SIOC ontology would allow for authoritative translation of `sioc:name`
values attached to `sioc:Users` into `foaf:nick` values attached to `foaf:Agents`.[16]

Applying the above chains to our dataset, we found 29,617 inferences; viz.:
29,373 `foaf:nick`, 216 `foaf:depiction`, 20 `foaf:mbox` and 8 `foaf:mbox_sha1sum`
statements respectively. Here it seems, the only practically convincing use-case
is the translation of `sioc:name` values into `foaf:nick` values, although perhaps
with the growth of online SIOC data, the above figures may begin to increase.

Complex keys. We examine one last use-case for the new OWL 2 constructs;
namely the `owl:hasKey` construct which is used to define a set of properties

---

[16] This translation seems a neat fit: the informal description of `foaf:nick` states that
it is for values "such as those use [sic.] in IRC chat, online accounts, and computer
logins" [4].

whose values together uniquely identify a member of the specified class. We foresee one possible use-case which again lies on the intersection of FOAF and SIOC: members of `foaf:OnlineAccount`, and thereby of `sioc:User`, are uniquely defined by the properties `foaf:accountName` and `foaf:accountServiceHomepage` together.

Applying the above key to our dataset, we found 4,576,310 non-reflexive `:sameAs` inferences (only includes inferences from application of **prp-key** and not of, e.g., **eq-trans**) mentioning 78,534 individuals, with the longest equivalence chain containing 723 individuals. Due to rare usage of URIs for `OnlineAccount` members, complex keys are the only solution currently available to uniquely identify and aggregate such resources.

4   Related Work

Several rule expressible non-standard OWL fragments; namely OWL-DLP [11], OWL$^-$ [7] (which is a slight extension of OWL-DLP), OWLPrime [25] and pD* [24]; have been defined in the literature and enable incomplete but sound RDFS and OWL Full inferences.

Some existing systems already implement a separation of TBox and ABox for scalable reasoning, where in particular, assertional data are stored in some RDBMS; e.g., Hawkeye [20] demonstrates reasoning over a 166m triple Web dataset – however, they use a prescribed TBox. Also, like us, they internally choose pivot identifiers to represent equivalent sets of individuals.

Work presented in [5] introduced the notion of an *authoritative description* which is very similar to ours; however, we provide much more extensive treatment of the issue, supporting a much more varied range of RDF(S)/OWL constructs.

One promising alternative to authoritative reasoning for the Web is the notion of "context-dependant" or "quarantined reasoning" introduced in [8], whereby inference results are only considered valid (quarantined) within the given context of a document.

5   Conclusion

In this paper, we have presented discussion relating to applying OWL 2 RL over Web data. In particular, we discussed a separation of terminological data from purely assertional data wherein terminological data represents a small fraction of an overall Web dataset and is the most frequently accessed during reasoning. We presented a categorisation of OWL 2 RL rules based on terminological/ assertional arity and discussed the implementational feasibility of said categories. We also discussed authoritative reasoning, which heeds the source of information when making inferences, thus countering unwanted third-party contributions. We identified those variable positions present in the OWL 2 RL/RDF rules which should be authoritatively restricted to counter-act ontology hijacking. Finally, motivated by a lack of Web reasoning discussion in the official specifications, we presented a number of Web use-cases for OWL 2 terms based on two popular Web ontologies: viz. SIOC and FOAF. Although some of the use-cases were not convincing when presented with a real Web dataset, we found some practical deployment for the `owl:propertyChainAxiom` and `owl:hasKey` constructs. In any case, our purview was limited to that of SIOC and FOAF, and we conclude that new OWL 2 terms may find more productive application in other/future Web ontologies.

References

1. Bojārs, U., Breslin, J.G.: SIOC Core Ontology Specification (January 2009), http://rdfs.org/sioc/spec/
2. Breslin, J.G., Harth, A., Bojars, U., Decker, S.: Towards semantically-interlinked online communities. In: Gómez-Pérez, A., Euzenat, J. (eds.) ESWC 2005. LNCS, vol. 3532, pp. 500–514. Springer, Heidelberg (2005)
3. Brickley, D., Guha, R.: RDF vocabulary description language 1.0: RDF Schema. W3C Recommendation (February 2004), http://www.w3.org/TR/rdf-schema/
4. Brickley, D., Miller, L.: FOAF Vocabulary Specification 0.91 (November 2007), http://xmlns.com/foaf/spec/
5. Cheng, G., Ge, W., Wu, H., Qu, Y.: Searching semantic web objects based on class hierarchies. In: Proceedings of Linked Data on the Web Workshop (2008)
6. de Bruijn, J.: RIF RDF and OWL Compatibility, W3C Working Draft (July 2009), http://www.w3.org/TR/rif-rdf-owl/
7. de Bruijn, J., Polleres, A., Lara, R., Fensel, D.: OWL⁻. Technical Report WSML d20.1v0.2 (2005)
8. Delbru, R., Polleres, A., Tummarello, G., Decker, S.: Context dependent reasoning for semantic documents in Sindice. In: Proc. of the 4th Int. Workshop on Scalable Semantic Web Knowledge Base Systems (SSWS 2008) (October 2008)
9. Ding, L., Finin, T.: Characterizing the Semantic Web on the Web. In: Cruz, I., Decker, S., Allemang, D., Preist, C., Schwabe, D., Mika, P., Uschold, M., Aroyo, L.M. (eds.) ISWC 2006. LNCS, vol. 4273, pp. 242–257. Springer, Heidelberg (2006)
10. Golbreich, C., Wallace, E.K.: OWL 2 New Features and Rationale. W3C Working Draft (2009), http://www.w3.org/TR/owl2-new-features/
11. Grosof, B., Horrocks, I., Volz, R., Decker, S.: Description logic programs: Combining logic programs with description logic. In: 13th International Conference on World Wide Web (2004)
12. Harth, A., Umbrich, J., Decker, S.: Multicrawler: A pipelined architecture for crawling and indexing semantic web data. In: Cruz, I., Decker, S., Allemang, D., Preist, C., Schwabe, D., Mika, P., Uschold, M., Aroyo, L.M. (eds.) ISWC 2006. LNCS, vol. 4273, pp. 258–271. Springer, Heidelberg (2006)
13. Hayes, P.: RDF semantics. W3C Recommendation (February 2004), http://www.w3.org/TR/rdf-mt/
14. Hogan, A., Harth, A., Decker, S.: Performing object consolidation on the semantic web data graph. In: I3: Identity, Identifiers, Identification Workshop (2007)
15. Hogan, A., Harth, A., Passant, A., Decker, S., Polleres, A.: Weaving the Pedantic Web. Technical report, DERI Galway (2009), http://www.deri.ie/fileadmin/documents/DERI-TR-2009-07-28.pdf

16. Hogan, A., Harth, A., Polleres, A.: Scalable Authoritative OWL Reasoning for the Web. Int. Journal on Semantic Web and Information Systems 5(2) (2009)
17. Lukácsy, G., Szeredi, P.: Efficient description logic reasoning in prolog: The dlog system. CoRR, abs/0904.0578 (2009)
18. Motik, B., Grau, B.C., Horrocks, I., Wu, Z., Fokoue, A., Lutz, C.: OWL 2 Web Ontology Language Profiles, W3C Candidate Recommendation (June 2009), http://www.w3.org/TR/owl2-profiles/
19. Motik, B., Patel-Schneider, P.F., Parsia, B.: OWL 2 Web Ontology Language Structural Specification and Functional-Style Syntax, W3C Candidate Recommendation (June 2009), http://www.w3.org/TR/owl2-syntax/
20. Pan, Z., Qasem, A., Kanitkar, S., Prabhakar, F., Heflin, J.: Hawkeye: A practical large scale demonstration of semantic web integration. In: Meersman, R., Tari, Z., Herrero, P. (eds.) OTM-WS 2007, Part II. LNCS, vol. 4806, pp. 1115–1124. Springer, Heidelberg (2007)
21. Schneider, M.: OWL 2 RDF-Based Semantics. W3C Candidate Recommendation (June 2009), http://www.w3.org/TR/owl2-rdf-based-semantics/
22. Smith, M., Horrocks, I., Krötzsch, M.: OWL 2 Web Ontology Language Conformance, W3C Candidate Recommendation (June 2009), http://www.w3.org/TR/owl2-conformance/
23. Smith, M.K., Welty, C., McGuinness, D.L.: OWL Web Ontology Language Guide. W3C Recommendation (February 2004), http://www.w3.org/TR/owl-guide/
24. ter Horst, H.J.: Completeness, decidability and complexity of entailment for RDF Schema and a semantic extension involving the OWL vocabulary. Journal of Web Semantics 3, 79–115 (2005)
25. Wu, Z., Eadon, G., Das, S., Chong, E.I., Kolovski, V., Annamalai, M., Srinivasan, J.: Implementing an Inference Engine for RDFS/OWL Constructs and User-Defined Rules in Oracle. In: 24th Int. Conf. on Data Engineering. IEEE, Los Alamitos (2008)

A   Rule Tables

In this Section, we provide Tables 1-9 for reference (which include all OWL 2 RL rules) presented in Turtle-like syntax; the default namespace refers to `owl:`. Rules are categorised according to increasing terminological/assertional antecedent arity; authoritative variable positions are denoted using bold-face.

Table 1. Rules with no antecedent

OWL2RL	SAOR	Consequent	Notes
\multicolumn{4}{c|}{$\mathcal{R}0$: no antecedent}			
prp-ap	-	?ap a :AnnotationProperty .	For each built-in annotation property
cls-thing	-	:Thing a :Class .	-
cls-nothing	-	:Nothing a :Class .	-
dt-type1	-	?dt a rdfs:Datatype .	For each built-in datatype
dt-type2	-	?l a ?dt .	For all ?l in the value space of datatype ?dt
dt-eq	-	$?l_1$:sameAs $?l_2$.	For all $?l_1$ and $?l_2$ with the same data value
dt-diff	-	$?l_1$:differentFrom $?l_2$.	For all $?l_1$ and $?l_2$ with different data values

Table 2. Only terminological antecedent patterns

$\mathcal{R}1$: only terminological patterns in antecedent			
OWL2RL	**SAOR**	**Antecedent** (terminological)	**Consequent**
cls-00	rdfc0	?c :oneOf (?x_1 ... ?x_n) .	?x_1 ... ?x_n a ?c .
scm-cls	-	?c a :Class .	?c rdfs:subClassOf ?c , :Thing ; :equivalentClass ?c . :Nothing rdfs:subClassOf ?c .
scm-sco	-	?c_1 rdfs:subClassOf ?c_2 . ?c_2 rdfs:subClassOf ?c_3 .	?c_1 rdfs:subClassOf ?c_3 .
scm-eqc1	-	?c_1 :equivalentClass ?c_2 .	?c_1 rdfs:subClassOf ?c_2 . ?c_2 rdfs:subClassOf ?c_1 .
scm-eqc2	-	?c_1 rdfs:subClassOf ?c_2 . ?c_2 rdfs:subClassOf ?c_1 .	?c_1 :equivalentClass ?c_2 .
scm-op	-	?p a :ObjectProperty .	?p rdfs:subPropertyOf ?p . ?p :equivalentProperty ?p .
scm-dp	-	?p a :DatatypeProperty .	?p rdfs:subPropertyOf ?p . ?p :equivalentProperty ?p .
scm-spo	-	?p_1 rdfs:subPropertyOf ?p_2 . ?p_2 rdfs:subPropertyOf ?p_3 .	?p_1 rdfs:subPropertyOf ?p_3 .
scm-eqp1	-	?p_1 :equivalentProperty ?p_2 .	?p_1 rdfs:subPropertyOf ?p_2 . ?p_2 rdfs:subPropertyOf ?p_1 .
scm-eqp2	-	?p_1 rdfs:subPropertyOf ?p_2 . ?p_2 rdfs:subPropertyOf ?p_1 .	?p_1 :equivalentProperty ?p_2 .
scm-dom1	-	?p rdfs:domain ?c_1 . ?c_1 rdfs:subClassOf ?c_2 .	?p rdfs:domain ?c_2 .
scm-dom2	-	?p_2 rdfs:domain ?c . ?p_1 rdfs:subPropertyOf ?p_2 .	?p_1 rdfs:domain ?c .
scm-rng1	-	?p rdfs:range ?c_1 . ?c_1 rdfs:subClassOf ?c_2 .	?p rdfs:range ?c_2 .
scm-rng2	-	?p_2 rdfs:range ?c . ?p_1 rdfs:subPropertyOf ?p_2 .	?p_1 rdfs:range ?c .
scm-hv	-	?c_1 :hasValue ?i ; :onProperty ?p_1 . ?c_2 :hasValue ?i ; :onProperty ?p_2 . ?p_1 rdfs:subPropertyOf ?p_2 .	?c_1 rdfs:subClassOf ?c_2 .
scm-svf1	-	?c_1 :someValuesFrom ?y_1 ; :onProperty ?p . ?c_2 :someValuesFrom ?y_2 ; :onProperty ?p . ?y_1 rdfs:subClassOf ?y_2 .	?c_1 rdfs:subClassOf ?c_2 .
scm-svf2	-	?c_1 :someValuesFrom ?y ; :onProperty ?p_1 . ?c_2 :someValuesFrom ?y ; :onProperty ?p_2 . ?p_1 rdfs:subPropertyOf ?p_2 .	?c_1 rdfs:subClassOf ?c_2 .
scm-avf1	-	?c_1 :allValuesFrom ?y_1 ; :onProperty ?p . ?c_2 :allValuesFrom ?y_2 ; :onProperty ?p . ?y_1 rdfs:subClassOf ?y_2 .	?c_1 rdfs:subClassOf ?c_2 .
scm-avf2	-	?c_1 :allValuesFrom ?y ; :onProperty ?p_1 . ?c_2 :allValuesFrom ?y ; :onProperty ?p_2 . ?p_1 rdfs:subPropertyOf ?p_2 .	?c_1 rdfs:subClassOf ?c_2 .
scm-int	-	?c :intersectionOf (?c_1 ... ?c_n) .	?c rdfs:subClassOf ?c_1...?c_n .
scm-uni	-	?c :unionOf (?c_1 ... ?c_n) .	?c_1...?c_n rdfs:subClassOf ?c .

Table 3. No terminological, but one assertional antecedent pattern

$\mathcal{R}2$: one assertional antecedent pattern				
OWL2RL	**SAOR**	**Antecedent**	**Consequent**	**Notes**
eq-ref	-	?s ?p ?o .	?s :sameAs ?s . ?p :sameAs ?p . ?o :sameAs ?o .	
eq-sym	rdfp6'	?x :sameAs ?y .	?y :sameAs ?x .	
cls-nothing2	-	?x a :Nothing .	false	
dt-not-type	-	?l a ?dt .	false	Where ?l is not in the value space of ?dt

Table 4. At least one terminological and exactly one assertional pattern

$\mathcal{R}3$: at least one terminological/only one assertional pattern in antecedent				
OWL2RL	**SAOR**	**Antecedent**		**Consequent**
		terminological	*assertional*	
prp-dom	rdfs2	?p rdfs:domain ?c .	?x ?p ?y .	?x a ?c .
prp-rng	rdfs3$'$	?p rdfs:range ?c .	?x ?p ?y .	?y a ?c .
prp-irp	-	?p a :IrreflexiveProperty .	?x ?p ?x .	false
prp-symp	rdfp3$'$	?p a :SymmetricProperty .	?x ?p ?y .	?y ?p ?x .
prp-spo1	rdfs7$'$	?p$_1$ rdfs:subPropertyOf ?p$_2$.	?x ?p$_1$?y .	?x ?p$_2$?y .
prp-eqp1	rdfp13a$'$	?p$_1$:equivalentProperty ?p$_2$.	?x ?p$_1$?y .	?x ?p$_2$?y .
prp-eqp2	rdfp13b$'$	?p$_1$:equivalentProperty ?p$_2$.	?x ?p$_2$?y .	?x ?p$_1$?y .
prp-inv1	rdfp8a$'$	?p$_1$:inverseOf ?p$_2$.	?x ?p$_1$?y .	?y ?p$_2$?x .
cls-int2	rdfc3a$'$	?c :intersectionOf (?c$_1$ ... ?c$_n$) .	?x a ?c .	?x a ?c$_1$...?c$_n$.
cls-uni	rdfc1$'$	?c :unionOf (?c$_1$ ... ?c$_i$ ... ?c$_n$) .	?x a ?c$_i$	?x a ?c .
cls-svf2	rdfp15$'$*	?x :someValuesFrom :Thing ; :onProperty ?p .	?u ?p ?v .	?u a ?x .
cls-hv1	rdfp14b$'$	?x :hasValue ?y ; :onProperty ?p .	?u a ?x .	?u ?p ?y .
cls-hv2	rdfp14a$'$	?x :hasValue ?y ; :onProperty ?p .	?u ?p ?y .	?u a ?x .
cax-sco	rdfs9	?c$_1$ rdfs:subClassOf ?c$_2$.	?x a ?c$_1$.	?x a ?c$_2$.
cax-eqc1	rdfp12a$'$	?c$_1$:equivalentClass ?c$_2$.	?x a ?c$_1$.	?x a ?c$_2$.
cax-eqc2	rdfp12b$'$	?c$_1$:equivalentClass ?c$_2$.	?x a ?c$_2$.	?x a ?c$_1$.

Table 5. No terminological, but multiple assertional patterns

$\mathcal{R}4$: no terminological pattern/multiple assertional patterns			
OWL2RL	**SAOR**	**Antecedent**	**Consequent**
eq-trans	rdfp7	?x :sameAs ?y . ?y :sameAs ?z .	?x :sameAs ?z .
eq-rep-s	rdfp11$'$*	?s :sameAs ?s$'$. ?s ?p ?o .	?s$'$?p ?o .
eq-rep-p	-	?p :sameAs ?p$'$. ?s ?p ?o .	?s ?p$'$?o .
eq-rep-o	rdfp11$''$*	?o :sameAs ?o$'$. ?s ?p ?o .	?s ?p ?o$'$.
eq-diff1	-	?x :sameAs ?y ; :differentFrom ?y .	false
prp-npa1	-	?x :sourceIndividual ?i$_1$; :assertionProperty ?p ; :targetIndividual ?i$_2$. ?i$_1$?p ?i$_2$.	false
prp-npa2	-	?x :sourceIndividual ?i$_1$; :assertionProperty ?p ; :targetValue ?lt . ?i$_1$?p ?lt .	false

Table 6. At least one terminological and mulitple assertional patterns

$\mathcal{R}5$: at least one terminological/multiple assertional patterns in antecedent				
OWL2RL	SAOR	Antecedent		Consequent
		terminological	*assertional*	
prp-fp	rdfp1$'$	?p a :FunctionalProperty .	?x ?p ?y$_1$, ?y$_2$.	?y$_1$:sameAs ?y$_2$.
prp-ifp	rdfp2	?p a :InverseFunctionalProperty .	?x$_1$?p ?z . ?x$_2$?p ?z .	?x$_1$:sameAs ?x$_2$.
prp-asyp	-	?p a :AsymmetricProperty .	?x ?p ?y . ?y ?p ?x .	false
prp-trp	rdfp4	?p a :TransitiveProperty .	?x ?p ?y . ?y ?p ?z .	?x ?p ?z .
prp-pdw	-	?p$_1$:disjointWith ?p$_2$.	?x ?p$_1$?y ; ?p$_2$?y .	false
prp-adp	-	?x a :AllDisjointProperties ; owl:members (?p$_1$ ... p$_n$) .	?u ?p$_i$?y ; ?p$_j$?y .	false
cls-com	-	?c$_1$:complementOf ?c$_2$.	?x a ?c$_1$, ?c$_2$.	false
cls-svf1	rdfp15$'$	?x :someValuesFrom ?y ; :onProperty ?p .	?u ?p ?v . ?v a ?y .	?u a ?x .
cls-avf	rdfp16$'$	?x :allValuesFrom ?y ; :onProperty ?p .	?u a ?x ; ?p ?v .	?v a ?y .
cls-maxc1	-	?x :maxCardinality 0 ; :onProperty ?p .	?u a ?x ; ?p ?y .	false
cls-maxc2	rdfc4b	?x :maxCardinality 1 ; :onProperty ?p .	?u a ?x ; ?p ?y$_1$, ?y$_2$.	?y$_1$:sameAs ?y$_2$.
cls-maxqc1	-	?x :maxQualifedCardinality 0 ; :onProperty ?p ; :onClass ?c .	?u a ?x ; ?p ?y . ?y a ?c .	false
cls-maxqc2	-	?x :maxQualifiedCardinality 0 ; :onProperty ?p ; :onClass :Thing .	?u a ?x ; ?p ?y .	false
cls-maxqc3	-	?x :maxQualifiedCardinality 1 ; :onProperty ?p ; :onClass ?c .	?u a ?x ; ?p ?y$_1$, ?y$_2$. ?y$_1$ a ?c . ?y$_2$ a ?c .	?y$_1$:sameAs ?y$_2$.
cls-maxqc4	-	?x :maxQualifiedCardinality 1 ; :onProperty ?p ; :onClass :Thing .	?u a ?x ; ?p ?y$_1$, ?y$_2$.	?y$_1$:sameAs ?y$_2$.
cax-dw	-	?c$_1$:disjointWith ?c$_2$.	?x a ?c1 , ?c2 .	false
cax-adc	-	?x a :AllDisjointClasses ; :members (?c$_1$ ... ?c$_n$) .	?z a ?c$_i$, ?c$_j$.	false

Table 7. No terminological, but a variable number of assertional patterns

$\mathcal{R}6$: no terminological pattern/variable assertional patterns			
OWL2RL	SAOR	Antecedent	Consequent
eq-diff2	-	?x a :AllDifferent ; :members (z$_1$ ... z$_n$) . ?z$_i$:sameAs ?z$_j$.	false
eq-diff3	-	?x a :AllDifferent ; :distinctMembers (z$_1$ ... z$_n$) . ?z$_i$:sameAs ?z$_j$.	false

Table 8. At least one terminological and variable assertional patterns

$\mathcal{R}7$: at least one terminological/variable assertional patterns in antecedent				
OWL2RL	SAOR	Antecedent		Consequent
		terminological	*assertional*	
prp-spo2	-	?p :propertyChainAxiom (?p$_1$ ... ?p$_n$) .	?u$_1$?p$_1$?u$_2$ . ... ?u$_n$?p$_n$?u$_{n+1}$.	?u$_1$?p ?u$_{n+1}$.
prp-key	-	?c :hasKey (?p$_1$ ... p$_n$) .	?x a ?c . ?x ?p$_1$?z$_1$ . ... ?x ?p$_n$?z$_n$. ?y a ?c . ?y ?p$_1$?z$_1$ . ... ?y ?p$_n$?z$_n$.	?x :sameAs ?y .
cls-int1	rdfc3c	?c :intersectionOf (?c$_1$ ... ?c$_n$) .	?y a ?c$_1$ ... ?c$_n$.	?y a ?c .

Table 9. Rules not in OWL 2 RL

		Rules not in OWL 2 RL			
		Antecedent		Consequent	$\mathcal{R}$
ID	SAOR	*terminological*	*assertional*		
cls-minc1	rdfc2	?x :minCardinality 1 ; :onProperty ?p .	?u ?p ?y .	?u a ?x .	$\mathcal{R}3$
cls-hs1	-	?x :hasSelf true ; :onProperty ?p .	?u ?a ?x .	?u ?p ?u .	$\mathcal{R}3$
cls-hs2	-	?x :hasSelf true ; :onProperty ?p .	?u ?p ?u .	?u a ?c .	$\mathcal{R}3$
cls-duni1	-	?x :disjointUnionOf ($?c_1$ ... $?c_i$ ... $?c_n$) .	?y a $?c_i$.	?y a ?x .	$\mathcal{R}3$
cls-duni2	-	?x :disjointUnionOf (... $?c_i$ ... $?c_j$ ...) .	?y a $?c_i$, $?c_j$.	false	$\mathcal{R}5$

Answer Sets in a Fuzzy Equilibrium Logic

Steven Schockaert[1], Jeroen Janssen[2], Dirk Vermeir[2], and Martine De Cock[1,3]

[1] Dept. of Applied Mathematics and Computer Science,
Ghent University, Belgium
{steven.schockaert,martine.decock}@ugent.be
[2] Dept. of Computer Science, Vrije Universiteit Brussel, Belgium
{jeroen.janssen,dvermeir}@vub.ac.be
[3] Institute of Technology, University of Washington, Tacoma, WA, USA
mdecock@u.washington.edu

Abstract. Since its introduction, answer set programming has been generalized in many directions, to cater to the needs of real-world applications. As one of the most general "classical" approaches, answer sets of arbitrary propositional theories can be defined as models in the equilibrium logic of Pearce. Fuzzy answer set programming, on the other hand, extends answer set programming with the capability of modeling continuous systems. In this paper, we combine the expressiveness of both approaches, and define answer sets of arbitrary fuzzy propositional theories as models in a fuzzification of equilibrium logic. We show that the resulting notion of answer set is compatible with existing definitions, when the syntactic restrictions of the corresponding approaches are met. We furthermore locate the complexity of the main reasoning tasks at the second level of the polynomial hierarchy. Finally, as an illustration of its modeling power, we show how fuzzy equilibrium logic can be used to find strong Nash equilibria.

1 Introduction

Answer set programming (ASP) is a widely used framework for non-monotonic reasoning [2], in which knowledge is represented as a set of rules of the form $\alpha \leftarrow \beta$. Intuitively, such a rule indicates that its head α should be assumed, whenever its body β is known to hold. Essentially, the expressions α and β are propositional expressions in negation-normal form, although two types of negation may occur in front of atoms, viz. strong negation $\neg$ and negation-as-failure not. The former corresponds to classical negation whereas the latter models lack of evidence, e.g. $a \leftarrow not\, b$ means that a should be assumed, unless it has been established that b holds. In contrast to other forms of logic programming such as traditional Prologs, ASP offers an elegant declarative semantics for both types of negations. In practice, an *answer set program* P, i.e. a set of rules, represents a problem instance, whose solutions correspond to the *answer sets* of the program. When no negations occur in P, the answer sets correspond to minimal models of the rules in P. The semantics of strong negation can be defined by treating $\neg a$ as an atom (which implies that an answer set can at the

A. Polleres and T. Swift (Eds.): RR 2009, LNCS 5837, pp. 135–149, 2009.
© Springer-Verlag Berlin Heidelberg 2009

same time contain a and $\neg a$, in which case it is inconsistent). The semantics of negation-as-failure is usually defined in terms of the Gelfond-Lifschitz reduct [9]; an interpretation I is then an answer set of P if it is an answer set of the reduct P^I, a particular program without negation-as-failure.

The syntax and semantics of answer set programs has been generalized in many directions. An interesting example is [23], where the notion of answer set is extended to arbitrary propositional theories, among others allowing programs with nested rules, or with occurrences of *not* in front of complex expressions. Answer sets are defined as models in a particular logic called *equilibrium logic*, whose most important characteristics are recalled in Section 2. Due to its generality, this theory has proven useful in defining the semantics of various practical extensions to ASP [8]. Moreover, due to the elegant definition of answer sets, in logical terms, equilibrium logic has also proven fundamental in obtaining important theoretical results on ASP [17].

Since classical ASP is based on boolean logic, it can only model discrete systems. *Fuzzy answer set programming* (FASP) is a quite different generalization of ASP, which allows to model continuous systems by using a (typically) infinite domain of truth values such as the unit interval $[0, 1]$ (e.g. [13,26,27]). In FASP, atoms correspond to gradations or intensities rather than boolean propositions, and operators from fuzzy logic are used to encode relationships between them. An obvious application of FASP is knowledge representation in the presence of vagueness. In this case, atoms correspond to statements whose truth is inherently gradual (e.g. "today is a rainy day"). Another example are optimization problems involving continuous quantities, e.g. a FASP program can be specified to assign papers to reviewers, by taking a continuous form of similarity between reviewers and authors into account to avoid conflicts of interest [14]. As a last example, FASP can be used as an elegant vehicle to specify quantities that are defined in terms of fixpoints, such as the well known PageRank measure [22]. FASP should not be confused with extensions to ASP that consider possibilistic [21] or probabilistic [18] uncertainty, even though to some extent, probabilistic uncertainty in ASP can be simulated in FASP [4]. Neither should FASP be confused with fuzzy logic based approaches to commonsense reasoning about continuous systems [11]. Essentially, fuzzy answer set programs specify a continuous function in a way that highlights causal relations between variables.

The aim of this paper is to combine the equilibrium logic approach from [23] with the ideas of FASP. To this end, we introduce a fuzzy equilibrium logic, and define a notion of answer sets which is analogous to answer sets in regular (or crisp) equilibrium logic. The resulting framework is useful in many ways. First, due to the generality of (fuzzy) equilibrium logic, many constructs from ASP that have not been generalized yet to FASP can easily be defined using fuzzy equilibrium logic, i.e. our approach increases the expressivity of FASP in a substantial way. Second, even restricted to (fuzzy) logic programs, the fuzzy equilibrium approach offers more flexibility than existing methods (e.g. regarding the choice of implicator, see Section 2.2). Third, answer sets in fuzzy equilibrium logic are defined in a way which is very different from existing approaches to FASP, which either generalize the Gelfond-Lifschitz definition [9], or generalize an equivalent definition based on unfounded sets [24]. Having different definitions of answer sets is desirable, as it may lead to different insights, implementations of different reasoners, or facilitate the derivation of new theoretical results [16]. While this is already true for classical ASP, it becomes even more important for FASP. Indeed, it is well known that "fuzzifying" concepts that are classically equivalent may lead to generalizations that are no longer equivalent. As an illustration, the definition of answer sets in FASP from [27], which is based on a generalization of unfounded sets, behaves quite different from the definition from [15], which is based on a generalization of the Gelfond-Lifschitz reduct.

The paper is structured as follows. In the next section, we familiarize the reader with the equilibrium logic of Pearce (Section 2.1) and with fuzzy answer set programming (Section 2.2). The main body of the paper is presented in Section 3 where we define fuzzy equilibrium models (Section 3.1), expose the relationship both with equilibrium models in the sense of Pearce and with answer sets in FASP (Section 3.2), and determine the computational complexity of main reasoning tasks (Section 3.3). In addition, Section 3.3 presents a geometric characterization of fuzzy equilibrium models in terms of polyhedra in a high-dimensional space. Next, in Section 4, as an application example, we show how the problem of finding strong Nash equilibria can be solved using fuzzy equilibrium logic. Finally, some connections with related work are discussed in Section 5. Proofs of the propositions in this paper are available as an online appendix[1].

2   Background

2.1   Equilibrium Logic

In [23], equilibrium logic is introduced by Pearce with the aim of extending the notion of answer set to general propositional theories. The formulation of this logic is based on the logic $N2$, which in turn corresponds to an extension of the logic of here-and-there with strong negation. Although these logics can be characterized axiomatically and algebraically, it will be sufficient for us to consider a model-theoretic characterization in terms of Kripke frames. In the following, let A be a set of atoms (or propositions), and let formulas be defined recursively as either atoms, or expressions of the form $\alpha \lor \beta$, $\alpha \land \beta$, $\alpha \to \beta$, $\neg \alpha$ or $not\, \alpha$, where α and β are formulas. An ($N2$-) valuation V assigns a truth value from $\{-1, 0, 1\}$ to formulas in two different worlds, h (or here) and t (or there), which are connected by the relation $\leq = \{(h, h), (t, t), (h, t)\}$; we write $V(w, \alpha)$ to denote the truth value of formula α in world w. Furthermore, we have the requirement that for each formula α, if $V(h, \alpha) \neq 0$ then $V(t, \alpha) = V(h, \alpha)$. Intuitively, -1 and 1 correspond to the classical truth degrees *false* and *true*, whereas 0 corresponds to *undecided*, i.e. our knowledge about the truth of formulas may be incomplete. The *there*-world corresponds to a refinement of the

---

[1] http://users.ugent.be/~sschocka/fuzzyEquilibriumProofs.pdf

knowledge in the *here*-world. Furthermore, the valuation of complex formulas is obtained from the valuation of atoms as follows ($w \in \{h, t\}$, α and β are formulas):

$$V(w, \neg\alpha) = -V(w, \alpha)$$
$$V(w, \alpha \wedge \beta) = \min(V(w, \alpha), V(w, \beta))$$
$$V(w, \alpha \vee \beta) = \max(V(w, \alpha), V(w, \beta))$$

$$V(w, \alpha \to \beta) = \begin{cases} 1 & \text{if } \forall w' \geq w \,.\, (V(w', \alpha) = 1) \Rightarrow (V(w', \beta) = 1) \\ -1 & \text{if } V(w, \alpha) = 1 \text{ and } V(w, \beta) = -1 \\ 0 & \text{otherwise} \end{cases}$$

$$V(w, not\,\alpha) = \begin{cases} 1 & \text{if } \forall w' \geq w \,.\, V(w', \alpha) < 1 \\ -1 & \text{if } V(w, \alpha) = 1 \\ 0 & \text{otherwise} \end{cases}$$

The valuation V is called an (N2-) model of a set of formulas Θ if for each $\alpha \in \Theta$, it holds that $V(h, \alpha) = V(t, \alpha) = 1$. Let Lit be the set of all literals, i.e. $Lit = A \cup \{\neg a | a \in A\}$. For a valuation V, let V_h and V_t be the set of literals that are *true* in worlds h and t:

$$V_h = \{l \in Lit | V(h, l) = 1\} \qquad V_t = \{l \in Lit | V(t, l) = 1\}$$

A model is called h-minimal if its *here* world is as little committing as possible, given its particular *there* world.

Definition 1. *[23] A model V of a set of formulas Θ is h-minimal if for every other model V' of Θ it holds that either $V_t \neq V'_t$ or $V_h \subseteq V'_h$.*

Note that minimality refers to the number of *literals* that are verified by a valuation, and not the number of *atoms* as for minimality in classical logics. Equilibrium models are h-minimal models whose valuation in h and t coincides.

Definition 2. *[23] A h-minimal model V of a set of formulas Θ is an equilibrium model if $V_h = V_t$.*

A set of formulas P is called a logic program when every formula α in P is of the form

$$l_1 \wedge ... \wedge l_m \wedge not\,l_{m+1} \wedge ... \wedge not\,l_n \to l_{n+1} \vee ... \vee l_s$$

where every l_i is either an atom or a strongly negated atom, i.e. $l_i \in Lit$. As the following proposition reveals, for logic programs the equilibrium models coincide with the answer sets. As common in logic programming, we will often reverse the direction of the implication arrow and write $\alpha \leftarrow \beta$ for $\beta \to \alpha$.

Proposition 1. *[23] Let P be a logic program and S a consistent set of literals (i.e. a and $\neg a$ cannot be both in S, for any atom in A). Then S is an answer set of P iff there is an equilibrium model V of P such that $S = V_t$.*

For clarity, we will sometimes talk about Pearce equilibrium models, to avoid confusion with the fuzzy equilibrium models introduced below.

Example 1. Let $\Theta = \{a \leftarrow not\, b, b \leftarrow not\, a\}$, then the model V defined by $V_t = V_h = \{a, b\}$ is not h-minimal, as witnessed by the model V' defined by $V'_t = \{a, b\}$ and $V'_h = \{\}$. Note that the absence of e.g. both a and $\neg a$ in V'_h implies that $V(h, a) = 0$. However, V' is not an equilibrium model because $V'_t \neq V'_h$. It is easy to see that the only equilibrium models are V'' and V''' defined by $V''_t = V''_h = \{a\}$ and $V'''_t = V'''_h = \{b\}$.

2.2   Fuzzy Answer Set Programs

The core idea of FASP is to replace the boolean truth values by truth values from an appropriate lattice $\mathcal{L} = (L, \leq_L)$, and to replace logical connectives by operators from some multi-valued logic. In this paper, we restrict our attention to fuzzy truth values, i.e. $\mathcal{L} = ([0, 1], \leq)$. Furthermore, we will focus on the variant of FASP introduced in [15]. In addition to the truth lattice $\mathcal{L}$, we then assume that a set $\mathcal{F} = \bigcup_n \mathcal{F}_n$ is given, where for each $n \in \mathbb{N}$, $\mathcal{F}_n$ is a finite set of *acceptable* $[0, 1]^n - [0, 1]$ functions. A function in $\mathcal{F}_n$ is called acceptable if for all $1 \leq i \leq n$ it is either monotonically increasing or decreasing in its ith argument. Note that such functions do not necessarily need to correspond to generalizations of the classical logical connectives, although this will often be the case in practice.

Logical conjunction is then, usually, generalized by a t-norm, i.e. a symmetric, associative, increasing $[0, 1]^2 - [0, 1]$ mapping $\otimes$ satisfying the boundary condition $1 \otimes u = u$ for all $u \in [0, 1]$. Similarly, logical disjunction is generalized by a t-conorm, i.e. a symmetric, associative, increasing $[0, 1]^2 - [0, 1]$ mapping $\oplus$ satisfying the boundary condition $0 \oplus u = u$ for all $u \in [0, 1]$. Typical examples of t-norms and t-conorms are

$$u \otimes_m v = \min(u, v) \quad u \otimes_p v = u \cdot v \qquad u \otimes_l v = \max(0, u + v - 1)$$
$$u \oplus_m v = \max(u, v) \quad u \oplus_p v = u + v - u \cdot v \quad u \oplus_l v = \min(1, u + v)$$

Negation is usually modeled as the complement w.r.t. 1, i.e. $\neg u = 1 - u$. To generalize the notion of implication, several strategies are commonly used. One possibility is to define an implicator $\rightarrow$ in terms of a negation operation $\neg$ and t-conorm $\oplus$, as $u \rightarrow v = \neg u \otimes v$. Such operators are called S-implicators. For example, the maximum $\oplus_m$ gives rise to the Kleene-Dienes implicator $\rightarrow_{kd}$ defined by $u \rightarrow_{kd} v = \max(1 - u, v)$. A less intuitive, but from a logical point of view often more interesting alternative is to define an implicator as the residuum of a left-continuous t-norm $\otimes$, i.e.

$$u \rightarrow v = \sup\{\lambda \in [0, 1] | u \otimes \lambda \leq v\}$$

Such operators are called residual implicators. The residual implicator corresponding with $\otimes_l$ is given by $u \rightarrow_l v = \min(1, 1 - u + v)$ and is often used because of its excellent logical properties; note for instance that $\rightarrow_l$ is at the same time a residual implicator and an S-implicator. The operators $\otimes_l$, $\oplus_l$ and $\rightarrow_l$ are called the Łukasiewicz t-norm, t-conorm and implicator.

Given the set $\mathcal{F}$, a fuzzy formula is defined recursively as either a constant from $[0,1]$, an atom from A, or the application of a function f from $\mathcal{F}_n$ to fuzzy formulas $e_1, ..., e_n$. An $A - [0,1]$ mapping I is called a (fuzzy) interpretation. Given an interpretation I, the (fuzzy) valuation $[\alpha]_I$ of a fuzzy formula is defined recursively as $[c]_I = c$, for any $c \in [0,1]$, $[a]_I = I(a)$ for any $a \in A$, and $[f(e_1, ..., e_n)]_I = f([e_1]_I, ..., [e_n]_I)$ for any $f \in \mathcal{F}_n$. A (fuzzy) rule is an expression of the form $\alpha \rightarrow a$ where the body α of the rule is a fuzzy formula, the head a of the rule is either an atom from A or a constant from $[0,1]$, and $\rightarrow$ is a residual implicator. Usually, in the context of logic programming, rules are written as $a \leftarrow \alpha$. When the head a is a constant, the rule is called a constraint.

A program P is then defined as a set of fuzzy rules. The program is called positive if P does not contain constraints and only applications of functions with increasing arguments. An interpretation I is called a (fuzzy) model of P if $I(r) = 1$ for every r in P.

Definition 3. *[15] The answer set of a positive program is defined as the (unique) minimal model of that program.*

To define answer sets of general programs, we also need the notion of negative and positive occurrence of an atom. An atom a occurs positively in the formula a. Furthermore, if a occurs positively (resp. negatively) in e_i and f is increasing (resp. decreasing) in its ith argument, then a occurs positively in the formula $f(e_1, ..., e_i, ..., e_n)$. Similarly, if a occurs positively (resp. negatively) in e_i and f is decreasing (resp. increasing) in its ith argument, then a occurs negatively in $f(e_1, ..., e_i, ..., e_n)$.

Definition 4. *[15] If P contains functions with decreasing partial mappings, but no constraints, a model I of P is an answer set of P iff it is an answer set of the program P^I, which is obtained from P by replacing all negative occurrences of a by its interpretation $I(a)$, for all atoms $a \in A$. Finally, if P contains constraints, then a model I of P is called an answer set of P iff it is an answer set of $P \setminus C$, where C is the set of constraints from P.*

Finally, a notion of approximate answer set can be defined by allowing that some rules are only satisfied to a certain degree. Specifically, let ρ be a mapping from rules to $[0,1]$. An interpretation I is then called a (fuzzy) ρ-model of Θ if $I(r) \geq \rho(r)$ for every rule r in Θ.

Definition 5. *[15] Let ρ be a mapping from rules to $[0,1]$, and let $P = \{r_1, ..., r_m\}$. A ρ-model I of Θ is called a ρ-answer set of P if it is the minimal ρ-model of $P^I \setminus C$, where C and P^I are as in Definition 4.*

Example 2. Consider the program P defined by $P = \{a \leftarrow_l not\, b, b \leftarrow_l not\, a\}$, then for every λ in $[0,1]$, the fuzzy interpretation I defined by $I(a) = \lambda$ and $I(b) = 1 - \lambda$ is an answer set. Indeed, I is easily seen to be the unique minimal model of $P^I = \{a \leftarrow_l 1 - (1 - \lambda), b \leftarrow_l 1 - \lambda\}$.

3 Fuzzy Equilibrium Logic

3.1 Definition

In the equilibrium logic of Pearce, a third truth value, 0, is used to define underspecified valuations. When moving from boolean to fuzzy truth degrees, underspecified valuations can be defined by assigning an interval of truth degrees to atoms, as opposed to a precise degree from $[0, 1]$. The restriction that the valuation in the *there*-world should be more specific than the valuation in the *here*-world then translates to the requirement that the interval assigned to an atom in t should be contained in the interval assigned to it in h. Thus we define a (fuzzy $N2$-) valuation V as a mapping from $W \times A$ to (possibly degenerate) subintervals of $[0, 1]$ such that $V(h, a) \supseteq V(t, a)$, where $W = \{h, t\}$ and A is a set of atoms as before. For $V(w, \alpha) = [u, v]$, we write $V^-(w, \alpha)$ to denote the lower bound u and $V^+(w, \alpha)$ to denote the upper bound v. We define the valuation of complex fuzzy formulas as follows ($w \in \{h, t\}$):

$$V(w, \neg\alpha) = [1 - V^+(w, \alpha), 1 - V^-(w, \alpha)]$$
$$V(w, \alpha \otimes \beta) = [V^-(w, \alpha) \otimes V^-(w, \beta), V^+(w, \alpha) \otimes V^+(w, \beta)]$$
$$V(w, \alpha \oplus \beta) = [V^-(w, \alpha) \oplus V^-(w, \beta), V^+(w, \alpha) \oplus V^+(w, \beta)]$$

To generalize the semantics of an implication $\alpha \to \beta$, note that $(V(w', \alpha) = 1) \Rightarrow (V(w', \beta) = 1)$ can be generalized as $V^-(w', \alpha) \to V^-(w', \beta) = 1$. Thus, $\min\left(V^-(h, \alpha) \to V^-(h, \beta), V^-(t, \alpha) \to V^-(t, \beta)\right)$ naturally emerges as the lower bound of the interval $V(h, \alpha \to \beta)$. Similarly, for the upper bound, note that the condition $V(w, \alpha) = 1$ and $V(w, \beta) = -1$ from equilibrium logic can be generalized as $(V^-(w, \alpha) \to V^+(w, \beta)) < 1$, which leads to $V^-(w, \alpha) \to V^+(w, \beta)$ as the upper bound of the interval $V(h, \alpha \to \beta)$. We obtain:

$$V(h, \alpha \to \beta) = [\min\left(V^-(h, \alpha) \to V^-(h, \beta), V^-(t, \alpha) \to V^-(t, \beta)\right),$$
$$V^-(h, \alpha) \to V^+(h, \beta)]$$
$$V(t, \alpha \to \beta) = [V^-(t, \alpha) \to V^-(t, \beta), V^-(t, \alpha) \to V^+(t, \beta)]$$

Finally, to generalize the semantics of *not*, note that $V(h, \alpha) \supseteq V(t, \alpha)$ for any fuzzy formula α. Indeed, this follows easily by structural induction, since the property is true by definition for atoms, and is conserved for each of the aforementioned operators. Therefore, we have that $\min(1 - V^-(h, \alpha), 1 - V^-(t, \alpha)) = 1 - V^-(t, \alpha)$. Following a similar strategy as for implications, we arrive at:

$$V(h, not\, \alpha) = [1 - V^-(t, \alpha), 1 - V^-(h, \alpha)]$$
$$V(t, not\, \alpha) = [1 - V^-(t, \alpha), 1 - V^-(t, \alpha)]$$

Above, we assumed that fuzzy formulas are composed of t-norms, t-conorms, implicators, as well as strong and weak negation. It is easy however to extend this definition to arbitrary functions; we omit the details. Also note that rules in fuzzy equilibrium logic are not necessarily implemented by residual implicators.

A valuation V is a (fuzzy $N2$-) model of a set of fuzzy formulas Θ if for every α in Θ, $V^-(h, \alpha) = 1$, which also implies $V^+(h, \alpha) = V^-(t, \alpha) = V^+(t, \alpha) = 1$. Analogous to models in crisp equilibrium logic, fuzzy equilibrium models are models which are in some sense minimal, and which assign the same value to literals in both worlds.

Definition 6. *A fuzzy $N2$-model V of a set of fuzzy formulas Θ is h-minimal if for every other fuzzy $N2$-model V' of Θ, it holds that either*

1. $V(t, a) \neq V'(t, a)$ *for some a in A; or*
2. $V'(h, a) \subseteq V(h, a)$ *for all a in A.*

Note that h-minimal fuzzy $N2$-models are those that are least committing, i.e. those whose valuation in the *here*-world corresponds to the largest possible interval.

Definition 7. *A h-minimal fuzzy $N2$-model V of a set of fuzzy formulas Θ is a fuzzy equilibrium model if $V(h, a) = V(t, a)$ for all a in A.*

Analogous to ρ-answer sets in FASP, we define the notion of fuzzy equilibrium ρ-model. First we define a (fuzzy $N2$-) ρ-model of a set of fuzzy formulas Θ as a valuation V satisfying $V^-(h, \alpha) \geq \rho(\alpha)$ for every α in Θ.

Definition 8. *Let ρ be a mapping from fuzzy formulas to $[0, 1]$. A fuzzy $N2$-ρ-model of a set of fuzzy formulas Θ is h-minimal if for every other fuzzy $N2$-ρ-model of Θ one of the two conditions from Definition 6 are satisfied. A h-minimal fuzzy $N2$-ρ-model V is a fuzzy equilibrium ρ-model if $V(h, a) = V(t, a)$ for all a in A.*

Example 3. Consider again the set of fuzzy formulas P from Example 2. A fuzzy $N2$-valuation V is a fuzzy $N2$-model of P iff

$$(V^-(h, a \leftarrow_l not\, b) = 1) \wedge (V^-(h, b \leftarrow_l not\, a) = 1)$$
$$\Leftrightarrow \min(V^-(h, a) \leftarrow_l V^-(h, not\, b), V^-(t, a) \leftarrow_l V^-(t, not\, b)) = 1$$
$$\wedge \min(V^-(h, b) \leftarrow_l V^-(h, not\, a), V^-(t, b) \leftarrow_l V^-(t, not\, a)) = 1$$
$$\Leftrightarrow V^-(h, a) \geq V^-(h, not\, b) \wedge V^-(t, a) \geq V^-(t, not\, b)$$
$$\wedge V^-(h, b) \geq V^-(h, not\, a) \wedge V^-(t, b) \geq V^-(t, not\, a)$$
$$\Leftrightarrow V^-(h, a) \geq 1 - V^-(t, b) \wedge V^-(t, a) \geq 1 - V^-(t, b)$$
$$\wedge V^-(h, b) \geq 1 - V^-(t, a) \wedge V^-(t, b) \geq 1 - V^-(t, a)$$
$$\Leftrightarrow V^-(h, a) \geq 1 - V^-(t, b) \wedge V^-(h, b) \geq 1 - V^-(t, a)$$

We find that for every λ in $[0, 1]$, the fuzzy $N2$-valuation V defined by $V(t, a) = V(h, a) = [\lambda, 1]$ and $V(t, b) = V(h, b) = [1 - \lambda, 1]$ is a fuzzy equilibrium model. Note that the fuzzy equilibrium models essentially correspond to the answer sets from Example 2. In Propositions 3 and 4 below, we will clarify this connection.

3.2   Relationship to Existing Frameworks

The connection between fuzzy equilibrium models and Pearce equilibrium models can be seen by observing that $[0,1]$ in fuzzy $N2$-valuations takes the role of 0 (*undecided*) in $N2$-valuations, whereas the degenerate intervals $[0,0]$ and $[1,1]$ take the role of respectively -1 (*false*) and 1 (*true*).

Proposition 2. *Let Θ_1 be a set of (crisp) formulas, and let Θ_2 be the set of fuzzy formulas obtained from Θ_1 by replacing classical conjunction, disjunction and implication by respectively $\otimes_m$, $\oplus_m$ and $\rightarrow_{kd}$. Furthermore, let V_1 be an $N2$-valuation, and let V_2 be the fuzzy $N2$-valuation obtained from V_1 by replacing -1, 0 and 1 by respectively $[0,0]$, $[0,1]$ and $[1,1]$. It holds that V_1 is a Pearce equilibrium model of Θ_1 iff V_2 is a fuzzy equilibrium model of Θ_2.*

The above proposition teaches us that Pearce equilibrium models of a set of formulas coincide with a particular subset of its fuzzy equilibrium models, i.e. those in which the valuation of atoms is restricted to $[0,0]$, $[0,1]$ and $[1,1]$. Note however, that translations of "crisp" logic programs can still have other fuzzy equilibrium models. This was already illustrated in Example 3.

We can also show that fuzzy equilibrium models generalize the notion of answer set from fuzzy answer set programs to arbitrary sets of fuzzy formulas.

Proposition 3. *Let P be a set of fuzzy rules, and let V be a fuzzy $N2$-valuation. Furthermore, let I be the fuzzy interpretation defined by $I(a) = V^-(t, a)$ for all $a \in A$. If V is a fuzzy equilibrium model of P, then I is an answer set of P. Similarly, if V is a fuzzy equilibrium ρ-model of P, for some $P - [0,1]$ mapping ρ, it holds that I is a ρ-answer set of P.*

Proposition 4. *Let P be a set of fuzzy rules, and let I be a fuzzy interpretation. Furthermore, let V be the fuzzy $N2$-valuation defined by $V(h, a) = [I(a), 1]$ and $V(t, a) = [I(a), 1]$ for all $a \in A$. If I is an answer set of P then V is a fuzzy equilibrium model of P. Similarly, if I is a ρ-answer set of P, for some $P - [0,1]$ mapping ρ, it holds that V is a fuzzy equilibrium ρ-model.*

3.3   Complexity and Geometrical Representation

To study aspects of reasoning with fuzzy equilibrium models, we limit the set of connectives in $\mathcal{F}$ to $\neg$, *not*, $\otimes_l$, $\oplus_l$, $\rightarrow_l$, $\otimes_m$, $\oplus_m$ and $\rightarrow_{kd}$, i.e. the connectives from Łukasiewicz logic, together with negation-as-failure. Let a linear relation over a set of variables X be defined as an expression of the form $a_1 x_1 + ... + a_n x_n \diamond b$, where $a_1, ..., a_n, b \in \mathbb{R}$, $x_1, ..., x_n$ are variables from X, and $\diamond \in \{\leq, =, \geq\}$. Furthermore, let a disjunctive linear relation (DLR) be defined as an expression of the form $\gamma_1 \vee ... \vee \gamma_m$, where each γ_i is a linear relation. It is not hard to see that, under the aforementioned restrictions on $\mathcal{F}$, the requirements that should be satisfied by a fuzzy $N2$-valuation V to be a model of a set of fuzzy formulas Θ can be written as a set Γ of DLRs, in which the variables correspond to expressions of the form $V^-(h, a)$, $V^+(h, a)$, $V^-(t, a)$ and $V^+(t, a)$; let us write these variables as a_h^-, a_h^+, a_t^- and a_t^+. For example, in the case of Example 3, we have $\Gamma = \{a_h^- + b_t^- \geq 1, b_h^- + a_t^- \geq 1\}$. Geometrically, solutions of Γ, i.e. instantiations of the variables verifying at least one disjunct of each DLR in Γ, correspond to points in the unit hypercube $[0,1]^n$, where $n = 4|A|$ is the number of variables in Γ. Moreover, the set of all solutions of Γ can be represented as the union of a finite number of polyhedra; recall that a polyhedron is the intersection of a finite number of half-spaces. Let us write $p(a_h^-)$ to denote the value of the coordinate of point p corresponding to variable a_h^- (and similar for a_h^+, a_t^- and a_t^+).

Now consider the set of points M that correspond to h-minimal fuzzy $N2$-models of Θ. Recall that a facet of a polyhedron is the intersection of the polyhedron with one of its bounding hyperplanes; a face of a polyhedron is defined recursively as either the polyhedron itself or one of the faces of its facets [1]. When the solution space of Γ corresponds to a single polyhedron, the h-minimal fuzzy $N2$-models correspond to the union of one or more faces of that polyhedron. More in particular, it is not hard to show that whenever an interior point of a face corresponds to a h-minimal fuzzy $N2$-model, then all points of that face correspond to h-minimal fuzzy $N2$-models.

Next, we consider the case where the solution space of Γ corresponds to more than one polyhedron. Let $\mathcal{D}(G)$ be the set of points that are dominated by a face G, i.e.

$$p \in \mathcal{D}(G) \Leftrightarrow \exists q \in G . \forall a \in A . p(a_h^-) \geq q(a_h^-) \wedge p(a_h^+) \leq q(a_h^+)$$

Clearly, the set $\mathcal{D}(G)$ corresponds to a polyhedron. In general, the set M of h-minimal fuzzy $N2$-models corresponds to a finite union of sets of the form $G \setminus \bigcup_i \mathcal{D}(H_i)$ where G is a face of one of the polyhedra, and each H_i is a face of one of the other polyhedra. The fuzzy equilibrium models of Θ are then geometrically characterized as the intersection of M with the polyhedron E defined by

$$p \in E \Leftrightarrow \forall a \in A . p(a_h^-) = p(a_t^-) \wedge p(a_h^+) = p(a_t^+)$$

Note that $M \cap E$ can be represented as the finite union of polyhedra, whose vertices are the intersection of $4|A|$ hyperplanes, each of which is defined in terms of small integer coefficients. Hence, if $M \cap E \neq \emptyset$, we can guess the coordinates of such a vertex in polynomial time on a non-deterministic Turing machine. In other words, if Θ has at least one fuzzy equilibrium model, we can guess a fuzzy equilibrium model V in NP. Moreover, we can verify that V is indeed a fuzzy equilibrium model as follows. First, note that we can verify in polynomial time that V is a fuzzy $N2$-model of Θ, and that $V(t,a) = V(h,a)$ for all a in A. Next, to verify that V is h-minimal, let Γ^V be the set of DLRs obtained from Γ by instantiating all variables of the form a_t^- and a_t^+ by respectively $V^-(t,a)$ and $V^+(t,a)$, i.e. we restrict our attention to points that correspond to fuzzy $N2$-models violating the first condition of Definition 6. The set of DLRs Γ^V can be converted to a mixed integer program Δ in polynomial time [12]. From the theory of mixed integer programming, we know that a point p minimizing

$\sum_a p(a_h^-) - \sum_a p(a_h^+)$ can be found in NP. Clearly V is h-minimal iff $\sum_a p(a_h^-) - \sum_a p(a_h^+) = \sum_a V^-(t,a) - \sum_a V^+(t,a)$. Recall that Σ_2^P is the set of problems that can be solved in NP with an NP oracle, i.e. $\Sigma_2^P = \mathrm{NP}^{\mathrm{NP}}$, while Π_2^P is the set of problems whose complement is in Σ_2^P. The preceding discussion then easily leads to the following results.

Proposition 5. *Let Θ be a set of fuzzy formulas. The problem of deciding whether Θ has a fuzzy equilibrium model is in Σ_2^P.*

Proposition 6. *Let Θ be a set of fuzzy formulas. The problem of deciding whether $V(t,a) \subseteq [\mu,\lambda]$, for $0 \leq \mu \leq \lambda \leq 1$ and $a \in A$, in at least one fuzzy equilibrium model V of Θ is in Σ_2^P.*

Proposition 7. *Let Θ be a set of fuzzy formulas. The problem of deciding whether $V(t,a) \subseteq [\mu,\lambda]$, for $0 \leq \mu \leq \lambda \leq 1$ and $a \in A$, in all fuzzy equilibrium models V of Θ is in Π_2^P.*

Hence, the main reasoning tasks are in the same complexity class as their counterparts in (disjunctive) answer set programming. We can also establish the following hardness results.

Proposition 8. *Let Θ be a set of fuzzy formulas. The problem of deciding whether Θ has a fuzzy equilibrium model is Σ_2^P-hard.*

Proposition 9. *Let Θ be a set of fuzzy formulas. The problem of deciding whether $V(t,a) \subseteq [\mu,\lambda]$, for $0 \leq \mu \leq \lambda \leq 1$ and $a \in A$, in at least one fuzzy equilibrium model V of Θ is Σ_2^P-hard.*

Proposition 10. *Let Θ be a set of fuzzy formulas. If $[\mu,\lambda] \neq [0,1]$, the problem of deciding whether $V(t,a) \subseteq [\mu,\lambda]$, for $0 \leq \mu \leq \lambda \leq 1$ and $a \in A$, in all fuzzy equilibrium models V of Θ is Π_2^P-hard.*

Note that the above discussion can be generalized from the Łukasiewicz connectives to any operator whose semantics can be defined in terms of a mixed integer program. This even holds when strict inequalities are allowed, hence we may also use operators such as the Gödel and Rescher implicators $\rightarrow_g$ and $\rightarrow_{rs}$, defined by

$$u \rightarrow_g v = \begin{cases} 1 & \text{if } u \leq v \\ v & \text{otherwise} \end{cases} \qquad u \rightarrow_{rs} v = \begin{cases} 1 & \text{if } u \leq v \\ 0 & \text{otherwise} \end{cases}$$

for all u and v in $[0,1]$.

4   Example

As an illustration of how fuzzy equilibrium logic can be used in the context of declarative problem solving, we present a technique to find strong Nash equilibria, a problem which is known to be Σ_2^P-complete [10]. Nash equilibria are one of the most fundamental notions from game theory. Assume that a finite set of players $p_1, ..., p_n$ is given, and for each player p_i a set of actions A_i. A choice of actions $(a_1, ..., a_n) \in A_1 \times ... \times A_n$ is called a global strategy. Furthermore, let μ_i be an $A_1 \times ... \times A_n - \mathbb{R}$ function representing the utility (or desirability) to player p_i of a certain global strategy. A global strategy $\mathcal{A} = (a_1, ..., a_n)$ is called a (pure) strong Nash equilibrium if there does not exist a non-empty $K \subseteq \{1, ..., n\}$ and a global strategy $\mathcal{A}' = (a'_1, ..., a'_n)$ such that

1. for all $i \notin K$ it holds that $a_i = a'_i$; and
2. $\mu_i(\mathcal{A}) < \mu_i(\mathcal{A}')$ for each i in K.

In other words, a global strategy is a strong Nash equilibrium if there cannot be a coalition of players that is able to (strictly) improve the current situation of each of its members, without help of others.

Often, the set of actions is assumed to be finite. Here we allow an infinite number of actions, which can be encoded as a vector $(a^1, ..., a^k)$, where $a^i \in [0, 1]$. Essentially, actions are thus specified by a finite number of continuous parameters. For simplicity, we will only consider the case $k = 1$, although the results below can easily be generalized to arbitrary values of k. Moreover, we assume that each utility function μ_i can be represented as a fuzzy formula. Without loss of generality, we can then assume that $\mu_i(a_1, ..., a_n) = 1 - U_i(a_1, ..., a_n; 1 - a_1, ..., 1 - a_n)$ where U_i is a function which is increasing in all of its arguments. In other words μ_i is maximized when U_i is minimized. Minimizing, rather than maximizing U_i makes the translation to fuzzy equilibrium logic easier and more intuitive.

Now we construct a set of fuzzy formulas Θ such that the fuzzy equilibrium models of Θ are exactly the strong Nash equilibria that correspond to the given utility functions. For each i in $\{1, ..., n\}$, Θ contains the following fuzzy formulas:

$$a_i \oplus_l \neg a_i \qquad\qquad c_i^- \oplus_m c_i^+ \qquad\qquad d_i^- \oplus_l d_i^+$$

$$e_i^+ \leftarrow_l (a_i \otimes_m c_i^-) \oplus_m (d_i^+ \otimes_m c_i^+) \qquad\qquad e_i^- \leftarrow_l (\neg a_i \otimes_m c_i^-) \oplus_m (d_i^- \otimes_m c_i^+)$$

Intuitively, on the first line a strong Nash equilibrium $(a_1, ..., a_n)$ is guessed. The use of the Lukasiewicz t-conorm ensures that in every fuzzy $N2$-model V, it holds that $V(h, a_i) = V(t, a_i) = [\lambda_i, \lambda_i]$ for some $\lambda_i \in [0, 1]$. To verify that we have indeed found a strong Nash equilibrium, a coalition is guessed, defined by the indices c_i^+ and c_i^-; the former intuitively means that player i belongs to the coalition, and the latter that i does not. Note that c_i^- and c_i^+ correspond to crisp properties. By using the maximum, $V(h, c_i^-)$ and $V(h, c_i^+)$ will either be $[0, 1]$ or $[1, 1]$ in all h-minimal fuzzy $N2$-models. Next, the strategies of the users in the coalition are guessed: e_i^+ corresponds to the new strategy for player i, whereas e_i^- corresponds intuitively to its negation (i.e. complement with 1). For players outside the coalition this value e_i^+ is simply a_i, whereas for players in the coalition e_i^+ corresponds to a new value d_i^+. To check whether the coalition and strategies that have been guessed provide a counterexample of $(a_1, ..., a_n)$ being a strong Nash equilibrium, the following fuzzy formulas are added:

$$w_i \leftarrow_l (U_i(e_1^+, ..., e_n^+; e_1^-, ..., e_n^-) \leftarrow_{rs} U_i(a_1, ..., a_n; \neg a_1, ..., \neg a_n))$$

$$w \leftarrow_l ((c_1^+ \otimes_m w_1) \oplus_m ... \oplus_m (c_n^+ \otimes_m w_n)) \oplus_m (c_1^- \otimes_m ... \otimes_m c_n^-)$$

Note that for x and y in $[0,1]$, $x \leftarrow_{rs} y = 1$ means that $x \geq y$. Thus, w_i intuitively means that player i does not improve his utility when going from $(a_1, ..., a_n)$ to $(e_1^+, ..., e_n^+)$, whereas w specifies whether this is the case for all users, and whether the coalition that was guessed is indeed non-empty. Finally, to ensure that all guesses of coalitions and corresponding strategies are tried, rather than just one, the following fuzzy formulas need to be added:

$$d_i^- \leftarrow_l w \qquad d_i^+ \leftarrow_l w \qquad c_i^- \leftarrow_l w$$

$$c_i^+ \leftarrow_l w \qquad w_i \leftarrow_l w \qquad 0 \leftarrow_l not\ w$$

Thus, when $(e_1^+, ..., e_n^+)$ does not correspond to a counterexample, all atoms different from a_i are intuitively made true. This ensures that any counterexample to $(a_1, ..., a_n)$ being a Nash equilibrium corresponds to a fuzzy $N2$-model that can only be h-minimal if there does not exist a single counterexample. A similar technique to find minimal structures is commonly used in disjunctive answer set programming; see [7] for examples on graph coloring, the traveling salesman problem and eigenvalue problems. We can show the following proposition.

Proposition 11. *A global strategy $(\lambda_1, ..., \lambda_n)$ is a strong Nash equilibrium iff Θ has a fuzzy equilibrium model V such that $V^-(t, a_i) = \lambda_i$.*

5   Related Work

A variety of approaches to multi-valued and fuzzy ASP have been proposed in recent years, which differ mainly in terms of the types of connectives they allow, the way in which they handle partial satisfaction of rules, and the truth lattices that are used [3,5,6,15,19,26,27]. Most approaches generalize either the fixpoint definition or the minimal model definition of answer sets (stable models), although [27] generalizes a definition in terms of unfounded sets. Regarding expressive power, typically only rules with literals in the head are considered. In the case of the Lukasiewicz connectives, this essentially corresponds to an NP-complete time complexity, as can be easily derived from the translation of FASP to mixed integer programming proposed in [13]. One exception is [19] which allows disjunctions of literals in the head of a rule to define a hybridization of FASP with fuzzy description logics. In a quite different context, multi-valued ASP with disjunctions in the head are used in [25], as a vehicle to deal with inconsistencies in classical ASP. To the best of our knowledge, the current paper presents the first approach to fuzzy answer set programming which is sufficiently general to cover arbitrary fuzzy propositional theories.

6   Concluding Remarks

We have proposed fuzzy equilibrium logic as a generalization of both equilibrium logic and fuzzy answer set programming. The precise connections with these latter two formalisms have been established, and the overall computational complexity of important reasoning tasks was shown to be Σ_2^P-complete, which is the same as for the original equilibrium logic as well as disjunctive ASP. We furthermore presented a geometrical characterization of fuzzy equilibrium models, which can be used to implement fuzzy equilibrium logic reasoners. In future work, we will examine the apparently close connection to bi-level mixed integer programming [20] with the aim of obtaining more efficient implementations. As illustrated for strong Nash equilibria, fuzzy equilibrium logic supports declarative problem solving for computationally demanding problems in a continuous domain. In addition to such practical uses, we mainly envision fuzzy equilibrium logic as a vehicle to facilitate the proofs of theoretical properties of fuzzy answer set programming (e.g. complexity of subfragments, strong equivalence, etc.).

Acknowledgments. Steven Schockaert was funded as a postdoctoral fellow of the Research Foundation – Flanders. Jeroen Janssen was funded by a joint Research Foundation – Flanders project.

References

1. Aguzzoli, S., Ciabattoni, A.: Finiteness in infinite-valued Łukasiewicz logic. Journal of Logic, Language, and Information 9, 5–29 (2000)
2. Baral, C.: Knowledge Representation, Reasoning and Declarative Problem Solving. Cambridge University Press, Cambridge (2003)
3. Damásio, C.V., Medina, J., Ojeda-Aciego, M.: Sorted multi-adjoint logic programs: Termination results and applications. In: Alferes, J.J., Leite, J. (eds.) JELIA 2004. LNCS (LNAI), vol. 3229, pp. 252–265. Springer, Heidelberg (2004)
4. Damásio, C.V., Pereira, L.M.: Hybrid probabilistic logic programs as residuated logic programs. In: Brewka, G., Moniz Pereira, L., Ojeda-Aciego, M., de Guzmán, I.P. (eds.) JELIA 2000. LNCS (LNAI), vol. 1919, pp. 57–72. Springer, Heidelberg (2000)
5. Damásio, C.V., Pereira, L.M.: Antitonic logic programs. In: Eiter, T., Faber, W., Truszczyński, M. (eds.) LPNMR 2001. LNCS (LNAI), vol. 2173, pp. 379–392. Springer, Heidelberg (2001)
6. Damasio, C.V., Pereira, L.M.: An encompassing framework for paraconsistent logic programs. Journal of Applied Logic 3, 67–95 (2003)
7. Eiter, T., Gottlob, G., Mannila, H.: Disjunctive datalog. ACM Transactions on Database Systems 22(3), 364–418 (1997)
8. Ferraris, P.: Answer sets for propositional theories. In: Baral, C., Greco, G., Leone, N., Terracina, G. (eds.) LPNMR 2005. LNCS (LNAI), vol. 3662, pp. 119–131. Springer, Heidelberg (2005)
9. Gelfond, M., Lifschitz, V.: The stable model semantics for logic programming. In: Proceedings of the Fifth International Conference and Symposium on Logic Programming, pp. 1081–1086 (1988)
10. Gottlob, G., Greco, G., Scarcello, F.: Pure Nash equilibria: hard and easy games. Journal of Artificial Intelligence Research 24, 357–406 (2005)
11. HadjAli, A., Dubois, D., Prade, H.: Qualitative reasoning based on fuzzy relative orders of magnitude. IEEE Transactions on Fuzzy Systems 11(1), 9–23 (2003)
12. Hähnle, R.: Many-valued logic and mixed integer programming. Annals of Mathematics and Artificial Intelligence 12, 231–264 (1994)

13. Janssen, J., Heymans, S., Vermeir, D., De Cock, M.: Compiling fuzzy answer set programs to fuzzy propositional theories. In: Garcia de la Banda, M., Pontelli, E. (eds.) ICLP 2008. LNCS, vol. 5366, pp. 362–376. Springer, Heidelberg (2008)

14. Janssen, J., Schockaert, S., Vermeir, D., De Cock, M.: General fuzzy answer set programming: The basic language (submitted)

15. Janssen, J., Schockaert, S., Vermeir, D., De Cock, M.: General fuzzy answer set programs. In: Proceedings of the 8th International Workshop on Fuzzy Logic and Applications (WILF), pp. 352–359 (2009)

16. Lifschitz, V.: Twelve definitions of a stable model. In: Garcia de la Banda, M., Pontelli, E. (eds.) ICLP 2008. LNCS, vol. 5366, pp. 37–51. Springer, Heidelberg (2008)

17. Lifschitz, V., Pearce, D., Valverde, A.: Strongly equivalent logic programs. ACM Trans. Comput. Logic 2(4), 526–541 (2001)

18. Lukasiewicz, T.: Probabilistic logic programming. In: Proceedings of the 13th European Conference on Artificial Intelligence (ECAI 1998), pp. 388–392 (1998)

19. Lukasiewicz, T., Straccia, U.: Tightly integrated fuzzy description logic programs under the answer set semantics for the semantic web. In: Marchiori, M., Pan, J.Z., de Marie, C.S. (eds.) RR 2007. LNCS, vol. 4524, pp. 289–298. Springer, Heidelberg (2007)

20. Moore, J., Bard, J.: The mixed integer linear bilevel programming problem. Operations Research 38(5), 911–921 (1990)

21. Nicolas, P., Garcia, L., Stéphan, I., Lefèvre, C.: Possibilistic uncertainty handling for answer set programming. Ann. Math. Artif. Intell. 47(1-2), 139–181 (2006)

22. Page, L., Brin, S., Motwani, R., Winograd, T.: The PageRank citation ranking: Bringing order to the web. Technical report, Stanford Digital Library Technologies Project (1998)

23. Pearce, D.: A new logical characterisation of stable models and answer sets. In: Dix, J., Przymusinski, T.C., Moniz Pereira, L. (eds.) NMELP 1996. LNCS, vol. 1216, pp. 57–70. Springer, Heidelberg (1997)

24. Saccá, D., Zaniolo, C.: Stable models and non-determinism in logic programs with negation. In: Proceedings of the ACM Symposium on Principles of Database Systems, pp. 205–217 (1990)

25. Sakama, C., Inoue, K.: Paraconsistent stable semantics for extended disjunctive programs. Journal of Logic and Computation 5, 265–285 (1995)

26. Straccia, U.: Annotated answer set programming. In: Proceedings of the 11th International Conference on Information Processing and Management of Uncertainty in Knowledge-Based Systems, IPMU 2006 (2006)

27. Van Nieuwenborgh, D., De Cock, M., Vermeir, D.: An introduction to fuzzy answer set programming. Annals of Mathematics and Artificial Intelligence 50(3-4), 363–388 (2007)

Belief Logic Programming
with Cyclic Dependencies

Hui Wan

State University of New York at Stony Brook
Stony Brook, NY 11794, USA

Abstract. Our previous work [26] introduced Belief Logic Programming (BLP), a novel form of quantitative logic programming with correlation of evidence. Unlike other quantitative approaches to logic programming, this new theory is able to provide accurate conclusions in the presence of uncertainty when the sources of information are not independent. However, the semantics defined in [26] is not sufficiently general—it does not allow cyclic dependencies among beliefs, which is a serious limitation of expressive power. This paper extends the semantics of BLP to allow cyclic dependencies. We show that the new semantics is backward compatible with the semantics for acyclic BLP and has the expected properties. The results are illustrated with examples of inference in a simple diagnostic expert system.

1 Introduction

Quantitative reasoning has been widely used for dealing with uncertainty and inconsistency in knowledge representation and management, and, more recently, on the Semantic Web [14]. Among the various forms of quantitative reasoning, quantitative logic programming is a very important one.

Based on the approaches of uncertainty deduction, Lakshmanan and Shiri [11] classified the proposed quantitative logic programming frameworks into *annotation-based* and *implication-based* as follows:

- In the annotation-based frameworks such as [2,9,10,16,15,17,24], a rule is of the form $A : f(\beta_1, ..., \beta_n)$:- $B_1 : \beta_1 \wedge ... \wedge B_n : \beta_n$ which asserts "the certainty of A is at least (or is in) $f(\beta_1, ..., \beta_n)$, whenever the certainty of B_i is at least (or is in) β_i, $1 \leq i \leq n$."
- In the implication-based frameworks such as [1,11,8,12,13,20,22,5], a rule is of the form $\beta : \quad A$:- $B_1 \wedge ... \wedge B_n$ which asserts " the certainty that $B_1 \wedge ... \wedge B_n$ implies A is at least (or is in) β."

In the annotation-based frameworks, when function f in every rule is a constant function, the certainty of the rule head does not depend on the certainty of atoms in the rule body. Recursive loops are not a problem in such cases. When function f is not a constant function, such as in [15], the certainty of the rule head depends on the certainty of atoms in the rule body. Recursive loops are looked on as feedback connections and may cause infinite feedback.

A. Polleres and T. Swift (Eds.): RR 2009, LNCS 5837, pp. 150–165, 2009.

Things are different when we look at implication-based frameworks. In some cases, such as [13,11], certainty values are assigned to atoms, rules are treated as constraints, and models of a program satisfy all the constraints in the program. Recursive loops do not present a problem in such cases. In other frameworks, such as [8,19,22], certainty values are assigned to possible worlds and certainty of the head of a rule cannot be computed until certainty of atoms in the rule body is established. Consequently recursive loops cause a problem: it becomes impossible to compute certainty of the atoms involved in a recursive loop. Some frameworks in this category [8,19] simply do not allow programs with recursive loops. Some others, such as [6,22], eliminate recursive loops by introducing time parameters into atoms involved in loops, either explicitly or implicitly.

In a vast open world like the Semantic Web, it is often impossible for an application to acquire complete information from one information source, thus combination and correlation of evidence from multiple sources are necessary. In our previous work [26], we introduced a novel form of implication-based quantitative reasoning, called *Belief Logic Programming* (BLP) [26]. BLP can take into account correlation of evidence obtained from different, but overlapping and, possibly, contradicting information sources. This makes BLP very suitable for Web reasoning. BLP's semantics is based on belief combination functions and is inspired by Dempster-Shafer theory of evidence [3,21]. In [26], we related BLP semantics to Dempster-Shafer theory and also showed the connection with certain forms of defeasible reasoning, such as Courteous Logic Programming [7] and, more generally, Logic Programming with Courteous Argumentation Theories (LPDA) [25]. [26] also provides a detailed motivation of BLP and the arguments showing limitations of the earlier logic programming approaches—those based on probability theory, Fuzzy Logic, Dempster-Shafer theory, and other approaches [1,2,9,10,11,13,15,17,18,19,24]. The same limitations apply to other frameworks of uncertainty reasoning in the Semantic Web, e.g., [4,23]. In this paper we will not rehash the detailed motivation of BLP but focus on how to deal with recursive loops in BLP.

Like other quantitative logic programming frameworks, BLP faces the same challenge from recursive loops, and the semantics defined in [26] was restricted to belief logic programs without cyclic dependency among atoms. In the present paper we extend the previous work to belief logic programs with cycles by adapting a novel approach: instead of introducing time parameters to eliminate loops, we analyze the program structure and discard the support from unwanted loop influence. We define a transformational semantics and a fixpoint semantics in which self-supported beliefs are discarded. We also show that the proposed semantics are reasonable and are backward compatible with the semantics defined in [26]. The query answering algorithms proposed in [27] can also be adapted to belief logic programs with cycles, but for reason of focus and space we will not address query answering in this paper.

The paper is organized as follows. Section 2 explains the problem in detail by a motivating example. Section 3 is an overview of the syntax and the semantics for *acyclic* BLP programs. In Section 4, we introduce a transformational semantics and a fixpoint semantics for general BLP programs, which may include cycles. Section 5 concludes the paper. Proofs can be found in [28].

2   Motivating Example

Suppose that the rate of false positive test results for a certain disease is 20%. Furthermore, suppose that the certainty that someone who had a contact with a contagious person will also contract that same disease is 60%.

Now, let us assume that the tests for two persons, p_1 and p_2, came back positive, but there is no evidence that p_1 and p_2 had a contact. An expert system might then diagnose both p_1 and p_2 as having contracted the disease with certainty 80%. Now, suppose that the test for two other persons, p_3 and p_4, came back positive and p_3 and p_4 are *known* to have had a contact. Common sense then suggests that p_3 and p_4 are more likely to have the disease compared with p_1 and p_2.

The support (or dependence) relation with regard to p_3 and p_4 is shown in Figure 1. We can see that there is a loop (a cyclic dependency) between "p_3 has disease" and "p_4 has disease".

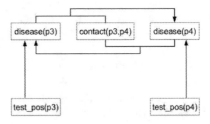

Fig. 1. Support Relation w.r.t. p_3 and p_4 in disease example

Due to the cyclic dependency, the belief in "p_3 has disease" pumps up the certainty of "p_4 has disease" and vice versa. In fact, this dependency is a self-supporting feedback loop, and if we keep combining evidence produced by this feedback loop, the belief in p_3 and p_4's diagnoses will end up close to 1. Clearly such inference is undesired. The problem, therefore, is: how can we discard loop influence when combining all the supporting evidence?

Our method, described in Section 4, identifies and discards such self-supporting feedback. After presenting the method, we will revisit the above example and show that the new method produces inference that is in accord with intuition.

3   Preliminaries

3.1   Syntax of BLP

A **belief logic program** (or a *blp*, for short) is a set of annotated rules. Each **annotated rule** has the following format:

$$[v, w]\ X\ \ \text{:-}\ \ Body$$

where X is a positive atom and *Body* is a conjunction of literals, i.e., a conjunction of atoms and negation of atoms.[1] An atom in BLP has the form $p(t_1, ..., t_n)$, where p is a predicate and t_i is a constant or a variable, $1 \leq i \leq n$. We will use capital letters to denote positive atoms, e.g., A, and a bar over such a letter will denote negation, e.g., $\overline{A}$. The annotation $[v, w]$ is called a **belief factor**, where v and w are real numbers such that $0 \leq v \leq w \leq 1$.

The informal meaning of the above rule is that if *Body* is true, then this rule supports X to the degree v and $\overline{X}$ to the degree $1 - w$. The difference, $w - v$, is the *information gap* (or the degree of ignorance) with regard to X.

An annotated rule of the form $[v, w]\ X\ \text{:-}\ true$ is called an **annotated fact**; it is often written simply as $[v, w]\ X$. In the remainder of this paper we will deal only with annotated rules and facts and refer to them simply as rules and facts.

Definition 1. *Given a blp P, an atom X is said to **depend on** an atom Y*

- *directly, if X is the head of a rule R and Y occurs in the body of R;*
- *indirectly, if X is dependent on Z, and Z depends on Y.* □

A blp P is said to be **cyclic** if there is an atom that depends on itself. Otherwise P is said to be **acyclic**. In [26], we required that in a blp there can be no circular dependency among atoms, i.e., [26] only considered acyclic blps. This is a serious limitation of express power. In this paper we will remove this restriction and allow circular dependency in Section 4.

3.2   Combination Functions

Definition 2. *Let D be the set of all belief factors, $\Phi : D \times D \to D$ is said to be a **belief combination function** if Φ is associative and commutative.* □

Due to the associativity of Φ, we can extend it from two arguments to nullary case, single argument, and three and more arguments: $\Phi() = [0, 1]$, $\Phi([v, w]) = [v, w]$, $\Phi([v_1, w_1], ..., [v_k, w_k]) = \Phi(\Phi([v_1, w_1], ..., [v_{k-1}, w_{k-1}]), [v_k, w_k])$. Note that the order of arguments in a belief combination function is immaterial, since such functions are commutative, so we often write such functions as functions on multisets of belief factors, e.g., $\Phi(\{[v_1, w_1], ..., [v_k, w_k]\})$.

Different types of beliefs might require different ways to combine them, so predicates in the same blp might be using different combination functions. Here are some popular combination functions:

- *Dempster's combination rule*:
  - $\Phi^{DS}([0, 0], [1, 1]) = [0, 1]$.

---

[1] In the BLP syntax and semantics in [26], rule bodies are Boolean combinations of literals. Since every blp can be transformed into an equivalent blp without body-disjunctions, as shown in [27], here we assume there is no disjunction in rule bodies.

- $\Phi^{DS}([v_1, w_1], [v_2, w_2]) - [v, w]$ if $\{[v_1, w_1], [v_2, w_2]\} \neq \{[0,0], [1,1]\}$, where
  $v = \frac{v_1 \cdot w_2 + v_2 \cdot w_1 - v_1 \cdot v_2}{K}$, $w = \frac{w_1 \cdot w_2}{K}$, and $K = 1 + v_1 \cdot w_2 + v_2 \cdot w_1 - v_1 - v_2$.
  In this case, $K \neq 0$ and thus v and w are well-defined.
- *Maximum:* $\Phi^{max}([v_1, w_1], [v_2, w_2]) = [max(v_1, v_2), max(w_1, w_2)]$.
- *Minimum:* $\Phi^{min}([v_1, w_1], [v_2, w_2]) = [min(v_1, v_2), min(w_1, w_2)]$.

3.3   Semantics of Acyclic BLP

Given a blp $\boldsymbol{P}$, the definitions of *Herbrand Universe* $U_{\boldsymbol{P}}$ and *Herbrand Base* $B_{\boldsymbol{P}}$ of $\boldsymbol{P}$ are the same as in the classical case. As usual in logic programming, the easiest way to define a semantics is by considering *ground* (i.e., variable-free) rules. We assume that each atom $X \in B_{\boldsymbol{P}}$ has an associated belief combination function, denoted Φ_X.[2] Intuitively, Φ_X is used to help determine the *combined* belief in X accorded by the rules in $\boldsymbol{P}$ that support X.

Definition 3. *A **truth valuation** over a set of atoms α is a mapping from α to $\{\mathbf{t}, \mathbf{f}, \mathbf{u}\}$. The set of all possible valuations over α is denoted as $\mathcal{T}Val(\alpha)$.*

*A **truth valuation** I for a blp $\boldsymbol{P}$ is a truth valuation over $B_{\boldsymbol{P}}$. Let $\mathcal{T}Val(\boldsymbol{P})$ denote the set of all the truth valuations for $\boldsymbol{P}$, so $\mathcal{T}Val(\boldsymbol{P}) = \mathcal{T}Val(B_{\boldsymbol{P}})$.* □

Definition 4. *A **support function** for a set of atoms α is a mapping m_α from $\mathcal{T}Val(\alpha)$ to $[0,1]$ such that $\sum_{I \in \mathcal{T}Val(\alpha)} m_\alpha(I) = 1$.*

*The atom-set α is called the **base** of m_α. A **support function** for a blp $\boldsymbol{P}$ is a mapping m from $\mathcal{T}Val(\boldsymbol{P})$ to $[0,1]$ such that $\sum_{I \in \mathcal{T}Val(\boldsymbol{P})} m(I) = 1$.* □

If α is a set of atoms, we will use $\mathcal{B}ool(\alpha)$ to denote the set of all Boolean formulas constructed out of these atoms (i.e., using $\wedge$, $\vee$, and negation).

Definition 5. *Given a truth valuation I over a set of atoms α and a formula $F \in \mathcal{B}ool(\alpha)$, $I(F)$ is defined as in Lukasiewicz's three-valued logic: $I(A \vee B) = max(I(A), I(B))$, $I(A \wedge B) = min(I(A), I(B))$, and $I(\overline{A}) = \neg I(A)$, where $\mathbf{f} < \mathbf{u} < \mathbf{t}$ and $\neg \mathbf{t} = \mathbf{f}$, $\neg \mathbf{f} = \mathbf{t}$, $\neg \mathbf{u} = \mathbf{u}$. We say that $I \models F$ if $I(F) = \mathbf{t}$.* □

Definition 6. *A mapping* `bel` $: \mathcal{B}ool(B_{\boldsymbol{P}}) \longrightarrow [0,1]$ *is said to be a **belief function** for $\boldsymbol{P}$ if there exists a support function m for $\boldsymbol{P}$, so that for all $F \in \mathcal{B}ool(B_{\boldsymbol{P}})$,* `bel`$(F) = \sum_{I \in \mathcal{T}Val(\boldsymbol{P}) \text{ such that } I \models F} m(I)$. □

Belief functions can be thought of as *interpretations* of belief logic programs. However, as usual in logic programming, we are interested not just in interpretations, but in models. We define the model of an acyclic blp next.

Definition 7. *Given an acyclic blp $\boldsymbol{P}$ and a truth valuation I, we define $\boldsymbol{P}$'s **reduct under** I to be $\boldsymbol{P}_I = \{R \mid R \in \boldsymbol{P}, I \models Body(R)\}$, where $Body(R)$ denotes the body of the rule R.*

*Let $\boldsymbol{P}(X)$ denote the set of rules in $\boldsymbol{P}$ with the atom X in the head. $\boldsymbol{P}$'s **reduct under** I **with** X **as head** is defined as $\boldsymbol{P}_I(X) = \boldsymbol{P}_I \cap \boldsymbol{P}(X)$. Thus, $\boldsymbol{P}_I(X)$ is simply that part of the reduct $\boldsymbol{P}_I$, which consists of the rules that have X as their head.* □

---

[2] Separate belief combination functions can be associated to different predicates or even ground atoms.

We now define a measure for the degree by which I is supported by $P(X)$.

Definition 8. *Given an acyclic blp P and a truth valuation I for P, for any $X \in B_P$, we define $s_P(I, X)$, called the P-support for X in I, as follows.*

1. *If $P_I(X) = \phi$, then*
   - *If $I(X) = \mathbf{t}$ or $I(X) = \mathbf{f}$, then $s_P(I, X) = 0$;*
   - *If $I(X) = \mathbf{u}$, then $s_P(I, X) = 1$.*
2. *If $P_I(X) = \{R_1, \ldots, R_n\}$, $n > 0$, let $[v, w]$ be the result of applying Φ_X to the belief factors of the rules $R_1, \ldots, R_n$. Then*
   - *If $I(X) = \mathbf{t}$, then $s_P(I, X) = v$;*
   - *If $I(X) = \mathbf{f}$, then $s_P(I, X) = 1 - w$;*
   - *If $I(X) = \mathbf{u}$, then $s_P(I, X) = w - v$.*   □

Informally, $I(X)$ represents what the truth valuation I believes about X. The above interval $[v, w]$ produced by the Φ_X represents the combined support accorded by the rule set $P_I(X)$ to that belief. $s_P(I, X)$ measures the degree by which a truth valuation I is supported by $P(X)$. If X is true in I, it is the combined belief in X supported by P given the truth valuation I. If X is false in I, $s_P(I, X)$ is the combined disbelief in X. Otherwise, it represents the combined information gap about X.

We now introduce the notion of P-support for I *as a whole*. It is defined as a cumulative P-support for all atoms in the Herbrand base.

Definition 9. *If I is a truth valuation for an acyclic blp P, then*

$$\hat{m}_P(I) = \prod_{X \in B_P} s_P(I, X)$$ □

Theorem 1. *For any acyclic blp P, $\sum_{I \in TVal(P)} \hat{m}_P(I) = 1$.*   □

This theorem is crucial, as it makes the following definition well-founded.

Definition 10. *The **model** of an acyclic blp P is the following belief function:*

$$\mathtt{model}(F) = \sum_{I \in TVal(P) \text{ such that } I \models F} \hat{m}_P(I), \quad \text{where } F \in \mathcal{B}ool(B_P).$$ □

In [26] we showed that model is a "correct" (and unique) belief functionthat one should expect: it provides each atom in the Herbrand base with precisely the right amount of support from all the applicable rules. Namely, if S is a suitable set of the rules that support A (see [26] for a precise formulation) then

$$\frac{\mathtt{model}\big(A \wedge \bigwedge_{R \in S} Body(R)\big)}{\mathtt{model}\big(\bigwedge_{R \in S} Body(R)\big)} = v \qquad \frac{\mathtt{model}\big(\overline{A} \wedge \bigwedge_{R \in S} Body(R)\big)}{\mathtt{model}\big(\bigwedge_{R \in S} Body(R)\big)} = 1 - w$$

where $[v, w] = \Phi_X(BF_S)$ and BF_S is the multiset of belief factors of rules in S.

4   Semantics for General BLP

We introduce some necessary notions first.

Definition 11. *Let P be a blp. B_P can always be partitioned into disjoint atom sets $C_1, \ldots, C_k$, such that two atoms are in the same set if and only if they depend on each other. We call $C_1, \ldots, C_k$ the* **atom cliques** *of P and use $clique(A)$ to denote the atom clique that contains atom A.*

A **clique ordering** *for P is a bijective function, $Order : \{C_1, \ldots, C_k\} \to \{1, \ldots, k\}$, such that, for any pair of atoms X and Y, X does not depend on Y if $order(clique(X)) < order(clique(Y))$.* ☐

We can safely infer that every atom clique in an acyclic blp has size 1, but not vice versa. Actually a blp can be cyclic even if there is only one atom in it – that atom, A, can depend on itself via the rule $[v, w]\ A$:- A.

4.1   Transformational Semantics for General BLP

In this section, we define a semantics for general (i.e., possibly cyclic) blps. We only consider *ground* blps in the semantics as usual. First, we transform a cyclic blp P into an acyclic blp P', which captures all the non-trivial and non-redundant belief derivations. Then the minimal model of P is defined to be the minimal model of the acyclic blp P'.

To simplify the description of the transformation, we assume that each rule, R, has a unique identifier, denoted ID_R — a new propositional constant.

Definition 12. *The* **dependency graph**, $\mathcal{H}$, *of a ground blp P is a directed bipartite graph whose nodes are partitioned into a set of* **atom nodes** *(a-nodes, for short) and* **rule nodes** *(r-nodes, for short). The nodes and edges are defined as follows:*

- *For each atom A in P, $\mathcal{H}$ has an a-node labeled A.*
- *For each rule R in P, $\mathcal{H}$ has an r-node labeled with proposition ID_R.*
- *For each rule R in P, an edge goes from the r-node labeled ID_R to the a-node labeled with R's head.*
- *For each rule R in P and each atom A that appears in R's body, an edge goes from the a-node labeled A to the r-node labeled ID_R.* ☐

The dependency graph $\mathcal{H}$ describes the dependency relation over B_P. Not only does $\mathcal{H}$ stores the information whether an atom A depends on another atom B, but also the structural information such as through which rules (or through which path) A depends on B. The structural information will be useful for us to split the undesired loop influence from the other supports.

Definition 13. *Let P be a blp and $\mathcal{H}$ be P's dependency graph. A directed graph $\mathcal{G}$ is called a* **partial-proof DAG of P for the atom A** *(pp-DAG for A, for short) if it has the following properties:*

1. *$\mathcal{G}$ is a maximal acyclic subgraph of $\mathcal{H}$ satisfying conditions 2-4, below.*
2. *Node A is the root of $\mathcal{G}$, i.e., every node in $\mathcal{G}$ is on a path leading to A.*

3. *Every a-node in $\mathcal{G}$ belongs to clique(A) and has exactly one child.*
4. *If an a-node D belongs to clique(A), and D's parent is in $\mathcal{G}$, then D itself is also in $\mathcal{G}$.* □

It is clear that an atom A can have more than one pp-DAGs. Each of them corresponds to a successful SLD-style derivation path for $? - A$, starting from outside of clique(A). The following example helps illustrate the observation.

Example 1. Returning to the example in Section 2. Suppose that the rate of false positive test results for a certain disease is 20%. The certainty that someone who had a contact with a contagious person will also contract the same disease is 60%. An expert system uses the following BLP rules to generate possible diagnosis.

$$[0.8, 1]\ disease(?X)\ \ :-\ \ test\_pos(?X).$$
$$[0.6, 1]\ disease(?X)\ \ :-\ \ contact(?X, ?Y) \wedge disease(?Y).$$

Suppose that the test for two persons, p_3 and p_4, came back positive and p_3 and p_4 are known to have had a contact. We get the following blp $\boldsymbol{P}_1$ after grounding.

$\boldsymbol{P}_1$
$\quad r_1 : [0.8, 1]\ disease(p_3)\ \ :-\ \ test\_pos(p_3).$
$\quad r_2 : [0.8, 1]\ disease(p_4)\ \ :-\ \ test\_pos(p_4).$
$\quad r_3 : [0.6, 1]\ disease(p_3)\ \ :-\ \ contact(p_3, p_4) \wedge disease(p_4).$
$\quad r_4 : [0.6, 1]\ disease(p_4)\ \ :-\ \ contact(p_4, p_3) \wedge disease(p_3).$
$\quad r_5 : \quad [1, 1]\ test\_pos(p_3).$
$\quad r_6 : \quad [1, 1]\ test\_pos(p_4).$
$\quad r_7 : \quad [1, 1]\ contact(p_3, p_4).$
$\quad r_8 : \quad [1, 1]\ contact(p_4, p_3).$

There are five atom cliques in $\boldsymbol{P}_1$: $\{test\_pos(p_3)\}$, $\{test\_pos(p_4)\}$, $\{contact(p_3, p_4)\}$, $\{contact(p_4, p_3)\}$ and $\{disease(p_3), disease(p_4)\}$. The dependency graph and the pp-DAGs are shown in Figure 2 and Figure 3, respectively. □

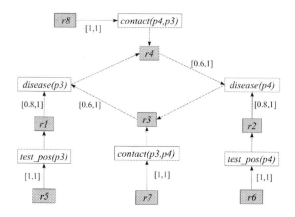

Fig. 2. The dependency graph for Example 1

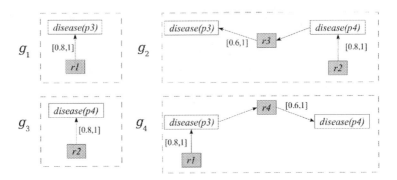

Fig. 3. The pp-DAGs in Example 1: g_1, g_2 are for $disease(p3)$, g_3, g_4 are for $disease(p4)$

Definition 14. *Let $\mathcal{G}$ be a pp-DAG of $\boldsymbol{P}$ for the atom A. Another pp-DAG $\mathcal{G}'$ of $\boldsymbol{P}$ is said to be a **child pp-DAG of $\mathcal{G}$** if:*

1. *$\mathcal{G}'$ is a subgraph of $\mathcal{G}$; and*
2. *$\mathcal{G}'$'s root, B, is a child of A's child in $\mathcal{G}$.* □

In Example 1, g_1 is a child pp-DAG of g_4, while g_3 is a child pp-DAG of g_2.

Now we are ready to define the transformation that converts cyclic blps to acyclic ones.

Definition 15. *Decyclification* *of $\boldsymbol{P}$, denoted $acyclic(\boldsymbol{P})$, is obtained from $\boldsymbol{P}$ as follows. Let $\mathcal{S}$ be the set of new atoms labeled with pp-DAGs of $\boldsymbol{P}$:*

$$\mathcal{S} = \{A^{\mathcal{G}} | A \in B_{\boldsymbol{P}} \text{ and } \mathcal{G} \text{ is a pp-DAG with root } A\}$$

For each rule $R \in \boldsymbol{P}$ of the form

$$[v,w]\ A_0 \quad :-\quad A_1,\ldots,A_k,\overline{A_{k+1}},\ldots,\overline{A_n},D_1,\ldots,D_l,\overline{D_{l+1}},\ldots,\overline{D_m}.$$

where $A_i \in clique(A_0), 1 \leq i \leq n$, $D_j \notin clique(A_0), 1 \leq j \leq m$, replace R with the rule

$$[v,w]\ A_0 \quad :-\ ID_R$$

where ID_R is the proposition that identifies R, plus, for every list $\mathcal{G}_0,\ldots,\mathcal{G}_n$ of pp-DAGs such that

- *$\mathcal{G}_i$ is a pp-DAG with the root A_i, $0 \leq i \leq n$, and*
- *ID_R is A_0's child in $\mathcal{G}_0$, and*
- *$\mathcal{G}_j$ is a child pp-DAG of $\mathcal{G}_0$, $1 \leq j \leq n$,*

we add the rules of the form

$$[v,w]\ A_0^{\mathcal{G}_0} \quad :-\quad A_1^{\mathcal{G}_1},\ldots,A_k^{\mathcal{G}_k},\overline{A_{k+1}^{\mathcal{G}_{k+1}}},\ldots,\overline{A_n^{\mathcal{G}_n}},D_1,\ldots,D_l,\overline{D_{l+1}},\ldots,\overline{D_m}.$$

$$[1,1]\ ID_R \quad :-\quad A_1^{\mathcal{G}_1},\ldots,A_k^{\mathcal{G}_k},\overline{A_{k+1}^{\mathcal{G}_{k+1}}},\ldots,\overline{A_n^{\mathcal{G}_n}},D_1,\ldots,D_l,\overline{D_{l+1}},\ldots,\overline{D_m}.$$ □

Intuitively, for each A, $A^{\mathcal{G}}$ is defined in such a way that its degree of belief is precisely that part of the belief in A, which is justified by the derivations that correspond to the pp-DAG $\mathcal{G}$. ID_R is defined in such a way that its degree of belief is the belief in R's body being derived without any loop influence. And the degree of belief in A_0 is obtained by combining the support to A_0 from all the ID_R's such that R has A_0 as head.

Note that in the resulting program, given any pair of atoms $A_i^{\mathcal{G}_i}$ and $A_j^{\mathcal{G}_j}$, $A_i^{\mathcal{G}_i}$ depends on $A_j^{\mathcal{G}_j}$ if and only if $\mathcal{G}_j$ is a child pp-DAG of $\mathcal{G}_i$. Thus, it is clear that the decyclification transformation eliminates all cycles.

Now we define the model of a general blp as follows.

Definition 16. *For a (possibly cyclic) blp $\boldsymbol{P}$, $\tilde{m}_{\boldsymbol{P}}$ is a support function for $\boldsymbol{P}$ such that for any $I \in T\mathcal{V}al(\boldsymbol{P})$,*

$$\tilde{m}_{\boldsymbol{P}}(I) = \sum_{I' \in T\mathcal{V}al(acyclic(\boldsymbol{P})),\, I'|_{B_{\boldsymbol{P}}}=I} \hat{m}_{acyclic(\boldsymbol{P})}(I')$$

*The **model** of $\boldsymbol{P}$ is a belief function for $\boldsymbol{P}$, $\mathrm{model}_c(\boldsymbol{P})$, such that for any formula F in $\mathcal{B}ool(B_{\boldsymbol{P}})$, $\mathrm{model}_c(\boldsymbol{P})(F) = \mathrm{model}\big(acyclic(\boldsymbol{P})\big)(F)$. (For acyclic blps, $\hat{m}$ and model are defined in Definitions 9 and 10.)* $\qquad\square$

In other words, the semantics for acyclic BLP can be applied on $acyclic(\boldsymbol{P})$ to compute the model of $\boldsymbol{P}$. In practice, the query answering algorithm in [27] can be used on $acyclic(\boldsymbol{P})$ to compute $\mathrm{model}_c(\boldsymbol{P})(F)$ for any F in $\mathcal{B}ool(B_{\boldsymbol{P}})$.

Example 2. (Example 1 continued.) Applying the decyclification transformation on $\boldsymbol{P}_1$, we get the following acyclic blp $\boldsymbol{P}_1'$.

$\boldsymbol{P}_1'$

$[0.8, 1]$ $disease(p_3)$:- r_1.
$[1, 1]$ r_1 :- $test\_pos(p_3)$.
$[0.8, 1]$ $disease(p_4)$:- r_2.
$[1, 1]$ r_2 :- $test\_pos(p_4)$.
$[0.6, 1]$ $disease(p_3)$:- r_3.
$[1, 1]$ r_3 :- $contact(p_3, p_4) \wedge disease^{g_2}(p_4)$.
$[0.6, 1]$ $disease(p_4)$:- r_4.
$[1, 1]$ r_4 :- $contact(p_4, p_3) \wedge disease^{g_1}(p_3)$.
$[0.8, 1]$ $disease^{g_1}(p_3)$:- $test\_pos(p_3)$.
$[0.8, 1]$ $disease^{g_2}(p_4)$:- $test\_pos(p_4)$.
$[0.6, 1]$ $disease^{g_3}(p_3)$:- $contact(p_3, p_4) \wedge disease^{g_2}(p_4)$.
$[0.6, 1]$ $disease^{g_4}(p_4)$:- $contact(p_4, p_3) \wedge disease^{g_1}(p_3)$.
$[1, 1]$ $test\_pos(p_3)$.
$[1, 1]$ $test\_pos(p_4)$.
$[1, 1]$ $contact(p_3, p_4)$.
$[1, 1]$ $contact(p_4, p_3)$.

If the combination function associated with $disease$ is Φ^{DS}, we get the following conclusions: $bel(disease(p_3)) = bel(disease(p_4)) = 0.896$. Note that the support for p_3 having the disease is greater than 0.8 because p_3 has positive test results *and* the prior contact with p_4 pumps up the confidence in the diagnosis. Note that the decyclification transformation eliminates the self-supporting feedback loop of $disease(p_3)$ and $disease(p_4)$. Otherwise, $bel(disease(p_3))$ and $bel(disease(p_4))$ would have ended up close to 1 via these self-supporting feedback loops. □

The following theorem shows that the semantics of general blps is an extension of the semantics for acyclic blps.

Theorem 2 (Backward Compatibility). *Let P be an acyclic blp, and $F \in \mathcal{B}ool(B_P)$. Then* $\tilde{m}_P = \hat{m}_P$ *and* $\mathtt{model}_c(P)(F) = \mathtt{model}(P)(F)$. □

The following theorem shows that defining the semantics of BLP through the decyclification is "reasonable" because it discards self-supported beliefs, i.e., belief in A produced by the rules that contain A in their bodies.

Theorem 3 (Self-support). *Let P be a (possibly cyclic) blp, and A an atom in B_P. Let P'_A be the blp obtained from P by deleting all the rules that contain A in their bodies. Then* $\mathtt{model}_c(P)(A) = \mathtt{model}_c(P'_A)(A)$. □

Example 3. (Example 2 continued.) Let P_2 be $P_1 - \{R_4\}$ where P_1 is the program in Example 1 and R_4 is the fourth rule of P_1. In the BLP semantics, P_2, P_1 and P'_1 (the decyclification of P_1, as shown in Example 2) give the same amount of support in $disease(p_3)$. □

4.2 Fixpoint Semantics and Modular Acyclicity

We now provide an alternative, fixpoint semantics for general blps, and show that the fixpoint semantics can be simplified for a special class of cyclic blps, called *modularly acyclic* blps.

First, for atom cliques we define some terms similar to those in Definition 7.

Definition 17. *Let P be a blp, I a truth valuation, and C an atom clique in the dependency graph of P. $P(C)$ is defined as the set of rules in P that has an atom from C in the head. P's reduct under I with respect to C, denoted $P_I(C)$,[3] is obtained from $P(C)$ by*

1. *Replace a rule body with false if it contains an atom $X \notin C$ and $I(X) \neq \mathbf{t}$.*
2. *Deleting every atom $X \notin C$ such that $I(X) = \mathbf{t}$.*
3. *If the combination Φ_A is such that $\forall v, w \ \Phi_A([v,w],[a,b]) = [a,b]$, and P has a fact of the form $[a,b] A$, then delete all the other rules with A in head.*

If $P_I(C)$ is acyclic, we say P is weakly cyclic with respect to I and C. □

Definition 18. *Suppose I is a truth valuation over a set of atoms β and $\alpha \subseteq \beta$. We define the restriction of I to α, denoted $I \mid_\alpha$, to be the truth valuation over α such that $\forall X \in \alpha$, $I \mid_\alpha (X) = I(X)$.*
    We write $I \mid_\alpha = \tau$ if $I(X) = \tau$ for all $X \in \alpha$, where $\tau \in \{t, f, u\}$. □

---

[3] Note that the definitions of $P(C)$ and $P_I(C)$ here is different from the definitions of $P(X)$ and $P_I(X)$ (in Definition 7): C is an atom clique, while X is an atom.

Let $I_\varnothing$ be an *empty* truth valuation $\varnothing \longrightarrow \{\mathbf{t}, \mathbf{f}, \mathbf{u}\}$, i.e., a trivial valuation with an empty domain. Since $I_\varnothing$ is the only truth valuation over $\varnothing$, it follows that $\mathcal{T}Val(\varnothing) = \{I_\varnothing\}$ and $\forall \alpha \forall I \in \mathcal{T}Val(\alpha) \ I \mid_\varnothing = I_\varnothing$. We also define a special support function $m_\varnothing : \mathcal{T}Val(\varnothing) \to [0,1]$ to be $m_\varnothing(I_\varnothing) = 1$; it is the only support function for $\varnothing$.

Next we define a $\hat{T}_{P,Ord}$ operator.

Definition 19. *Let P be a blp with n atom cliques and a clique ordering Ord. Let $\mathcal{C}_1, \ldots, \mathcal{C}_n$ be the atom cliques of P, such that $Ord(\mathcal{C}_i) = i, 1 \le i \le n$, and let $\alpha_0 = \varnothing$, $\alpha_i = \mathcal{C}_1 \cup \cdots \cup \mathcal{C}_i, 1 \le i \le n$.*

Given a support function m for $\alpha_k, 0 \le k < n$, $\hat{T}_{P,Ord}(m)$ is a support function for α_{k+1} such that for every truth valuation $I \in \mathcal{T}Val(\alpha_{k+1})$,

$$\hat{T}_{P,Ord}(m)(I) = m(I \mid_{\alpha_k}) \cdot \tilde{m}_Q(I \mid_{\mathcal{C}_{k+1}}) \tag{1}$$

where $Q = P_{I \mid_{\alpha_k}}(\mathcal{C}_{k+1})$. □

Theorem 4 (Equivalence of fixpoint and transformational semantics)
Let P be a blp with n atom cliques and Ord a clique ordering of P. Beginning with $m_0 = m_\varnothing$, let m_k be $\hat{T}_{P,Ord}^{\uparrow k}(m_0)$, $k = 0, 1, \ldots, n$. The support function m_n coincides with $\tilde{m}_P$. □

The above theorem shows that the fixpoint semantics does not depend on the choice of the clique ordering in P and that this semantics coincides with the transformational semantics of Section 4.

Next, we will show that the computation in (1) can be simplified for a special class of cyclic blps.

Definition 20. *Let P be a blp with n atom cliques and a clique ordering Ord. Let $\mathcal{C}_1, \ldots, \mathcal{C}_n$ be the atom cliques of P, such that $Ord(\mathcal{C}_i) = i, 1 \le i \le n$, and let $\alpha_0 = \varnothing$, $\alpha_i = \mathcal{C}_1 \cup \cdots \cup \mathcal{C}_i, 1 \le i \le n$. Also let $m_k = \hat{T}_{P,Ord}^{\uparrow k}(m_\varnothing)$, $k = 0, 1, \ldots, n$.*

*P is **modularly acyclic** if for every $0 \le k \le n-1$ and for every $I \in \mathcal{T}Val(\alpha_k)$ such that $m_k(I) \ne 0$, P is weakly cyclic with respect to I and $\mathcal{C}_{k+1}$.* □

If P is modularly acyclic, it follows from Theorem 2 that $\tilde{m}_Q(I \mid_{\mathcal{C}_{k+1}})$ in (1) is equivalent to $\hat{m}_Q(I \mid_{\mathcal{C}_{k+1}})$, where Q is $P_{I \mid_{\alpha_k}}(\mathcal{C}_{k+1})$, as defined in Definition 19. (Indeed, it follows from Definition 20 that Q is acyclic if P is modularly acyclic.) Since the computation of $\hat{m}$ does not involve decyclification, the computation of the model of a modularly acyclic blp can be greatly simplified.

Proposition 1. *Let P be a blp. If in every cycle in P (i.e., in every cycle in the dependency graph of P), there is a rule R such that some atom in R's body is not in the head of any rule, then P is modularly acyclic.* □

Example 4. Let us return to the diagnosis case for p_1 and p_2 in Section 2. Suppose that the test for two persons, p_1 and p_2, came back positive, but there is no evidence that p_1 and p_2 had contact. The resulting blp P_3 is

$$[0.8, 1]\ disease(p_1)\ :-\ test\_pos(p_1).$$
$$[0.8, 1]\ disease(p_2)\ :-\ test\_pos(p_2).$$
$$[0.6, 1]\ disease(p_1)\ :-\ contact(p_1, p_2) \wedge disease(p_2).$$
$$[0.6, 1]\ disease(p_2)\ :-\ contact(p_2, p_1) \wedge disease(p_1).$$
$$[1, 1]\ test\_pos(p_1).$$
$$[1, 1]\ test\_pos(p_2).$$

with atom cliques $C_1 = \{test\_pos(p_1)\}$, $C_2 = \{test\_pos(p_2)\}$, $C_3 = \{contact(p_1, p_2)\}$, $C_4 = \{contact(p_2, p_1)\}$, $C_5 = \{disease(p_1), disease(p_2)\}$.

Since $contact(p_1, p_2)$ and $contact(p_2, p_1)$ are not supported by any rule, it follows from Proposition 1 that $\boldsymbol{P}_3$ is modularly acyclic. The belief in $disease(p_1)$ is 0.8 and so is the belief in $disease(p_2)$. □

More interestingly, a blp can be modularly acyclic even when the condition in Proposition 1 is not satisfied, as shown in the following example.

Example 5. Consider a blp $\boldsymbol{P}_4$

$[0.8, 1]\ a\ :-\ d.$	$[1, 1]\ d.$
$[0.8, 1]\ b\ :-\ e.$	$[1, 1]\ e.$
$[0.6, 1]\ a\ :-\ b \wedge c_1.$	$[0.5, 0.5]\ c_1.$
$[0.6, 1]\ b\ :-\ a \wedge c_2.$	$[1, 1]\ c_2\ :-\ \overline{c_1}.$

According to Definition 20, $\boldsymbol{P}_4$ is modularly acyclic. The underlying intuition is as follows. The last rule is the only rule that supports c_2, so we know that c_2 and c_1 can not both be true in a truth valuation. Consequently, the rule $[0.6, 1]\ a\ :-\ b \wedge c_1$ and the rule $[0.6, 1]\ b\ :-\ a \wedge c_2$ do not both fire in a truth valuation. So, the cycle is "weak" and this program is modularly acyclic. □

4.3   Discussion

In this section, we will contrast our method with an alternative approach of eliminating cycles by adding time parameters, which is utilized in [6] and implicitly in [22]. Adopting a similar methodology, the program of Example 1 can be transformed to the following blp:

$$[0.8, 1]\ disease(p_3, T)\ :-\ test\_pos(p_3).$$
$$[0.8, 1]\ disease(p_4, T)\ :-\ test\_pos(p_4).$$
$$[0.6, 1]\ disease(p_3, T)\ :-\ contact(p_3, p_4) \wedge disease(p_4, T-1).$$
$$[0.6, 1]\ disease(p_4, T)\ :-\ contact(p_4, p_3) \wedge disease(p_3, T-1).$$
$$[1, 1]\ disease(p_3, T)\ :-\ disease(p_3, T-1).$$
$$[1, 1]\ disease(p_4, T)\ :-\ disease(p_4, T-1).$$

$[1, 1]\ test\_pos(p_3).$	$[1, 1]\ contact(p_3, p_4).$
$[1, 1]\ test\_pos(p_4).$	$[1, 1]\ contact(p_4, p_3).$

As a consequence of adding time parameters, the fifth and sixth rules must be added to ensure consistency. It is also worth noting that the first two rules assert that the test results provide support for diagnoses at any time point.

It is not difficult to observe the differences between the above transformed program and P_1' in Example 2 by our approach. One critical question in the time parameter methodology in [6] is, for a query $q(.)$, at which time point t does $q(., t)$ yield the correct answer. In [22], this problem is avoided by choosing the stationary state. However, this is based on a restriction that only stationary dynamic Bayesian networks can be modeled. Another assumption in the time parameter methodology is that an atom without time parameter takes the same value all the time. In this particular example, such an assumption translates to that p_3 and p_4 are having contact at all the time points. Obviously, this assumption may not hold in all applications. Our approach avoids the above problems by providing an alternative method to eliminate cycles.

It is worth noting that, the fact that we do not use the time parameter methodology to eliminate cycles *does not* mean that we do not allow time parameters. In the applications where time parameters are appropriate and feedbacks over time are desirable, time parameters may also be encoded into blps, e.g., $[0.9, 1] \ p(X, T) \ :- \ p(X, T - 1)$.

5   Conclusions

In [26] we introduced a novel logic theory, Belief Logic Programming, for reasoning with uncertainty, which can correlate structural information contained in derivation paths for beliefs. In this paper we extended the previous work to cyclic BLP by defining a transformational and a fixpoint semantics in which self-supported belief is discarded. We also showed that the proposed semantics are backward compatible with the semantics for acyclic BLP [26] and has the expected properties.

For future work, we plan to lift the decyclification transformation to non-ground level and extend the query answering algorithm proposed in [27] to cyclic BLP.

Acknowledgement

This work is part of the SILK (Semantic Inference on Large Knowledge) project sponsored by Vulcan, Inc. The author thanks Michael Kifer for the insightful discussions.

References

1. Baldwin, J.F.: Evidential support logic programming. Fuzzy Sets and Systems 24(1), 1–26 (1987)
2. Dekhtyar, A., Subrahmanian, V.S.: Hybrid probabilistic programs. J. of Logic Programming 43, 391–405 (1997)
3. Dempster, A.P.: Upper and lower probabilities induced by a multi-valued mapping. Ann. Mathematical Statistics 38 (1967)

4. Ding, Z., Peng, Y., Pan, R.: BayesOWL: Uncertainty modeling in semantic web ontologies. Studies in Fuzziness and Soft Computing 204, 3–29 (2006)
5. Van Emden, M.H.: Quantitative deduction and its fixpoint theory. J. of Logic Programming 3(1), 37–53 (1986)
6. Glesner, S., Koller, D.: Constructing flexible dynamic belief networks from first-order probabilistic knowledge bases. In: Froidevaux, C., Kohlas, J. (eds.) ECSQARU 1995. LNCS, vol. 946, pp. 217–226. Springer, Heidelberg (1995)
7. Grosof, B.N.: A courteous compiler from generalized courteous logic programs to ordinary logic programs. Technical Report Supplementary Update Follow-On to RC 21472, IBM (July 1999)
8. Kersting, K., De Raedt, L.: Bayesian logic programs. Technical report, Albert-Ludwigs University at Freiburg (2001)
9. Kifer, M., Li, A.: On the semantics of rule-based expert systems with uncertainty. In: Gyssens, M., Van Gucht, D., Paredaens, J. (eds.) ICDT 1988. LNCS, vol. 326, pp. 102–117. Springer, Heidelberg (1988)
10. Kifer, M., Subrahmanian, V.S.: Theory of generalized annotated logic programming and its applications. J. of Logic Programming 12(3,4), 335–367 (1992)
11. Lakshmanan, L.V.S., Shiri, N.: A parametric approach to deductive databases with uncertainty. IEEE Trans. on Knowledge and Data Engineering 13(4), 554–570 (2001)
12. Lukasiewicz, T.: Probabilistic logic programming under inheritance with overriding. In: Annual Conf. on Uncertainty in Artificial Intelligence (UAI 2001), pp. 329–336. Morgan Kaufmann Publishers, San Francisco (2001)
13. Lukasiewicz, T.: Probabilistic logic programming with conditional constraints. ACM Trans. on Computational Logic 2(3), 289–339 (2001)
14. Lukasiewicz, T., Straccia, U.: Managing uncertainty and vagueness in description logics for the semantic web. Journal of Web Semantics 6(4), 291–308 (2008)
15. Ng, R.T.: Reasoning with uncertainty in deductive databases and logic programs. Intl. Journal of Uncertainty, Fuzziness and Knowledge-Based Systems 5(3), 261–316 (1997)
16. Ng, R.T., Subrahmanian, V.S.: Probabilistic logic programming. Information and Computation 101(2), 150–201 (1992)
17. Ng, R.T., Subrahmanian, V.S.: A semantical framework for supporting subjective probabilities in deductive databases. J. of Automated Reasoning 10(2), 191–235 (1993)
18. Poole, D.: The Independent Choice Logic and Beyond. In: De Raedt, L., Frasconi, P., Kersting, K., Muggleton, S.H. (eds.) Probabilistic Inductive Logic Programming. LNCS (LNAI), vol. 4911, pp. 222–243. Springer, Heidelberg (2008)
19. De Raedt, L., Kersting, K.: Probabilistic inductive logic programming. In: De Raedt, L., Frasconi, P., Kersting, K., Muggleton, S.H. (eds.) Probabilistic Inductive Logic Programming. LNCS (LNAI), vol. 4911, pp. 1–27. Springer, Heidelberg (2008)
20. Richardson, M., Domingos, P.: Markov logic networks. Machine Learning 62(1-2), 107–136 (2006)
21. Shafer, G.: A Mathematical Theory of Evidence. Princeton University Press, Princeton (1976)
22. Shen, Y.: Reasoning with recursive loops under the plp framework. ACM Trans. Comput. Logic 9(4), 1–31 (2008)
23. Stoilos, G., Stamou, G., Tzouvaras, V., Pan, J.Z., Horrocks, I.: Fuzzy OWL: Uncertainty and the semantic web. In: Proc. of the International Workshop on OWL: Experiences and Directions (2005)

24. Subrahmanian, V.S.: On the semantics of quantitative logic programs. In: SLP, pp. 173–182 (1987)
25. Wan, H., Grosof, B.N., Kifer, M., Fodor, P., Liang, S.: Logic programming with defaults and argumentation theories. In: Hill, P.M., Warren, D.S. (eds.) ICLP 2009. LNCS, vol. 5649, pp. 432–448. Springer, Heidelberg (2009)
26. Wan, H., Kifer, M.: Belief logic programming: Uncertainty reasoning with correlation of evidence. In: Erdem, E., Li, F., Schaub, T. (eds.) LPNMR 2009. LNCS, vol. 5753, pp. 316–328. Springer, Heidelberg (2009)
27. Wan, H., Kifer, M.: Query answering in belief logic programming. In: Godo, L., Pugliese, A. (eds.) SUM 2009. LNCS, vol. 5785, pp. 268–281. Springer, Heidelberg (2009)
28. Wan, H., Kifer, M.: Technical report: Belief logic programming. Technical report, Stony Brook University (2009), http://www.cs.sunysb.edu/~hwan/BLP_TR.html

A Minimal Deductive System for General Fuzzy RDF

Umberto Straccia

Istituto di Scienza e Tecnologie dell'Informazione (ISTI - CNR), Pisa, Italy
straccia@isti.cnr.it

Abstract. It is well-known that crisp RDF is not suitable to represent vague information. Fuzzy RDF variants are emerging to overcome this limitations. In this work we provide, under a very general semantics, a deductive system for a salient fragment of fuzzy RDF. We then also show how we may compute the top-k answers of the union of conjunctive queries in which answers may be scored by means of a scoring function.

1 Introduction

RDF [17] has become a quite popular Semantic Web representation formalism. The basic ingredients are triples of the form (s, p, o), such as $(tom, likes, tomato)$, stating that subject s has property p with value o.

However, under the classical semantics, RDF cannot represent vague information and, to this purpose, some *Fuzzy RDF* variants have been proposed [12,13,14,21,22]: essentially they allow to state that a triple is true to some degree, *e.g.*, $(tom, likes, tomato)$ is true to degree at least 0.9.

Our main goal of this study is to provide, under a very general semantics, a minimal deductive system for fuzzy RDF, along the lines described by [15]. The advantage is that, (i) (unlike [12,13,14,22]) we abstract from the underlying XML representation; (ii) the semantics is quite general, *i.e.* is based on a t-norm [9]; (iii) we get a clear insight of the supported inference mechanism; and (iv) we concentrate on the main ingredients of RDF from a reasoning point of view. We then also address the query answering problem and show how effectively we may compute the top-k answers of the union of conjunctive queries in which answers may be scored by means of a scoring function.

Related work: The most general work so far and closest to our work is [14], to which respect we provide additionally a more general semantics, correctness and completeness and complexity results, add the notion of top-k answers of the union of conjunctive queries in which answers may be scored by means of a scoring function, and show how to compute the top-k answers. Another related work is [21], which allows to annotate triples with truth values taken from a finite partial order, while we rely on $[0, 1]$ instead[1]. However, we provide some desired inference capabilities not provided by [21], *e.g.*, from "a sport car is a fast car to degree 0.8" and "a fast car is an expensive car to degree

---

[1] But we can extend the truth space to other truth-spaces as well, provided that we extend the t-norm and residuated implication accordingly [9].

A. Polleres and T. Swift (Eds.): RR 2009, LNCS 5837, pp. 166–181, 2009.

0.9" we may infer that "a sport car is an expensive car to degree 0.72" (under product t-norm). Essentially, [21] does not provide a truth combination function to propagate the truth in such inferences, while we consider a t-norm instead. Additionally, as for [14], the top-k retrieval problem for the union of conjunctive queries is not addressed.

In the next section, we recall the main aspects of classical RDF as described in [15], which we extend then to the fuzzy case.

2   Preliminaries

For the sake of our purposes, we will rely on a minimal, but significant RDF fragment, called ρdf [15], that covers the essential features of RDF. According to [15], ρdf (read rho-df, the ρ from restricted rdf) is defined as the following subset of the RDFS vocabulary:

$$\rho\mathsf{df} = \{\mathsf{sp}, \mathsf{sc}, \mathsf{type}, \mathsf{dom}, \mathsf{range}\} \ .$$

Informally, the meaning of a triple (s, p, o) with $p \in \rho$df is:

- (p, sp, q) means that property p is a *sub property* of property q;
- (c, sc, d) means that class c is a *sub class* of class d;
- (a, type, b) means that a is of *type* b;
- (p, dom, c) means that the *domain* of property p is c;
- (p, range, c) means that the *range* of property p is c.

Syntax. Assume pairwise disjoint alphabets **U** (*RDF URI references*), **B** (*Blank nodes*), and **L** (*Literals*). Through the paper we assume **U**, **B**, and **L** fixed, and for simplicity we will denote unions of these sets simply concatenating their names. We call elements in **UBL** *terms* (denoted t), and elements in **B** *variables* (denoted x)[2].

An *RDF triple* (or *RDF atom*) is a triple $(s, p, o) \in$ **UBL** $\times$ **U** $\times$ **UBL**. In this tuple, s is the *subject*, p is the *predicate*, and o is the *object*. An *RDF graph* (or simply a graph, or *RDF Knowledge Base*) is a set of RDF triples τ. A subgraph is a subset of a graph. The *universe* of a graph G, denoted by $universe(G)$ is the set of elements in **UBL** that occur in the triples of G. The *vocabulary* of G, denoted by $voc(G)$ is the set $universe(G) \cap$ **UL**. A graph is *ground* if it has no blank nodes, *i.e.* variables.

In what follows we will need some technical notions. A *variable assignment* is a function $\mu :$ **UBL** $\to$ **UBL** preserving URIs and literals, *i.e.*, $\mu(t) = t$, for all $t \in$ **UL**. Given a graph G, we define $\mu(G) = \{(\mu(s), \mu(p), \mu(o)) \mid (s, p, o) \in G\}$. We speak of a variable assignment μ from G_1 to G_2, and write $\mu : G_1 \to G_2$, if μ is such that $\mu(G_1) \subseteq G_2$.

Semantics. An *RDF interpretation* $\mathcal{I}$ over a vocabulary V is a tuple

$$\mathcal{I} = \langle \Delta_R, \Delta_P, \Delta_C, \Delta_L, P[\![\cdot]\!], C[\![\cdot]\!], \cdot^{\mathcal{I}} \rangle ,$$

where $\Delta_R, \Delta_P, \Delta_C, \Delta_L$ are the interpretations domains of $\mathcal{I}$, and $P[\![\cdot]\!], C[\![\cdot]\!], \cdot^{\mathcal{I}}$ are the interpretation functions of $\mathcal{I}$. They have to satisfy:

1. Δ_R is a nonempty set of resources, called the domain or universe of $\mathcal{I}$;
2. Δ_P is a set of property names (not necessarily disjoint from Δ_R);

---

[2] All symbols may have upper or lower script.

3. $\Delta_C \subseteq \Delta_R$ is a distinguished subset of Δ_R identifying if a resource denotes a class of resources;
4. $\Delta_L \subseteq \Delta_R$, the set of literal values, Δ_L contains all plain literals in $\mathbf{L} \cap V$;
5. $P[\![\cdot]\!]$ maps each property name $p \in \Delta_P$ into a subset $P[\![p]\!] \subseteq \Delta_R \times \Delta_R$, *i.e.* assigns an extension to each property name;
6. $C[\![\cdot]\!]$ maps each class $c \in \Delta_C$ into a subset $C[\![c]\!] \subseteq \Delta_R$, *i.e.* assigns a set of resources to every resource denoting a class;
7. $\cdot^{\mathcal{I}}$ maps each $t \in \mathbf{UL} \cap V$ into a value $t^{\mathcal{I}} \in \Delta_R \cup \Delta_P$, *i.e.* assigns a resource or a property name to each element of $\mathbf{UL}$ in V, and such that $\cdot^{\mathcal{I}}$ is the identity for plain literals and assigns an element in Δ_R to elements in $\mathbf{L}$;
8. $\cdot^{\mathcal{I}}$ maps each variable $x \in \mathbf{B}$ into a value $x^{\mathcal{I}} \in \Delta_R$, *i.e.* assigns a resource to each variable in $\mathbf{B}$.

The notion entailment is defined using the idea of *satisfaction* of a graph under certain interpretation. Intuitively a ground triple (s, p, o) in an RDF graph G will be true under the interpretation $\mathcal{I}$ if p is interpreted as a property name, s and o are interpreted as resources, and the interpretation of the pair (s, o) belongs to the extension of the property assigned to p.

In RDF, blank nodes, *i.e.* variables, work as existential variables. Intuitively the triple (x, p, o) with $x \in \mathbf{B}$ would be true under $\mathcal{I}$ if there exists a resource s such that (s, p, o) is true under $\mathcal{I}$.

Now, let G be a graph. An interpretation $\mathcal{I}$ is a *model* of G, denoted $\mathcal{I} \models G$, iff $\mathcal{I}$ is an interpretation over the vocabulary ρdf $\cup$ *universe*(G) that satisfies the following conditions:

Simple:
    1. for each $(s, p, o) \in G$, $p^{\mathcal{I}} \in \Delta_P$ and $(s^{\mathcal{I}}, o^{\mathcal{I}}) \in P[\![p^{\mathcal{I}}]\!]$;

Subproperty:
    1. $P[\![\mathsf{sp}^{\mathcal{I}}]\!]$ is transitive over Δ_P;
    2. if $(p, q) \in P[\![\mathsf{sp}^{\mathcal{I}}]\!]$ then $p, q \in \Delta_P$ and $P[\![p]\!] \subseteq P[\![q]\!]$;

Subclass:
    1. $P[\![\mathsf{sc}^{\mathcal{I}}]\!]$ is transitive over Δ_C;
    2. if $(c, d) \in P[\![\mathsf{sc}^{\mathcal{I}}]\!]$ then $c, d \in \Delta_C$ and $C[\![c]\!] \subseteq C[\![d]\!]$;

Typing I:
    1. $x \in C[\![c]\!]$ iff $(x, c) \in P[\![\mathsf{type}^{\mathcal{I}}]\!]$;
    2. if $(p, c) \in P[\![\mathsf{dom}^{\mathcal{I}}]\!]$ and $(x, y) \in P[\![p]\!]$ then $x \in C[\![c]\!]$;
    3. if $(p, c) \in P[\![\mathsf{range}^{\mathcal{I}}]\!]$ and $(x, y) \in P[\![p]\!]$ then $y \in C[\![c]\!]$;

Typing II:
    1. For each $e \in \rho$df, $e^{\mathcal{I}} \in \Delta_P$
    2. if $(p, c) \in P[\![\mathsf{dom}^{\mathcal{I}}]\!]$ then $p \in \Delta_P$ and $c \in \Delta_C$
    3. if $(p, c) \in P[\![\mathsf{range}^{\mathcal{I}}]\!]$ then $p \in \Delta_P$ and $c \in \Delta_C$
    4. if $(x, c) \in P[\![\mathsf{type}^{\mathcal{I}}]\!]$ then $c \in \Delta_C$

We define G *entails* H under ρdf, denoted $G \models H$, iff every model under ρdf of G is also a model under ρdf of H.

Please note that in [15], $P[\![\mathsf{sp}^{\mathcal{I}}]\!]$ (resp. $C[\![\mathsf{sc}^{\mathcal{I}}]\!]$) besides being required to be transitive over Δ_P (resp. Δ_C), is also *reflexive* over Δ_P (resp. Δ_C). We omit this requirement and, thus, do not support inferences such as $G \models (a, \mathsf{sp}, a)$ and $G \models (a, \mathsf{sc}, a)$, which anyway are of marginal interest (see [15] for a more in depth discussion on this issue).

Deductive system. In what follows, we recall the sound and complete deductive system for the fragment of RDF presented in [15]. The system is arranged in groups of rules that captures the semantic conditions of models. In every rule, A, B, C, X, and Y are meta-variables representing elements in **UBL**. An instantiation of a rule is a uniform replacement of the metavariables occurring in the triples of the rule by elements of **UBL**, such that all the triples obtained after the replacement are well formed RDF triples. The rules are as follows:

1. Simple:

   (a) $\frac{G}{G'}$ for a map $\mu : G' \rightarrow G$ (b) $\frac{G}{G'}$ for $G' \subseteq G$

2. Subproperty:

   (a) $\frac{(A,\mathrm{sp},B),(B,\mathrm{sp},C)}{(A,\mathrm{sp},C)}$ (b) $\frac{(A,\mathrm{sp},B),(X,A,Y)}{(X,B,Y)}$

3. Subclass:

   (a) $\frac{(A,\mathrm{sc},B),(B,\mathrm{sc},C)}{(A,\mathrm{sc},C)}$ (b) $\frac{(A,\mathrm{sc},B),(X,\mathrm{type},A)}{(X,\mathrm{type},B)}$

4. Typing:

   (a) $\frac{(A,\mathrm{dom},B),(X,A,Y)}{(X,\mathrm{type},B)}$ (b) $\frac{(A,\mathrm{range},B),(X,A,Y)}{(Y,\mathrm{type},B)}$

5. Implicit Typing:

   (a) $\frac{(A,\mathrm{dom},B),(C,\mathrm{sp},A),(X,C,Y)}{(X,\mathrm{type},B)}$ (b) $\frac{(A,\mathrm{range},B),(C,\mathrm{sp},A),(X,C,Y)}{(Y,\mathrm{type},B)}$

A *proof* is defined in the usual way. Let G and H be graphs. Then $G \vdash H$ iff there is a sequence of graphs $P_1, \ldots, P_k$ with $P_1 = G$ and $P_k = H$, and for each j ($2 \leqslant j \leqslant k$) one of the following holds:

1. there exists a map $\mu : P_j \rightarrow P_{j-1}$ (rule (1a));
2. $P_j \subseteq P_{j-1}$ (rule (1b));
3. there is an instantiation $\frac{R}{R'}$ of one of the rules (2)(5), such that $R \subseteq P_{j-1}$ and $P_j = P_{j-1} \cup R'$.

The sequence of rules used at each step (plus its instantiation or map), is called a *proof* of H from G.

Proposition 1 (Soundness and completeness [15]). *The proof system* $\vdash$ *is sound and complete for* $\models$, *that is,* $G \vdash H$ *iff* $G \models H$.

Let G be a ground graph and τ be a gound triple. The *closure* of G is defined as

$$cl(G) = \{\tau \mid \tau \text{ ground and } G \vdash \tau\}.$$

Note that the size of the closure of G is $\mathcal{O}(|G|^2)$ and, thus a naive method to answer whether $G \models \tau$ consists in computing $cl(G)$ and check whether τ is included in $cl(G)$ [15]. [15] provides also an alternative method to test $G \models \tau$ that runs in time $\mathcal{O}(|G| \log |G|)$.

Query Answering. For the sake of our purpose, we get inspired by [6][3] and we assume that a RDF graph G is *ground* and *closed*, *i.e.*, G is closed under the application of the rules (2)-(5). Then a *conjunctive query* is a Datalog-like rule of the form

$$q(\mathbf{x}) \leftarrow \exists \mathbf{y}.\tau_1, \ldots, \tau_n$$

where $n \geqslant 1$, $\tau_1, \ldots, \tau_n$ are triples, $\mathbf{x}$ is a vector of variables occurring in $\tau_1, \ldots, \tau_n$, called the *distinguished variables*, $\mathbf{y}$ are so-called *non-distinguished variables* and are distinct from the variables in $\mathbf{x}$, each variable occurring in τ_i is either a distinguished variable or a non-distinguished variable. If clear from the context, we may omit the exitential quantification $\exists \mathbf{y}$. For instance, the query

$$q(x,y) \leftarrow (x, creates, y), (x, \mathsf{type}, Flemish), (y, exhibited, Uffizi)$$

has intended meaning to retrieve all the artifacts x created by Flemish artists y, being exhibited at Uffizi Gallery.

We will also write a query as

$$q(\mathbf{x}) \leftarrow \exists \mathbf{y}.\varphi(\mathbf{x}, \mathbf{y}) \,,$$

where $\varphi(\mathbf{x}, \mathbf{y})$ is $\tau_1, \ldots, \tau_n$. Furthermore, $q(\mathbf{x})$ is called the *head* of the query, while $\exists \mathbf{y}.\varphi(\mathbf{x}, \mathbf{y})$ is called the *body* of the query.

Finally, a *disjunctive query* (or, *union of conjunctive queries*) $\mathbf{q}$ is, as usual, a finite set of conjunctive queries in which all the rules have the same head.

Given a graph G, a query $q(\mathbf{x}) \leftarrow \exists \mathbf{y}.\varphi(\mathbf{x}, \mathbf{y})$, and a vector $\mathbf{t}$ of terms in $\mathbf{UL}$, we say that $q(\mathbf{t})$ is *entailed* by G, denoted $G \models q(\mathbf{t})$, iff in any model $\mathcal{I}$ of G, there is a vector $\mathbf{t}'$ of terms in $\mathbf{UL}$ such that $\mathcal{I}$ is a model of $\varphi(\mathbf{t}, \mathbf{t}')$. If $G \models q(\mathbf{t})$ then $\mathbf{t}$ is called an *answer* to q. For a disjunctive query $\mathbf{q} = \{q_1, \ldots, q_m\}$, we say that $\mathbf{q}(\mathbf{t})$ is *entailed* by G, denoted $G \models \mathbf{q}(\mathbf{t})$, iff $G \models q_i(\mathbf{t})$ for some $q_i \in \mathbf{q}$. The *answer set* of $\mathbf{q}$ w.r.t. G is defined as

$$ans(G, \mathbf{q}) = \{\mathbf{t} \mid G \models \mathbf{q}(\mathbf{t})\} \,.$$

A simple method to determine $ans(G, \mathbf{q})$ is as follows. Compute the closure $cl(G)$ of G and store it into a database, *e.g.*, using the method [1]. It is easily verified that any disjunctive query can be mapped into union of SQL queries over the underlying database schema. Hence, $ans(G, \mathbf{q})$ is determined by issuing these SQL queries to the database.

3   Fuzzy RDF

We now present fuzzy RDF in its general form, by extending [12,13,14]. To do so and to make the paper self-contained, we first recall basic notions of mathematical fuzzy logic [9].

3.1   Preliminaries: Mathematical Fuzzy Logic

In mathematical fuzzy logics, the convention prescribing that a statement is either true or false is changed and is a matter of degree taken from a *truth space* S, usually $[0, 1]$ (in that case we speak about *Mathematical Fuzzy Logic* [9]) or $\{\frac{0}{n}, \frac{1}{n}, \ldots, \frac{n}{n}\}$ for an

---

[3] Note that [15] does not address conjunctive query answering.

integer $n \geqslant 1$. Often $\mathcal{S}$ may also be a complete lattice or a bilattice [3,5] (often used in logic programming [4]). In the sequel, we assume $\mathcal{S} = [0, 1]$. This degree is called *degree of truth* of the statement ϕ in the interpretation $\mathcal{I}$.

In the illustrative fuzzy logic that we consider in this section, *fuzzy statements* have the form $\phi[n]$, where $n \in [0, 1]$ [8,9] and ϕ is a statement, which encodes that the degree of truth of ϕ is *at least* n. For example, $ripe\_tomato[0.9]$ says that we have a rather ripe tomato (the degree of truth of $ripe\_tomato$ is at least 0.9). Semantically, a *fuzzy interpretation* $\mathcal{I}$ maps each basic statement p_i into $[0, 1]$ and is then extended inductively to all statements as follows:

$$\begin{aligned}
\mathcal{I}(\phi \wedge \psi) &= \mathcal{I}(\phi) \otimes \mathcal{I}(\psi) & \mathcal{I}(\phi \vee \psi) &= \mathcal{I}(\phi) \oplus \mathcal{I}(\psi), \\
\mathcal{I}(\phi \rightarrow \psi) &= \mathcal{I}(\phi) \Rightarrow \mathcal{I}(\psi) & \mathcal{I}(\neg\phi) &= \ominus \mathcal{I}(\phi), \\
\mathcal{I}(\exists x.\phi(x)) &= \sup_{c \in \Delta^{\mathcal{I}}} \mathcal{I}(\phi(c)) & \mathcal{I}(\forall x.\phi(x)) &= \inf_{c \in \Delta^{\mathcal{I}}} \mathcal{I}(\phi(c))
\end{aligned} \tag{1}$$

where $\otimes$, $\oplus$, $\Rightarrow$, and $\ominus$ are so-called *combination functions*, namely, *triangular norms* (or *t-norms*), *triangular co-norms* (or *s-norms*), *implication functions*, and *negation functions*, respectively, which extend the classical Boolean conjunction, disjunction, implication, and negation, respectively, to the fuzzy case.

Several t-norms, s-norms, implication functions, and negation functions have been given in the literature. An important aspect of such functions is that they satisfy some properties that one expects to hold for the connectives; see Tables 1 and 2. Note that in Table 1, the two properties Tautology and Contradiction follow from Identity, Commutativity, and Monotonicity. Usually, the implication function $\Rightarrow$ is defined as *r-implication*, that is, $a \Rightarrow b = \sup \{c \mid a \otimes c \leqslant b\}$.

Some t-norms, s-norms, implication functions, and negation functions of various fuzzy logics are shown in Table 3 [9]. In fuzzy logic, one usually distinguishes three different logics, namely, Łukasiewicz, Gödel, and Product logic; the popular Zadeh logic is a sublogic of Łukasiewicz logic as, $min(x, y) = x \wedge (x \Rightarrow y)$ and $\max(x, y) = (x \Rightarrow y) \Rightarrow y$. Some salient properties of these logics are shown in Table 4. For more properties, see especially [9,16]. Note also, that a fuzzy logic having all properties shown in Table 4, collapses to boolean logic, *i.e.* the truth-set can be $\{0, 1\}$ only. Also note that the importance of these three logics is due the fact that any t-norm can be obtained as a combination of Łukasiewicz, Gödel, and Product t-norm. The implication $x \Rightarrow y = \max(1 - x, y)$ is called *Kleene-Dienes implication* in the fuzzy logic literature. Note that we have the following inferences: Let $a \geqslant n$ and $a \Rightarrow b \geqslant m$. Then, under Kleene-Dienes implication, we infer that "if $n > 1 - m$ then $b \geqslant m$". More importantly, under an r-implication relative to a t-norm $\otimes$, we have that

$$\text{from } a \geqslant n \text{ and } a \Rightarrow b \geqslant m, \text{ we infer } b \geqslant n \otimes m . \tag{2}$$

To see this, as $a \geqslant n$ and $a \Rightarrow b = \sup \{c \mid a \otimes c \leqslant b\} = \bar{c} \geqslant m$ it follows that $b \geqslant a \otimes \bar{c} \geqslant n \otimes m$. In a similar way, under an r-implication relative to a t-norm $\otimes$, we have that

$$\text{from } a \Rightarrow b \geqslant n \text{ and } b \Rightarrow c \geqslant m, \text{ we infer that } a \Rightarrow c \geqslant n \otimes m . \tag{3}$$

As we will see later on, these are the main inference patterns we will rely on in this paper.

Table 1. Properties for t-norms and s-norms

Axiom Name	T-norm	S-norm
Tautology / Contradiction	$a \otimes 0 = 0$	$a \oplus 1 = 1$
Identity	$a \otimes 1 = a$	$a \oplus 0 = a$
Commutativity	$a \otimes b = b \otimes a$	$a \oplus b = b \oplus a$
Associativity	$(a \otimes b) \otimes c = a \otimes (b \otimes c)$	$(a \oplus b) \oplus c = a \oplus (b \oplus c)$
Monotonicity	if $b \leqslant c$, then $a \otimes b \leqslant a \otimes c$	if $b \leqslant c$, then $a \oplus b \leqslant a \oplus c$

Table 2. Properties for implication and negation functions

Axiom Name	Implication Function	Negation Function
Tautology / Contradiction	$0 \Rightarrow b = 1,\ a \Rightarrow 1 = 1,\ 1 \Rightarrow 0 = 0$	$\ominus 0 = 1,\ \ominus 1 = 0$
Antitonicity	if $a \leqslant b$, then $a \Rightarrow c \geqslant b \Rightarrow c$	if $a \leqslant b$, then $\ominus a \geqslant \ominus b$
Monotonicity	if $b \leqslant c$, then $a \Rightarrow b \leqslant a \Rightarrow c$	

Table 3. Combination functions of various fuzzy logics

	Łukasiewicz Logic	Gödel Logic	Product Logic	Zadeh Logic
$a \otimes b$	$\max(a + b - 1, 0)$	$\min(a, b)$	$a \cdot b$	$\min(a, b)$
$a \oplus b$	$\min(a + b, 1)$	$\max(a, b)$	$a + b - a \cdot b$	$\max(a, b)$
$a \Rightarrow b$	$\min(1 - a + b, 1)$	$\begin{cases} 1 & \text{if } a \leqslant b \\ b & \text{otherwise} \end{cases}$	$\min(1, b/a)$	$\max(1 - a, b)$
$\ominus a$	$1 - a$	$\begin{cases} 1 & \text{if } a = 0 \\ 0 & \text{otherwise} \end{cases}$	$\begin{cases} 1 & \text{if } a = 0 \\ 0 & \text{otherwise} \end{cases}$	$1 - a$

Table 4. Some additional properties of combination functions of various fuzzy logics

Property	Łukasiewicz Logic	Gödel Logic	Product Logic	Zadeh Logic
$x \otimes \ominus x = 0$	+	+	+	−
$x \oplus \ominus x = 1$	+	−	−	−
$x \otimes x = x$	−	+	−	+
$x \oplus x = x$	−	+	−	+
$\ominus \ominus x = x$	+	−	−	+
$x \Rightarrow y = \ominus x \oplus y$	+	−	−	+
$\ominus (x \Rightarrow y) = x \otimes \ominus y$	+	−	−	+
$\ominus (x \otimes y) = \ominus x \oplus \ominus y$	+	+	+	+
$\ominus (x \oplus y) = \ominus x \otimes \ominus y$	+	+	+	+

Note that implication functions and t-norms are also used to define the degree of subsumption between fuzzy sets and the composition of two (binary) fuzzy relations. A *fuzzy set* R over a countable crisp set X is a function $R \colon X \to [0, 1]$. The *degree of subsumption* between two fuzzy sets A and B, denoted $A \sqsubseteq B$, is defined as

$$\inf_{x \in X} A(x) \Rightarrow B(x) , \tag{4}$$

where $\Rightarrow$ is an implication function. Note that in First-Order-Logic terms, A is a subclass of B may be seen as the formula

$$\forall x . A(x) \Rightarrow B(x) ,$$

and, as in fuzzy logic $\forall$ is the inf, we get equation above. Together with (2), these are the two major notions we need later on.

Note that if $A(x) \leqslant B(x)$, for all $x \in [0, 1]$, then $A \sqsubseteq B$ evaluates to 1. Of course, $A \sqsubseteq B$ may evaluate to a value $v \in (0, 1)$ as well. A (binary) *fuzzy relation* R over two countable crisp sets X and Y is a function $R \colon X \times Y \to [0, 1]$. The *composition* of two fuzzy relations $R_1 \colon X \times Y \to [0, 1]$ and $R_2 \colon Y \times Z \to [0, 1]$ is defined as $(R_1 \circ R_2)(x, z) = \sup_{y \in Y} R_1(x, y) \otimes R_2(y, z)$. A fuzzy relation R is *transitive* iff $R(x, z) \geqslant (R \circ R)(x, z)$.

A fuzzy interpretation $\mathcal{I}$ *satisfies* a fuzzy statement $\phi[n]$ or $\mathcal{I}$ is a *model* of $\phi[n]$, denoted $\mathcal{I} \models \phi[n]$, iff $\mathcal{I}(\phi) \geqslant n$. The notions of satisfiability and logical consequence are defined in the standard way. We say $\phi[n]$ is a *tight logical consequence* of a set of fuzzy statements KB iff n is the infimum of $\mathcal{I}(\phi)$ subject to all models $\mathcal{I}$ of KB. Notice that the latter is equivalent to $n = \sup \{ r \mid KB \models \phi[r] \}$. We refer the reader to [7,8,9] for reasoning algorithms for fuzzy propositional and First-Order Logics.

3.2   Generalized Fuzzy RDF

We are now ready to extend the notions introduced in the previous section to fuzzy RDF. We start with the syntax and then define the semantics.

Syntax. A *fuzzy RDF triple* is an expression $\tau[n]$, where τ is a triple and $n \in [0, 1]$. The intended semantics is that the degree of truth of τ is not less than n. For instance, $(audiTT, \text{type}, SportsCar)[0.8]$ is a fuzzy triple, intending that AudiTT is almost a sport car. In a fuzzy triple $\tau[n]$, the truth value n may be omitted and, in that case, the value $n = 1$ is assumed. A *fuzzy RDF graph* $\tilde{G}$ (or simply a fuzzy graph, or *fuzzy RDF Knowledge Base*) is a set of fuzzy RDF triples $\tilde{\tau}$. The notions of *universe* of a graph $\tilde{G}$, the *vocabulary* of $\tilde{G}$, *ground* graph and *variable assignment* are as for the crisp case. Without loss of generality we may assume that there are not two fuzzy triples $\tau[n]$ and $\tau[m]$ in a fuzzy graph $\tilde{G}$. If this is the case, we may just remove the fuzzy triple with the lower score.

Semantics. The fuzzy semantics is derived directly from the crisp one, where the extension functions are no longer sets, but functions assigning a truth in $[0, 1]$. So, let $\otimes$ be a t-norm and let $\Rightarrow$ be its r-implication. A *fuzzy RDF interpretation* $\mathcal{I}$ over a vocabulary V is a tuple

$$\mathcal{I} = \langle \Delta_R, \Delta_P, \Delta_C, \Delta_L, P[\![\cdot]\!], C[\![\cdot]\!], \cdot^{\mathcal{I}} \rangle,$$

where $\Delta_R, \Delta_P, \Delta_C, \Delta_L$ are the interpretations domains of $\mathcal{I}$, and $P[\![\cdot]\!], C[\![\cdot]\!], \cdot^{\mathcal{I}}$ are the interpretation functions of $\mathcal{I}$. They have to satisfy:

1. Δ_R is a nonempty set of resources, called the domain or universe of $\mathcal{I}$;
2. Δ_P is a set of property names (not necessarily disjoint from Δ_R);
3. $\Delta_C \subseteq \Delta_R$ is a distinguished subset of Δ_R identifying if a resource denotes a class of resources;
4. $\Delta_L \subseteq \Delta_R$, the set of literal values, Δ_L contains all plain literals in $\mathbf{L} \cap V$;
5. $P[\![\cdot]\!]$ maps each property name $p \in \Delta_P$ into a partial function $P[\![p]\!] : \Delta_R \times \Delta_R \to [0, 1]$, *i.e.* assigns a degree to each pair of resources, denoting the degree of being the pair an instance of the property p;

6. $C[\![\cdot]\!]$ maps each class $c \in \Delta_C$ into a partial function $C[\![c]\!] : \Delta_R \to [0,1]$, *i.e.* assigns a degree to every resource, denoting the degree of of the resource being an instance of class c;

7. $\cdot^{\mathcal{I}}$ maps each $t \in \mathbf{UL} \cap V$ into a value $t^{\mathcal{I}} \in \Delta_R \cup \Delta_P$, *i.e.* assigns a resource or a property name to each element of $\mathbf{UL}$ in V, and such that $\cdot^{\mathcal{I}}$ is the identity for plain literals and assigns an element in Δ_R to elements in $\mathbf{L}$;

8. $\cdot^{\mathcal{I}}$ maps each variable $x \in \mathbf{B}$ into a value $x^{\mathcal{I}} \in \Delta_R$, *i.e.* assigns a resource to each variable in $\mathbf{B}$.

Note that the only difference so far relies on points 5. and 6., in which the extension function become now fuzzy membership functions. Note also that $C[\![\cdot]\!]$ (resp. $P[\![\cdot]\!]$) is a *partial* function and, thus, is not defined on all arguments. Alternatively, we may define it to be a total function. We use the former formulation to distinguish the case where a tuple **t** may be an answer to a query, even though the score is 0, from the case where a tuple is not retrieved, since it does not satisfy the query conditions. In particular, if a triple does not belong to a fuzzy graph, then its truth is assumed to be undefined, while if $C[\![\cdot]\!]$ (resp. $P[\![\cdot]\!]$) is total, then its truth of this triple would be 0, which is a small though fundamental difference. Please note that both [14,21] rely on total interpretations. We prefer the partial semantics approach as we believe it is better suited for applications, as it is more "database-like" in query answering. For instance, suppose we are looking for a second-hand car, which is cheap and not too old, where cheap and old are functions of the price and age, respectively, and the cheapness and oldness scores are aggregated via weighted linear combination. Then under total semantics one may retrieve a car with non zero score, despite its age is unknown (the tuple relating the car to its age is not in the graph and, thus, the degree of oldness is 0, but there may be a tuple dictating the price of the car), while under partial semantics, this car will not be retrieved (as it would happen for a top-k database engine or using *e.g.* SPARQL [18] in which the scoring component of the query is omitted).

The notion entailment is defined using the idea of *satisfaction* of a graph under certain interpretation. Intuitively a ground fuzzy triple $(s, p, o)[n]$ in a fuzzy RDF graph $\tilde{G}$ will be satisfied under the interpretation $\mathcal{I}$ if p is interpreted as a property name, s and o are interpreted as resources, and the interpretation of the pair (s, o) belongs to the extension of the property assigned to p to degree not less than n.

Now, let $\tilde{G}$ be a fuzzy graph over ρdf. A fuzzy interpretation $\mathcal{I}$ is a *model* of $\tilde{G}$ under ρdf, denoted $\mathcal{I} \models \tilde{G}$, iff $\mathcal{I}$ is a fuzzy interpretation over the vocabulary ρdf $\cup$ *universe*$(\tilde{G})$ that satisfies the following conditions:

Simple:

1. for each $(s, p, o)[n] \in G$, $p^{\mathcal{I}} \in \Delta_P$ and $P[\![p^{\mathcal{I}}]\!](s^{\mathcal{I}}, o^{\mathcal{I}}) \geqslant n$;

Subproperty:

1. $P[\![\mathsf{sp}^{\mathcal{I}}]\!]$ is transitive over Δ_P;
2. if $P[\![\mathsf{sp}^{\mathcal{I}}]\!](p, q)$ is defined then $p, q \in \Delta_P$ and

$$P[\![\mathsf{sp}^{\mathcal{I}}]\!](p, q) = \inf_{(x,y) \in \Delta_R \times \Delta_R} P[\![p]\!](x, y) \Rightarrow P[\![q]\!](x, y) \, ;$$

Subclass:

1. $P[\![\mathsf{sc}^{\mathcal{I}}]\!]$ is transitive over Δ_C;

2. if $P[\![sc^\mathcal{I}]\!](c,d)$ is defined then $c, d \in \Delta_C$ and

$$P[\![sc^\mathcal{I}]\!](c,d) = \inf_{x \in \Delta_R} C[\![c]\!](x) \Rightarrow C[\![d]\!](x) \ ;$$

Typing I:

1. $C[\![c]\!](x) = P[\![type^\mathcal{I}]\!](x,c)$;
2. if $P[\![dom^\mathcal{I}]\!](p,c)$ is defined then

$$P[\![dom^\mathcal{I}]\!](p,c) = \inf_{(x,y) \in \Delta_R \times \Delta_R} P[\![p]\!](x,y) \Rightarrow C[\![c]\!](x) \ ;$$

3. if $P[\![range^\mathcal{I}]\!](p,c)$ is defined then

$$P[\![range^\mathcal{I}]\!](p,c) = \inf_{(x,y) \in \Delta_R \times \Delta_R} P[\![p]\!](x,y) \Rightarrow C[\![c]\!](y) \ ;$$

Typing II:

1. For each $e \in \rho df$, $e^\mathcal{I} \in \Delta_P$
2. if $P[\![dom^\mathcal{I}]\!](p,c)$ is defined then $p \in \Delta_P$ and $c \in \Delta_C$
3. if $P[\![range^\mathcal{I}]\!](p,c)$ is defined then $p \in \Delta_P$ and $c \in \Delta_C$
4. if $P[\![type^\mathcal{I}]\!](x,c)$ is defined then $c \in \Delta_C$

Some explanations about the above definitions are in place. To do so, let us keep in mind Eq. (4). At first, let us explain condition 2 of the subclass condition. In the crisp case if c is a sub-class of d then we impose that $C[\![c]\!] \subseteq C[\![d]\!]$. The fuzzyfication of this subsumption condition yields the degree of subsumption and, thus, using Eq. (4), we get immediately

$$P[\![sc^\mathcal{I}]\!](c,d) = \inf_{x \in \Delta_R} C[\![c]\!](x) \Rightarrow C[\![d]\!](x) \ .$$

i.e., $P[\![sc^\mathcal{I}]\!](c,d)$ is evaluated as the degree of subsumption between class c and class d. In First-Order-Logic terms, we recall that c is a sub-class of d may be seen as the formula

$$\forall x.c(x) \Rightarrow d(x) \ ,$$

and, thus, as in fuzzy logic $\forall$ is the inf, we get equation above. The argument for the sub-property condition is similar. Concerning condition 2 of Typing I, we may write the condition that property p has domain c in First-Order-Logic as

$$\forall x \forall y.p(x,y) \Rightarrow c(x) \ ,$$

which then gives us immediately the condition

$$P[\![dom^\mathcal{I}]\!](p,c) = \inf_{(x,y) \in \Delta_R \times \Delta_R} P[\![p]\!](x,y) \Rightarrow C[\![c]\!](x) \ .$$

The argument for condition 3 of Typing I is similar. We define $\tilde{G}$ *entails* $\tilde{H}$ under ρdf, denoted $\tilde{G} \models \tilde{H}$, iff every fuzzy model under ρdf of $\tilde{G}$ is also a model under ρdf of $\tilde{H}$.

As for the crisp case, it can be shown that any fuzzy graph is *consistent*, *i.e.* has a model.

Proposition 2 (Consistency). *Any fuzzy RDF graph has a model.*

Therefore, unlike [21], we do not have to care about consistency checking.

Deductive system. In what follows, we present a sound and complete deductive system for our fuzzy RDF fragment. As we will see, it is an extension of the one we have seen for the crisp case. Indeed, for each crisp rule (except for group 1, which remains identical) there is a fuzzy analogue. The rules are as follows[4]:

1. Simple:

   (a) $\frac{\tilde{G}}{\tilde{G}'}$ for a map $\mu : \tilde{G}' \to \tilde{G}$ (b) $\frac{G}{G'}$ for $\tilde{G}' \subseteq \tilde{G}$

2. Subproperty:

   (a) $\dfrac{(A, \mathsf{sp}, B)[n],(B, \mathsf{sp}, C)[m]}{(A, \mathsf{sp}, C)[n \otimes m]}$ (b) $\dfrac{(A, \mathsf{sp}, B)[n],(X, A, Y)[m]}{(X, B, Y)[n \otimes m]}$

3. Subclass:

   (a) $\dfrac{(A, \mathsf{sc}, B)[n],(B, \mathsf{sc}, C)[m]}{(A, \mathsf{sc}, C)[n \otimes m]}$ (b) $\dfrac{(A, \mathsf{sc}, B)[n],(X, \mathsf{type}, A)[m]}{(X, \mathsf{type}, B)[n \otimes m]}$

4. Typing:

   (a) $\dfrac{(A, \mathsf{dom}, B)[n],(X, A, Y)[m]}{(X, \mathsf{type}, B)[n \otimes m]}$ (b) $\dfrac{(A, \mathsf{range}, B)[n],(X, A, Y)[m]}{(Y, \mathsf{type}, B)[n \otimes m]}$

5. Implicit Typing:

   (a) $\dfrac{(A, \mathsf{dom}, B)[n],(C, \mathsf{sp}, A)[m],(X, C, Y)[r]}{(X, \mathsf{type}, B)[n \otimes m \otimes r]}$

   (b) $\dfrac{(A, \mathsf{range}, B)[n],(C, \mathsf{sp}, A)[m],(X, C, Y)[r]}{(Y, \mathsf{type}, B)[n \otimes m \otimes r]}$

It suffices to explain the rules of the sub-class category, as all the rules of categories 2-5 follow the same schema. To do so, consider inference schemas (2) and (3).
   Consider the rule
$$\frac{(A, \mathsf{sc}, B)[n], (B, \mathsf{sc}, C)[m]}{(A, \mathsf{sc}, C)[n \otimes m]} .$$

Let us show that the rule is sound, *i.e.* for a fuzzy interpretation $\mathcal{I}$, if $\mathcal{I} \models (A, \mathsf{sc}, B)[n]$ and $\mathcal{I} \models (B, \mathsf{sc}, C)[m]$ then $\mathcal{I} \models (A, \mathsf{sc}, C)[n \otimes m]$. Indeed,

1. As $P[\![\mathsf{sc}^{\mathcal{I}}]\!]$ is transitive over Δ_C, we have that

$$P[\![\mathsf{sc}^{\mathcal{I}}]\!](A^{\mathcal{I}}, C^{\mathcal{I}}) \geqslant P[\![\mathsf{sc}^{\mathcal{I}}]\!](A^{\mathcal{I}}, B^{\mathcal{I}}) \otimes P[\![\mathsf{sc}^{\mathcal{I}}]\!](B^{\mathcal{I}}, C^{\mathcal{I}}) .$$

2. As $\mathcal{I} \models (A, \mathsf{sc}, B)[n]$, it follows that

$$P[\![\mathsf{sc}^{\mathcal{I}}]\!](A^{\mathcal{I}}, B^{\mathcal{I}}) = \inf_{x \in \Delta_R} C[\![A^{\mathcal{I}}]\!](x) \Rightarrow C[\![B^{\mathcal{I}}]\!](x) \geqslant n ;$$

3. As $\mathcal{I} \models (B, \mathsf{sc}, C)[m]$, it follows that

$$P[\![\mathsf{sc}^{\mathcal{I}}]\!](B^{\mathcal{I}}, C^{\mathcal{I}}) = \inf_{x \in \Delta_R} C[\![B^{\mathcal{I}}]\!](x) \Rightarrow C[\![C^{\mathcal{I}}]\!](x) \geqslant m ;$$

---

[4] An excerpt of them has been provided in [19].

4. From 1-3, it follows immediately that

$$P[\![sc^{\mathcal{I}}]\!](A^{\mathcal{I}}, C^{\mathcal{I}}) \geqslant n \otimes m$$

and, thus $\mathcal{I} \models (A, sc, C)[n \otimes m]$.

Next, let us show that

$$\frac{(A, sc, B)[n], (X, type, A)[m]}{(X, type, B)[n \otimes m]}$$

is correct.

1. As $\mathcal{I} \models (A, sc, B)[n]$, it follows that

$$P[\![sc^{\mathcal{I}}]\!](A^{\mathcal{I}}, B^{\mathcal{I}}) = \inf_{x \in \Delta_R} C[\![A^{\mathcal{I}}]\!](x) \Rightarrow C[\![B^{\mathcal{I}}]\!](x) \geqslant n \; ;$$

2. As $\mathcal{I} \models (X, type, A)[m]$, it follows that

$$P[\![type^{\mathcal{I}}]\!](X^{\mathcal{I}}, A^{\mathcal{I}}) \geqslant m \; ;$$

3. But $C[\![A^{\mathcal{I}}]\!](X^{\mathcal{I}}) = P[\![type^{\mathcal{I}}]\!](X^{\mathcal{I}}, A^{\mathcal{I}})$ and, thus, from 2. we get

$$C[\![A^{\mathcal{I}}]\!](X^{\mathcal{I}}) \geqslant m \; .$$

4. Consider $X^{\mathcal{I}} \in \Delta_R$. As 1. holds for all $x \in \Delta_R$, we have that

$$C[\![A^{\mathcal{I}}]\!](X^{\mathcal{I}}) \Rightarrow C[\![B^{\mathcal{I}}]\!](x) \geqslant n \; .$$

5. By schema (2) and point 3. and 4., we get

$$C[\![B^{\mathcal{I}}]\!]((X^{\mathcal{I}}) \geqslant n \otimes m \; .$$

6. But, $P[\![type^{\mathcal{I}}]\!](X^{\mathcal{I}}, B^{\mathcal{I}}) = C[\![B^{\mathcal{I}}]\!](X^{\mathcal{I}})$ and, thus, by 5. we get

$$P[\![type^{\mathcal{I}}]\!](X^{\mathcal{I}}, B^{\mathcal{I}}) \geqslant n \otimes m$$

and, thus $\mathcal{I} \models (X, type, B)[n \otimes m]$.

The notion of proof is as for the crisp case and we have:

Proposition 3 (Soundness and completeness). *For fuzzy RDF, the proof system $\vdash$ is sound and complete for $\models$, that is, $\tilde{G} \vdash \tilde{H}$ iff $\tilde{G} \models \tilde{H}$.*

Query Answering. We extend the notion of conjunctive query to the case in which a scoring function can be specified to score the answers similarly as in [11] (see also [20]).

For the sake of our purpose, we assume that a fuzzy RDF graph $\tilde{G}$ is *ground* and *closed, i.e.,* $\tilde{G}$ is closed under the application of the rules (2)-(5). Then a *fuzzy conjunctive query* extends a crisp query and is of the form

$$q(\mathbf{x})[s] \leftarrow \exists \mathbf{y}. \tau_1[s_1], \ldots, \tau_n[s_n], s = f(s_1, \ldots, s_n, p_1(\mathbf{z}_1), \ldots, p_h(\mathbf{z}_h))$$

where additionally

1. $\mathbf{z_i}$ are tuples of terms in **UL** or variables in **x** or **y**;

2. p_j is an n_j-ary *fuzzy predicate* assigning to each n_j-ary tuple $\mathbf{t}_j$ in $\mathbf{UL}$ a *score* $p_j(\mathbf{t}_j) \in [0,1]_m$. Such predicates are called *expensive predicates* in [2] as the score is not pre-computed off-line, but is computed on query execution. We require that an n-ary fuzzy predicate p is *safe*, that is, there is not an m-ary fuzzy predicate p' such that $m < n$ and $p = p'$. Informally, all parameters are needed in the definition of p;

3. f is a *scoring* function $f \colon ([0,1])^{n+h} \to [0,1]$, which combines the scores s_i of the n triples and the h fuzzy predicates into an overall *score* to be assigned to the rule head. We assume that f is *monotone*, that is, for each $\mathbf{v}, \mathbf{v}' \in ([0,1])^{n+h}$ such that $\mathbf{v} \leqslant \mathbf{v}'$, it holds $f(\mathbf{v}) \leqslant f(\mathbf{v}')$, where $(v_1, \dots, v_{n+h}) \leqslant (v'_1, \dots, v'_{n+h})$ iff $v_i \leqslant v'_i$ for all i;

4. the scoring variables s and s_i are distinct from those in $\mathbf{x}$ and $\mathbf{y}$ and s is distinct from each s_i.

We may omit s_i and in that case $s_i = 1$ is assumed. $s = f(s_1, \dots, s_n, p_1(\mathbf{z}_1), \dots, p_h(\mathbf{z}_h))$ is called the *scoring atom*. We may also omit the scoring atom and in that case $s = 1$ is assumed. For instance, the query

$$q(x)[s] \leftarrow (x, \mathsf{type}, SportsCar)[s_1], (x, hasPrice, y), s = s_1 \cdot cheap(y)$$

where *e.g.* $cheap(p) = \max(0, 1 - \frac{p}{12000})$, has intended meaning to retrieve all cheap sports car. Any answer is scored according to the product of being cheap and a sports car.

The notion of disjunctive query is as for the crisp case. We will also write a query as

$$q(\mathbf{x})[s] \leftarrow \exists \mathbf{y}.\varphi(\mathbf{x}, \mathbf{y})[s] \ ,$$

where $\varphi(\mathbf{x}, \mathbf{y})$ is $\tau_1[s_1], \dots, \tau_n[s_n]$, $s = f(\mathbf{s}, p_1(\mathbf{z}_1), \dots, p_h(\mathbf{z}_h))$ and $\mathbf{s} = \langle s_1, \dots, s_n \rangle$.

Consider a fuzzy graph $\tilde{G}$, a query $q(\mathbf{x})[s] \leftarrow \exists \mathbf{y}.\varphi(\mathbf{x}, \mathbf{y})[s]$, a vector $\mathbf{t}$ of terms in $\mathbf{UL}$ and $s \in [0,1]$. We say that $q(\mathbf{t})[s]$ is *entailed* by $\tilde{G}$, denoted $\tilde{G} \models q(\mathbf{t})[s]$, iff in any model $\mathcal{I}$ of $\tilde{G}$, there is a vector $\mathbf{t}'$ of terms in $\mathbf{UL}$, a vector $\mathbf{s}$ of scores in $[0,1]$ such that $\mathcal{I}$ is a model of $\varphi(\mathbf{t}, \mathbf{t}')[s]$ (the scoring atom is satisfied iff s is the value of the evaluation of the score combination function). The definition is extended to a disjunctive query as for the crisp case.

We say that s is *tight* iff $s = \sup\{s' \mid \tilde{G} \models q(\mathbf{t})[s']\}$. If $\tilde{G} \models q(\mathbf{t})[s]$ and s is tight then $\mathbf{t}[s]$ is called an *answer* to $\mathbf{q}$ w.r.t. $\tilde{G}$. The *answer set*, $ans(\tilde{G}, \mathbf{q})$ of $\mathbf{q}$ w.r.t. $\tilde{G}$, is defined as the set of answers to $\mathbf{q}$ w.r.t. $\tilde{G}$.

As now each answer to a query has a degree of truth (*i.e.* score), the basic inference problem that is of interest in is the top-k retrieval problem, formulated as follows.

Top-k Retrieval. Given a fuzzy graph $\tilde{G}$, and a disjunctive query $\mathbf{q}$, retrieve k answers $\mathbf{t}[s]$ with maximal scores and rank them in decreasing order relative to the score s, denoted

$$ans_k(\tilde{G}, \mathbf{q}) = Top_k \ ans(\tilde{G}, \mathbf{q}) \ .$$

Next, we describe a method to determine $ans_k(\tilde{G}, \mathbf{q})$. So, let $\tilde{G}$ be a ground fuzzy graph. similarly to the crisp case, the *closure* of $\tilde{G}$ is defined as

$$cl(\tilde{G}) = \{\tau[n] \mid \tau \text{ ground}, \tilde{G} \vdash \tau[n] \text{ and } n \text{ is tight}\}^5.$$

---

[5] Note that rule (Simple a) is not required to compute the closure.

Note that by definition of $cl(\tilde{G})$, there cannot bet two fuzzy triples $\tau[n]$ and $\tau[m]$ in $cl(\tilde{G})$ such that $n < m$. The closure $cl(\tilde{G})$ can be computed by repeatedly applying the fuzzy inference rules together with the *redundancy elimination rule* below:

– Redundancy Elimination Rule (RER):

$$\frac{\tau[n], \ \tau[m]}{\text{remove } \tau[n]} \text{ if } n < m$$

Essentially, each time we generate a tuple τ, we keep the one involving τ with highest degree. This rule is necessary in order to guarantee the termination of the closure computation in case of cyclic graphs, such as *e.g.*

$$(A, \mathsf{sc}, B)[n], \ \ (B, \mathsf{sc}, C)[n], \ \ (C, \mathsf{sc}, A)[n]$$

where the t-norm is *e.g.* product. Without (RER), we may generate an infinite sequence of fuzzy triples $(A, \mathsf{sc}, A)[n^{3k}]$ ($k = 1, 2, \ldots$) and, thus, do not terminate. Please note that with (RER), only $(X, \mathsf{sc}, X)[n^3] \in cl(\tilde{G})$, where $X \in \{A, B, C\}$.

Now, under the above closure computation, we have, as for the crisp case [15]:

Proposition 4 (Size of Closure)

1. *The size of the closure of $\tilde{G}$ is $\mathcal{O}(|\tilde{G}|^2)$.*
2. *The size of the closure of $\tilde{G}$ is in the worst case no smaller than $\Omega(|\tilde{G}|^2)$.*

Therefore, a method to determine $ans_k(\tilde{G}, \mathbf{q})$ is as follows.

1. Compute the closure $cl(\tilde{G})$ of $\tilde{G}$ and store it into a database that supports top-k retrieval (*e.g.*, RankSQL [10][6]).
2. It can be verified that any fuzzy disjunctive query can be mapped into union of top-k SQL queries [10] over the underlying database schema.
3. Hence, $ans_k(\tilde{G}, \mathbf{q})$ is determined by issuing these top-k SQL queries to the database.

4 Summary and Outlook

We have presented fuzzy RDF under a generalized semantics based on t-norms and its r-implication. We provided a minimal deductive system, top-k fuzzy disjunctive queries and showed how these can be answered by relying on the closure computation and state of the art top-k database engines. An implementation is under development, where fuzzy triples are stored as RDF triples using reification and, thus, no change to RDF is required. We follow the method [1] to store the closure in the database.

Concerning future research: so far, we considered the closure of the graph (of quadratic size) to be stored into a database and then we submit top-k SQL queries to it. While one may think of extending the method described in [15] to check entailment

---

[6] But, *e.g.*, Postgres http://www.postgresql.org/,
MonetDB http://monetdb.cwi.nl/
may work as well.

for ground fuzzy tuples in $\mathcal{O}(|\tilde{G}| \log |\tilde{G}|)$, it remains to see whether similar methods exist to determine the top-k answers. To this end we are looking to the techniques developed for top-k query answering in fuzzy logic programming (see, *e.g.* [11,20]).

Another topic concerns the extension and mapping of SPARQL to fuzzy disjunctive queries.

References

1. Abadi, D.J., Marcus, A., Madden, S., Hollenbach, K.: Sw-store: a vertically partitioned DBMS for semantic web data management. VLDB J. 18(2), 385–406 (2009)
2. Chang, K.C.-C., won Hwang, S.: Minimal probing: Supporting expensive predicates for top-k queries. In: SIGMOD Conference, pp. 346–357 (2002)
3. Fitting, M.C.: Bilattices are nice things. In: Conference on Self-Reference, Copenhagen, Denmark (2002)
4. Fitting, M.C.: Fixpoint semantics for logic programming - a survey. Theoretical Computer Science 21(3), 25–51 (2002)
5. Ginsberg, M.L.: Multi-valued logics: a uniform approach to reasoning in artificial intelligence. Computational Intelligence 4, 265–316 (1988)
6. Gutierrez, C., Hurtado, C., Mendelzon, A.O.: Foundations of semantic web databases. In: Proceedings of the 23rd ACM SIGMOD-SIGACT-SIGART symposium on Principles of database systems (PODS 2004). ACM Press, New York (2004)
7. Hähnle, R.: Many-valued logics and mixed integer programming. Annals of Mathematics and Artificial Intelligence 3, 4(12), 231–264 (1994)
8. Hähnle, R.: Advanced many-valued logics. In: Gabbay, D.M., Guenthner, F. (eds.) Handbook of Philosophical Logic, 2nd edn., vol. 2. Kluwer, Dordrecht (2001)
9. Hájek, P.: Metamathematics of Fuzzy Logic. Kluwer, Dordrecht (1998)
10. Li, C., Chang, K.C.-C., Ilyas, I.F., Song, S.: RankSQL: query algebra and optimization for relational top-k queries. In: Proceedings of the 2005 ACM SIGMOD International Conference on Management of Data (SIGMOD 2005), pp. 131–142. ACM Press, New York (2005)
11. Lukasiewicz, T., Straccia, U.: Top-k retrieval in description logic programs under vagueness for the semantic web. In: Prade, H., Subrahmanian, V.S. (eds.) SUM 2007. LNCS (LNAI), vol. 4772, pp. 16–30. Springer, Heidelberg (2007)
12. Mazzieri, M.: A fuzzy RDF semantics to represent trust metadata. In: Proceedings of the 1st Italian Semantic Web Workshop: Semantic Web Applications and Perspectives, SWAP 2004 (2004)
13. Mazzieri, M., Dragoni, A.F.: A fuzzy semantics for semantic web languages. In: Proceedings of the ISWC Workshop on Uncertainty Reasoning for the Semantic Web (URSW 2005). CEUR Workshop Proceedings (2005)
14. Mazzieri, M., Dragoni, A.F.: A fuzzy semantics for the resource description framework. In: da Costa, P.C.G., d'Amato, C., Fanizzi, N., Laskey, K.B., Laskey, K.J., Lukasiewicz, T., Nickles, M., Pool, M. (eds.) URSW 2005 - 2007. LNCS (LNAI), vol. 5327, pp. 244–261. Springer, Heidelberg (2008)
15. Muñoz, S., Pérez, J., Gutierrez, C.: Minimal deductive systems for RDF. In: Franconi, E., Kifer, M., May, W. (eds.) ESWC 2007. LNCS, vol. 4519, pp. 53–67. Springer, Heidelberg (2007)
16. Novák, V.: Which logic is the real fuzzy logic? Fuzzy Sets and Systems, 635–641 (2005)
17. RDF, http://www.w3.org/RDF/
18. SPARQL, http://www.w3.org/TR/rdf-sparql-query/

19. Straccia, U.: Basic concepts and techniques for managing uncertainty and vagueness in semantic web languages. In: Reasoning Web, 4th International Summer School (2007) (invited Lecture)
20. Straccia, U.: Managing uncertainty and vagueness in description logics, logic programs and description logic programs. In: Baroglio, C., Bonatti, P.A., Małuszyński, J., Marchiori, M., Polleres, A., Schaffert, S. (eds.) Reasoning Web. LNCS, vol. 5224, pp. 54–103. Springer, Heidelberg (2008)
21. Udrea, O., Recupero, D.R., Subrahmanian, V.S.: Annotated RDF. In: Sure, Y., Domingue, J. (eds.) ESWC 2006. LNCS, vol. 4011, pp. 487–501. Springer, Heidelberg (2006)
22. Vaneková, V., Bella, J., Gurský, P., Horváth, T.: Fuzzy rdf in the semantic web: Deduction and induction. In: Proceedings of Workshop on Data Analysis, WDA 2005 (2005)

An Efficient Method for Computing Alignment Diagnoses

Christian Meilicke and Heiner Stuckenschmidt

Computer Science Institute, University of Mannheim,
B6,26 68159 Mannheim, Germany

Abstract. Formal, logic-based semantics have long been neglected in ontology matching. As a result, almost all matching systems produce incoherent alignments of ontologies. In this paper we propose a new method for repairing such incoherent alignments that extends previous work on this subject. We describe our approach within the theory of diagnosis and introduce the notion of a local optimal diagnosis. We argue that computing a local optimal diagnosis is a reasonable choice for resolving alignment incoherence and suggest an efficient algorithm. This algorithm partially exploits incomplete reasoning techniques to increase runtime performance. Nevertheless, the completeness and optimality of the solution is still preserved. Finally, we test our approach in an experimental study and discuss results with respect to runtime and diagnostic quality[1].

1 Introduction

It has widely been acknowledged that logical semantics and reasoning are the basis of intelligent applications on the semantic web. This is underlined by the design of standard languages, like the Web Ontology Language (OWL), which have a clearly defined logical semantics. Contrary to this, in the area of ontology matching the use of logical semantics as a guiding principle has long been neglected. Existing matching systems are primarily based on lexical and heuristic methods [2] that often result in alignments that contain logical contradictions. At first glimpse some systems seem to be an exception, for example ASMOV and S-Match. ASMOV [5] has become a successful participant of the OAEI over the last years. One of its constituents is a semantic verification component used to filter out conflicting correspondence. In particular, a comprehensive set of pattern is applied to detect certain kind of conflicts. However, ASMOV lacks a well defined alignment semantics and notions as correctness or completeness are thus not applicable. The S-Match system [4], on the contrary, employs sound and complete reasoning procedures. Nevertheless, the underlying semantics is restricted to propositional logic due to the fact that ontologies are interpreted as tree-like structures. S-Match can thus not guarantee to generate a coherent alignment between expressive OWL-ontologies. We have already argued that the problem of generating coherent alignments can best be solved by applying principles of diagnostic reasoning [11]. In this paper, we extend previous work on this topic in different directions.

---

[1] An extended version of this paper is available as technical report at
http://webrum.uni-mannheim.de/math/lski/matching/lod/

A. Polleres and T. Swift (Eds.): RR 2009, LNCS 5837, pp. 182–196, 2009.
© Springer-Verlag Berlin Heidelberg 2009

- We define the general notion of a reductionistic alignment semantics and introduce a natural interpretation as concrete specification. Contrary to previous work, we support different alignment semantics within our framework.
- As extension of our previous work we do not only cover concept correspondences but additionally support correspondences between properties.
- We describe the problem of repairing incoherent alignments in terms of Reiters theory of diagnosis [14] and introduce the notion of a local optimal diagnosis.
- We present an algorithm for constructing a local optimal diagnosis - based on the algorithm described in [12] - and show how this algorithm can be enhanced by partially exploiting efficient but incomplete reasoning methods.
- We report on several experiments concerned with both the diagnostic quality as well as the runtime of both algorithms.

In Section 2 we define our terminology and introduce some definitions centered around the the notion of alignment incoherence. In Section 3 we argue that repairing an incoherent alignment can be understood as diagnosis task. In particular, we introduce the notion of a local optimal diagnosis. In Section 4 we briefly introduce different reasoning techniques and algorithms exploiting these reasoning techniques in order to compute a local optimal diagnosis. These algorithms are applied on different datasets in Section 5 where we also discuss the results and compare them against other approaches. In Section 6 we end with a short summary and some concluding remarks.

2   Preliminaries

The task of aligning two ontologies $\mathcal{O}_1$ and $\mathcal{O}_2$ (sets of axioms) can be understood as detecting links between elements of $\mathcal{O}_1$ and $\mathcal{O}_2$. These links are referred to as correspondences and express a semantic relation. According to Euzenat and Shvaiko [2] we define a correspondence as follows and introduce an alignment as set of correspondences.

Definition 1 (Correspondence and Alignment). *Given ontologies $\mathcal{O}_1$ and $\mathcal{O}_2$, let Q be a function that defines sets of matchable elements $Q(\mathcal{O}_1)$ and $Q(\mathcal{O}_2)$. A correspondence between $\mathcal{O}_1$ and $\mathcal{O}_2$ is a 4-tuple $\langle e, e', r, n \rangle$ such that $e \in Q(\mathcal{O}_1)$ and $e' \in Q(\mathcal{O}_2)$, r is a semantic relation, and $n \in [0, 1]$ is a confidence value. An alignment $\mathcal{A}$ between $\mathcal{O}_1$ and $\mathcal{O}_2$ is a set of correspondences between $\mathcal{O}_1$ and $\mathcal{O}_2$.*

Our approach is applicable to alignments between ontologies represented in Description Logics, e.g. to alignments between OWL-DL ontologies. In this work the matchable elements $Q(\mathcal{O})$ are restricted to be atomic concepts or atomic properties. Further r is a semantic relation expressing equivalence or subsumption. We use the symbols $\overset{\leftrightarrow}{\equiv}$, $\overset{\leftrightarrow}{\sqsubseteq}$ and $\overset{\leftrightarrow}{\sqsupseteq}$ to refer to these relations. The semantics of these symbols has not yet been specified, although we might have a rough idea about their interpretation. The confidence value n describes the trust in the correctness of a correspondence. Given a correspondence c, we use $conf(c) = n$ to refer to the confidence of c. Additionally, we require that in an alignment $\mathcal{A}$ there exist no $c \neq c' \in \mathcal{A}$ such that $conf(c) = conf(c')$. We know that most matching systems will not fullfill this requirement. Another source of evidence has to decide which correspondence should be annoted with higher confidence. Thus, we avoid an explicit treatment of different total orderings derivable from the partial order of confidence values[2]. In the following we frequently need to talk about concepts or properties of an ontology $\mathcal{O}_i$. We use prefix notation $i\#e$ to uniquely determine that an entity e belongs to the signature of $\mathcal{O}_i$.

A concept $i\#C$ is defined to be unsatisfiable iff all models of $\mathcal{O}_i$ interpret $i\#C$ as empty set. We use the notion of unsatisfiability in a wider sense and define it with respect to both concepts and properties.

Definition 2 (Unsatisfiability). *A concept or property $i\#e$ is unsatisfiable in ontology $\mathcal{O}_i$, iff for all models $\mathcal{I}$ of $\mathcal{O}_i$ we have $i\#e^{\mathcal{I}} = \emptyset$. Otherwise $i\#e$ is satisfiable in $\mathcal{O}_i$.*

Usually, an ontology is referred to as incoherent whenever it contains an atomic unsatisfiable concept. We define ontology incoherence as follows.

Definition 3 (Ontology Incoherence). *An ontology $\mathcal{O}$ is incoherent iff there exists an atomic unsatisfiable concept or property in $\mathcal{O}$. Otherwise $\mathcal{O}$ is coherent.*

There are two ways to introduce the notion of alignment incoherence. The first approach requires a *specific model-theoretic alignment semantics*. Distributed Description Logics (DDL)[1] is an example for such a specific semantics, which we focused on in previous work [11]. The second approach, already sketched in [7], is based on *interpreting an alignment as a set of axioms X in a merged ontology*. Given an alignment $\mathcal{A}$ between $\mathcal{O}_1$ and $\mathcal{O}_2$, the (in)coherence of $\mathcal{A}$ is reduced to the (in)coherence $\mathcal{O}_1 \cup \mathcal{O}_2 \cup X$. We refer to such a semantics as reductionistic alignment semantics.

Definition 4 (Reductionistic Semantics). *Given an alignment $\mathcal{A}$ between ontologies $\mathcal{O}_1$ and $\mathcal{O}_2$. A reductionistic alignment semantics $\mathcal{S} = \langle ext, trans \rangle$ is a pair of functions where ext maps an ontology to a set of axioms (extension function) and trans maps an alignment to a set of axioms (translation function).*

Considering its role in the context of a merged ontology, it becomes clear how to apply such a reductionistic alignment semantics, abbreviated as alignment semantics in the following.

Definition 5 (Merged ontology). *Given an alignment $\mathcal{A}$ between ontologies $\mathcal{O}_1$ and $\mathcal{O}_2$ and an alignment semantics $\mathcal{S} = \langle ext, trans \rangle$. The merged ontology is defined as $\mathcal{O}_1 \cup_{\mathcal{A}}^{\mathcal{S}} \mathcal{O}_2 = \mathcal{O}_1 \cup \mathcal{O}_2 \cup ext(\mathcal{O}_1) \cup ext(\mathcal{O}_2) \cup trans(\mathcal{A})$.*

The merged ontology is merely a technical means to treat different semantics within a similar framework. Based on this framework we apply the definition of ontology incoherence in the context of a merged ontology resulting in the notion of alignment incoherence.

Definition 6 (Alignment Incoherence). *Given an alignment $\mathcal{A}$ between ontologies $\mathcal{O}_1$ and $\mathcal{O}_2$ and an alignment semantics $\mathcal{S}$. $\mathcal{A}$ is incoherent with respect to $\mathcal{O}_1$ and $\mathcal{O}_2$ according to $\mathcal{S}$, iff there exists an atomic concept or property $i\#C$ with $i \in \{1,2\}$ that is satisfiable in $\mathcal{O}_i$ and unsatisfiable in $\mathcal{O}_1 \cup_{\mathcal{A}}^{\mathcal{S}} \mathcal{O}_2$. Otherwise $\mathcal{A}$ is coherent.*

---

[2] For the experiments reported on in Section 5 we derived a total order - given correspondences with the same confidence value - from the lexicographical ordering of the URIs of the matched entities. Experiments with different orderings resulted in insignificant differences.

We now introduce an example of a reductionistic alignment semantics, primarily defined in [7] and [8] with respect to a less general framework.

Definition 7 (Natural Semantics). *Given an alignment $\mathcal{A}$ and an ontology $\mathcal{O}$. The natural semantics $\mathcal{S}_n = \langle ext_n, trans_n \rangle$ is defined by a specification of its components $ext_n(\mathcal{O}) \mapsto \emptyset$ and $trans_n(\mathcal{A}) \mapsto \{t_n(c) | c \in \mathcal{A}\}$ where t_n is defined as*

$$t_n(c) \mapsto \begin{cases} 1\#e \equiv 2\#e' & \text{if } r = \overset{\leftrightarrow}{\equiv} \\ 1\#e \sqsubseteq 2\#e' & \text{if } r = \overset{\leftrightarrow}{\sqsubseteq} \\ 1\#e \sqsupseteq 2\#e' & \text{if } r = \overset{\leftrightarrow}{\sqsupseteq} \end{cases}$$

The natural alignment semantics consists of an empty extension function *ext* and a translation function *trans* that maps correspondences one-to-one to axioms. It can be seen as self-evident and straightforward way to interpret correspondences as axioms.

An example for an alignment semantics with $ext(\mathcal{O}) \neq \emptyset$ is given by DDL. DDL is a formalism for supporting distributed reasoning based on a semantics where each ontology is interpreted within its own domain interrelated via bridge rules. Nevertheless, it is also possible to reduce DDL to ordinary DL [1]. As a result we obtain a reductionistic alignment semantics where the extension function maps $\mathcal{O}_1$ and $\mathcal{O}_2$ to a non empty set of additional axioms while the translation function differs significantly from the translation function of the natural semantics.

3   Problem Statement

In this section we show that the problem of debugging alignments can be understood as diagnostic problem and characterize a certain type of diagnosis. Throughout the remaining parts we use $\mathcal{A}$ to refer to an alignment, we use $\mathcal{O}$ with or without subscript to refer to an ontology, and $\mathcal{S}$ to refer to some reductionistic alignment semantics.

In ontology debugging a minimal incoherency preserving sub-TBox (MIPS) $\mathcal{M} \subseteq \mathcal{O}$ is an incoherent set of axioms while any proper subset $\mathcal{M}' \subset \mathcal{M}$ is coherent [15]. The same notion can be applied to the field of alignment debugging where we have to consider sets of correspondences instead of axioms.

Definition 8 (MIPS Alignment). *$\mathcal{M} \subseteq \mathcal{A}$ is a minimal incoherence preserving sub-alignment (MIPS alignment), iff $\mathcal{M}$ is incoherent with respect to $\mathcal{O}_1$ and $\mathcal{O}_2$ and there exists no $\mathcal{M}' \subset \mathcal{M}$ such that $\mathcal{M}'$ is coherent with respect to $\mathcal{O}_1$ and $\mathcal{O}_2$. The collection of all MIPS alignments is referred to as $MIPS_\mathcal{S}(\mathcal{A}, \mathcal{O}_1, \mathcal{O}_2)$.*

As already indicated in [11], the problem of debugging an incoherent alignment can be understood in terms of Reiters theory of diagnosis [14]. Reiter describes a diagnostic problem in terms of a system and its components. The need for a diagnosis arises, when the observed system behavior differs from the expected behaviour. According to Reiter, the diagnostic problem is to determine a set of those system components which, when assumed to be functioning abnormally, explain the discrepancy between observed and correct behaviour. If this set of components is minimal, it is referred to as diagnosis Δ. In our context a system is a tuple $\langle \mathcal{A}, \mathcal{O}_1, \mathcal{O}_2, \mathcal{S} \rangle$. The discrepancies between observed and correct behaviour are the terminological entities that were satisfiable in $\mathcal{O}_1$ and $\mathcal{O}_2$ and have become unsatisfiable in $\mathcal{O}_1 \cup_{\mathcal{A}}^{S} \mathcal{O}_2$. The components of the system are the axioms of $\mathcal{O}_1$ and $\mathcal{O}_2$ as well as the correspondences of $\mathcal{A}$. Nevertheless, with respect to alignment debugging the set of possibly erroneous components is restricted to the correspondences of $\mathcal{A}$. We conclude, that an alignment diagnosis should be defined as a minimal set $\Delta \subseteq \mathcal{A}$ such that $\mathcal{A} \setminus \Delta$ is coherent.

Definition 9 (Alignment Diagnosis). $\Delta \subseteq \mathcal{A}$ *is a diagnosis for* $\mathcal{A}$ *with respect to* $\mathcal{O}_1$ *and* $\mathcal{O}_2$ *iff* $\mathcal{A} \setminus \Delta$ *is coherent with respect to* $\mathcal{O}_1$ *and* $\mathcal{O}_2$ *and for each* $\Delta' \subset \Delta$ *the alignment* $\mathcal{A} \setminus \Delta'$ *is incoherent with respect to* $\mathcal{O}_1$ *and* $\mathcal{O}_2$.

Reiter argues that a diagnosis is a minimal hitting set over the set of all minimal conflict sets. A minimal conflict set in the general theory of diagnosis is equivalent to a MIPS in the context of diagnosing ontology alignments. A diagnosis for an incoherent alignment $\mathcal{A}$ is thus a minimal hitting set for $MIPS_S(\mathcal{A}, \mathcal{O}_1, \mathcal{O}_2)$.

Proposition 1 (Diagnosis and Minimal Hitting Set). *Given an alignment* $\mathcal{A}$ *between ontologies* $\mathcal{O}_1$ *and* $\mathcal{O}_2$. $\Delta \subseteq \mathcal{A}$ *is a diagnosis for* $\mathcal{A}$ *with respect to* $\mathcal{O}_1$ *and* $\mathcal{O}_2$, *iff* Δ *is a minimal hitting set for* $MIPS_S(\mathcal{A}, \mathcal{O}_1, \mathcal{O}_2)$.

Proposition 1 is a special case of corollary 4.5 in [14] where an accordant proof is given. In general there exist many different diagnosis for an incoherent alignment. Reiter proposes the hitting set tree algorithm for enumerating all minimal hitting sets. With respect to our problem we will not be able to compute a complete hitting set tree for large matching problems. Instead of that we focus on a specific type of diagnosis explained by discussing the example alignments $\mathcal{A}_I$ to $\mathcal{A}_{IV}$ depicted in Figure 1.

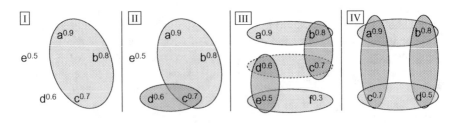

Fig. 1. Four examples for an alignment and its MIPS alignments. Correspondences are denoted by letters $a, b, \ldots$, their confidence values are specified in upper script.

$\mathcal{A}_I$ is an alignment that contains only one MIPS $\mathcal{M} = \{a, b, c\}$. Thus, there are exactly three diagnosis $\{a\}$, $\{b\}$ and $\{c\}$. Taking the confidence values into account, the most reasonable choice for fixing the incoherence is obviously the removal of the 'weakest correspondence' in $\mathcal{M}$, namely $argmin_{x \in \mathcal{M}} conf(x)$. Therefore, we prefer $\Delta = \{c\}$ as diagnosis. Does the naive strategy to remove the correspondence with lowest confidence from each MIPS always result in a diagnosis? $\mathcal{A}_{II}$ disproves this assumption. Following the naive approach we would remove both c and d, although, it is sufficient to remove c. The following recursive definition introduces the notion of an accused correspondence to cope with this problem.

Definition 10 (Accused Correspondence). *A correspondence $c \in \mathcal{A}$ is accused by $\mathcal{A}$ with respect to $\mathcal{O}_1$ and $\mathcal{O}_2$, iff there exists some $\mathcal{M} \in MIPS_{\mathcal{S}}(\mathcal{A}, \mathcal{O}_2, \mathcal{O}_2)$ with $c \in \mathcal{M}$ such that for all $c' \in \mathcal{M} \setminus \{c\}$ it holds that (1) conf(c') $>$ conf(c) and (2) c' is not accused by $\mathcal{A}$ with respect to $\mathcal{O}_1$ and $\mathcal{O}_2$.*

We have chosen the term 'accused correspondence' because the correspondence with lowest confidence in a MIPS alignment $\mathcal{M}$ is 'accused' to cause the problem. This charge will be rebuted if one of the other correspondences in $\mathcal{M}$ is already accused due to the existence of another MIPS alignment. We can apply this definition on the example alignment $\mathcal{A}_{II}$. Correspondence c is an accused correspondence, while correspondence d is not accused due to condidtion (2) in Definition 10. Obviously, the removal of the accused correspondence seems to be the most reasonable decision. In particular, it can be shown by induction that the set of accused correspondences is a diagnosis. Due to the lack of space we have to refer the reader to [9] where an accordant proof is given.

Proposition 2. *The alignment $\Delta \subseteq \mathcal{A}$ which consists of all correspondences accused by $\mathcal{A}$ with respect to $\mathcal{O}_1$ and $\mathcal{O}_2$ is a diagnosis for $\mathcal{A}$ with respect to $\mathcal{O}_1$ and $\mathcal{O}_2$.*

The set of accused correspondences is defined in a way where the whole collection $MIPS_{\mathcal{S}}(\mathcal{A}, \mathcal{O}_1, \mathcal{O}_2)$ is not taken into account from a global point of view. At the same time each removal decision seems to be the optimal choice with respect to the MIPS under discussion. Therefore, it is referred to as local optimal diagnosis in the following.

Definition 11 (Local Optimal Diagnosis). *A diagnosis Δ such that all $c \in \Delta$ are accused by $\mathcal{A}$ with respect to $\mathcal{O}_1$ and $\mathcal{O}_2$ is referred to as local optimal diagnosis.*

For the third alignment depicted in Figure 1 the set $\Delta = \{b, d, f\}$ is a local optimal diagnosis. The effects of a local removal decision can have strong effects on the whole diagnosis. One of the MIPS of $\mathcal{A}'$ is depicted with dashed lines. Suppose that we would not know this MIPS. As a result we would compute $\Delta = \{b, e\}$ as diagnosis. This small example indicates that each decision might have effects on a chain of consequent decisions. Thus, we need to construct an algorithm that is complete with respect to the detection of incoherence, because missing out a reason for incoherence might have significant effects on the whole diagnosis.

  We discussed examples where the removal of the accused correspondences is a reasonable choice, nevertheless, it is disputable whether a local optimal diagnosis is the best choice among all diagnosis. Instead of comparing confidences within a MIPS, it is e.g. also possible to aggregate (e.g. sum up) the confidences of Δ as proposed in [7]. In our framework we would refer to such a diagnosis as a global optimal diagnosis. The fourth alignment $\mathcal{A}_{IV}$ is an example where local optimal diagnosis Δ_L and global optimal diagnosis Δ_G differ, in particular we have $\Delta_L = \{b, c\}$ and global optimal diagnosis $\Delta_G = \{a, d\}$. We will see in Section 4 that a local optimal diagnosis can be computed in polynomial time (leaving aside the complexity of the reasoning involved). Opposed to this, we have to solve the weighted variant of the hitting set problem to construct a global optimal solution, which is known to be a NP-complete problem [3]. The experimental results presented in Section 5 will also show that the removal of a local optimal diagnosis has positive effects on the quality of the alignment.

4 Algorithms

A straightforward way to check the coherence of an alignment can be described as follows. We have to iterate over the atomic entities $i\#e_{i\in\{1,2\}}$ of both $\mathcal{O}_1$ and $\mathcal{O}_2$ each time checking whether $i\#e$ is unsatisfiable in $\mathcal{O}_1 \cup_{\mathcal{A}}^{\mathcal{S}} \mathcal{O}_2$ and satisfiable in $\mathcal{O}_i$. The (un)satisfiability of a property $i\#R$ is decided via checking the (un)satisfiability of $\exists i\#R.\top$. Given a coherent alignment $\mathcal{A}$, we have to iterate over all atomic entities to conclude that $\mathcal{A}$ is coherent. If $\mathcal{A}$ is incoherent we can stop until we detect a first unsatisfiable class. Alternatively, we might also completely classify $\mathcal{O}_1 \cup_{\mathcal{A}}^{\mathcal{S}} \mathcal{O}_2$ and ask the reasoner for unsatisfiable classes. In the following we refer to the application of such a strategy by the procedure call ISCOHERENTALIGNMENT($\mathcal{A}, \mathcal{O}_1, \mathcal{O}_2$).

There exists an approach to decide the coherence *for most dual-element alignments* which outperforms ISCOHERENTALIGNMENT by far. This approach and its application requires to introduce the notion of a conflict pair. A conflict pair is an incoherent subset of an alignment that contains exactly two correspondences. Moreover, it turns out that most elements in $MIPS_{\mathcal{S}_n}(\mathcal{A}, \mathcal{O}_1, \mathcal{O}_2)$ are conflict pairs of a certain type. We believe that there exists a pattern based reasoning method for each alignment semantics that detects a (large) fraction of all conflict pairs within an alignment. We present such a reasoning method for the natural semantics $\mathcal{S}_n$ and argue finally how to develop a similar method for other alignment semantics using the example of DDL.

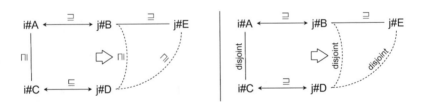

Fig. 2. Subsumption and disjointness propagation pattern. Arrows represent correspondences, solid lines represent axioms or entailed statements in $\mathcal{O}_i$ resp. $\mathcal{O}_j$, and dashed lines represent statements entailed by the merged ontology. Figure taken from [10], where these patterns have been used to support manual mapping revision.

First, we focus on the pattern depicted on the left of Figure 2. Given correspondences $\langle i\#A, j\#B, \overleftrightarrow{\sqsupseteq}, n \rangle$ and $\langle i\#C, j\#D, \overleftrightarrow{\sqsubseteq}, n' \rangle$ as well as axiom $i\#A \sqsubseteq i\#C$ we can conclude that $\mathcal{O}_i \cup_{\mathcal{A}}^{\mathcal{S}_n} \mathcal{O}_j \models j\#B \sqsubseteq j\#D$ and thus $\mathcal{O}_i \cup_{\mathcal{A}}^{\mathcal{S}_n} \mathcal{O}_j \models j\#E \sqsubseteq j\#D$ for each subconcept $j\#E$ of $j\#B$. Now we have $\mathcal{O}_i \cup_{\mathcal{A}}^{\mathcal{S}_n} \mathcal{O}_j \models \bot \sqsupseteq j\#E$ whenever $\mathcal{O}_j$ entails the disjointness of $j\#E$ and $j\#D$. In such a case we detected a conflict pair given the satisfiability of $j\#E$ in $\mathcal{O}_j$. The disjointness propagation pattern works similar. We abstain from a detailed description and refer the reader to the presentation in Figure 2. If we combine both patterns and check their occurrence in all possible combinations given a pair of correspondences, we end up with a sound but incomplete algorithm for deciding the incoherence of an alignment that contains exactly two correspondences. We will refer to this algorithm as POSSIBLYCOHERENT($c_1, c_2, \mathcal{O}_i, \mathcal{O}_j$) with $c_1, c_2 \in \mathcal{A}$.

$\mathcal{S}_n$ might in general induce complex interdependences between $\mathcal{A}$, $\mathcal{O}_1$ and $\mathcal{O}_2$. Therefore, neither are all conflict pairs detectable by the pattern-based approach, nor are all MIPS conflict pairs.

We extend our algorithms (respectively the described pattern) to correspondences between properties by replacing $i\#A$ by $\exists i\#A.\top$ in case that $i\#A$ is a property (the same for $i\#B$, $i\#C$, and $i\#D$). This allows us to consider dependencies between domain restrictions and the subsumption hierarchy within our pattern based reasoning approach. The patterns depicted in Figure 2 are specific to the natural semantics $\mathcal{S}_n$. Similar patterns very likely exist for any reductionistic alignment semantics. For DLL e.g. it is possible to construct corresponding patterns easily. The subsumption propagation pattern is a specific case of (and in particular inspired by) the general propagation rule used within the tableau algorithm proposed in [16], while the disjointness propagation pattern does not hold in DDL.

In the following we need to enumerate the correspondences of an alignment to access elements or subsets of the alignment by index or range. Thus, we sometimes treat an alignment $\mathcal{A}$ as a field using a notation $\mathcal{A}[i]$ to refer to the i-th element of $\mathcal{A}$ and $\mathcal{A}[j \ldots k]$ to refer to $\{\mathcal{A}[i] \in \mathcal{A} \mid j \leq i \leq k\}$. For the sake of convenience we use $\mathcal{A}[\ldots k]$ to refer to $\mathcal{A}[0 \ldots k]$, similar we use $\mathcal{A}[j \ldots]$ to refer to $\mathcal{A}[j \ldots |A| - 1]$. Further, let the index of an alignment start at 0.

Algorithm 1

BRUTEFORCELOD($\mathcal{A}, \mathcal{O}_1, \mathcal{O}_2$)

```
 1: if IsCOHERENTALIGNMENT(A, O₁, O₂) then
 2:    return ∅
 3: else
 4:    ▷ sort A descending according to confidence values
 5:    A' ← ∅
 6:    for all c ∈ A do
 7:       if IsCOHERENTALIGNMENT(A' ∪ {c}, O₁, O₂) then
 8:          A' ← A' ∪ {c}
 9:       end if
10:    end for
11:    return A \ A'
12: end if
```

We already argued that the set of accused correspondences forms a special kind of diagnosis referred to as local optimal diagnosis. Algorithm 1, which has been proposed in [12], is an iterative procedure that computes such a diagnosis. First, we check the coherence of $\mathcal{A}$ and return $\emptyset$ as diagnosis for a coherent alignment. Given $\mathcal{A}$'s incoherence, we have to order $\mathcal{A}$ by descending confidence values. Then an empty alignment $\mathcal{A}'$ is step by step extended by adding correspondences $c \in \mathcal{A}$. Whenever $\mathcal{A}' \cup c$ becomes incoherent, which is decided by reasoning in the merged ontology, c is not added. Finally, we end up with a local optimal diagnosis $\mathcal{A} \setminus \mathcal{A}'$.

Proposition 3. BRUTEFORCELOD($\mathcal{A}, \mathcal{O}_1, \mathcal{O}_2$) *is a local optimal diagnosis for $\mathcal{A}$ with respect to $\mathcal{O}_1$ and $\mathcal{O}_2$.*

Algorithm 1 is completely built on reasoning in the merged ontology and does not exploit efficient reasoning techniques. A more efficient algorithm requires to solve the following problem. Given an incoherent alignment $\mathcal{A}$ ordered descending according to its confidences, we want to find an index i such that $\mathcal{A}[\ldots i-1]$ is coherent and $\mathcal{A}[\ldots i]$ is incoherent. Obviously, a binary search can be used to detect this index. The accordant algorithm, referred to as SEARCHINDEXOFACCUSEDCORRESPONDENCE($\mathcal{A}, \mathcal{O}_1, \mathcal{O}_2$), starts with an index m that splits the incoherent alignment $\mathcal{A}$ in two parts of equal size. Let now i be the index we are searching for. If $\mathcal{A}[\ldots m]$ is coherent we know that $i > m$, otherwise $i \leq m$. Based on this observation we can start a binary search which finally requires $log_2(|\mathcal{A}|)$ iterations to terminate.

Algorithm 2

EFFICIENTLOD($\mathcal{A}, \mathcal{O}_1, \mathcal{O}_2$)

```
 1:  ▷ sort A descending according to confidence values
 2:  A' ← A, k ← 0
 3:  loop
 4:      for i ← k to |A'| − 1 do
 5:          for j ← 0 to i − 1 do
 6:              if not POSSIBLYCOHERENT(A'[j], A'[i], O₁, O₂) then
 7:                  A' ← A' \ {A'[i]}
 8:                  i ← i − 1  ▷ adjust i to continue with next element of A'
 9:                  break  ▷ exit inner for-loop
10:              end if
11:          end for
12:      end for
13:      k ← SEARCHINDEXOFACCUSEDCORRESPONDENCE(A', O₁, O₂)
14:      if k = NIL then
15:          return A \ A'
16:      end if
17:      ▷ let k* be the counterpart of k adjusted for A such that A[k*] = A'[k]
18:      A' ← A'[. . . k − 1] ∪ A[k* + 1 . . .]
19:  end loop
```

We are now prepared to construct an efficient algorithm to compute a local optimal diagnosis (LOD) (Algorithm 2). First we have to sort the input alignment $\mathcal{A}$, prepare a copy $\mathcal{A}'$ of $\mathcal{A}$, and init an index $k = 0$. Variable k works as a separator between the part of $\mathcal{A}'$ that has already been processed successfully and the part of $\mathcal{A}'$ that has not yet been processed or has not been processed successfully. More precisely, it holds that $\mathcal{A}[\ldots k^*] \setminus \mathcal{A}'[\ldots k]$ is a LOD for $\mathcal{A}[\ldots k^*]$ where k^* is an index such that $\mathcal{A}'[k] = \mathcal{A}[k^*]$. Within the main loop we have two nested loops. These are used to check whether correspondence $\mathcal{A}'[i]_{i \geq k}$ possibly conflicts with one of $\mathcal{A}'[j]_{j < i}$. In case a conflict has been detected, $\mathcal{A}'[i]$ is removed from $\mathcal{A}'$. Notice that this approach would directly result in a LOD if both (1) all $\mathcal{M} \in MIPS_S(\mathcal{A}, \mathcal{O}_1, \mathcal{O}_2)$ were conflict pairs, and all conflict pairs were detectable by procedure POSSIBLYCOHERENT. Obviously, these assumptions are not correct and thus we have to search for an index k such that $\mathcal{A}[\ldots k^*] \setminus \mathcal{A}'[\ldots k]$ is a LOD for $\mathcal{A}[\ldots k^*]$. Index k is determined by the binary search presented above. If no such index could be detected, we know that $\mathcal{A} \setminus \mathcal{A}'$ is a LOD (line 14-16). Otherwise, the value of $\mathcal{A}'$ is readjusted to the union of $\mathcal{A}'[\ldots k-1]$, which can be understood as the validated part of $\mathcal{A}'$, and $\mathcal{A}[k^* + 1 \ldots]$, which is the remaining part of $\mathcal{A}$ to be processed in the next iteration. $\mathcal{A}'[k]$ is removed from $\mathcal{A}'$ and thus becomes a part of the diagnosis returned finally.

Proposition 4. EFFICIENTLOD($\mathcal{A}, \mathcal{O}_1, \mathcal{O}_2$) *is a local optimal diagnosis for $\mathcal{A}$ with respect to $\mathcal{O}_1$ and $\mathcal{O}_2$.*

Suppose now that Δ' is a LOD for a subset of $MIPS_{\mathcal{S}}(\mathcal{A}, \mathcal{O}_1, \mathcal{O}_2)$, namely those that are detected by our pattern based reasoning approach, while Δ is the LOD for the complete set $MIPS_{\mathcal{S}}(\mathcal{A}, \mathcal{O}_1, \mathcal{O}_2)$. The correctness of Proposition 4 is based on the fact, that Δ' can be split in a correct and an incorrect part. The correspondence where the correct part ends is exactly the correspondences that is detected by the binary search. Due to the stable ordering, the correct part can be extended over several iterations until we finally end up with a complete and correct local optimal diagnosis Δ.

5   Experiments

Our experiments are based on datasets used within two subtracks of the Ontology Alignment Evaluation Initiative (OAEI). These tracks are the benchmark track about the domain of publications and the conference track. In opposite to the other OAEI tracks, the reference alignments of these tracks are open available.

The benchmark dataset consists of an ontology #101 and alignments to a set of artificial variations #1xx to #2xx. Furthermore, there are reference alignments to four real ontologies known as #301 to #304. We have chosen these four ontologies for our experiments to avoid any interdependencies between the specifics of the artificial test sets and our approach. For our experimental study we had to apply some minor modifications. Neither ontology #101 nor ontologies #301 to #304 contain disjointness axioms; even a highly incorrect alignment cannot introduce any incoherences. Therefore, we decided to extend ontology #101 by disjointness axioms between sibling classes. In the 2008 evaluation 8 matching systems submitted results to the benchmark track that were annotated with confidence values. In the following we refer to this dataset as B_{08}^d.

Our second dataset is based on the conference dataset. In 2008 for the first time reference alignments between five ontologies (= 10 alignments) have been used as part of the official OAEI evaluation. We had to reduce this set four ontologies (= 6 alignments), since one ontology, namely the IASTED ontology, resulted in reasoning problems when merging this ontology with one of the other ontologies. In particular, the runtime behaviour of our algorithms was strongly affected by underlying reasoning problems with IASTED. Unfortunately, only three systems participated in the conference track in 2008, only two of them distinguishing between different degrees of confidence. Therefore, we also used the submissions to the 2007 campaign were we also had two matching systems producing meaningful confidence values. We refer to the resulting dataset as C_{07}, respectively C_{08}. Disjointness is modeled in this dataset incompletely depending on the specific ontology. Thus, we decided to apply our approach to the official OAEI dataset as well as to a dataset enriched with obvious disjointness statements between sibling concepts. These disjointness statements have been manually added as part of the work reported in [12]. The resulting datasets are referred to as C_{07}^d, respectively C_{08}^d.

In our experiments we used the reasoner Pellet [17], in particular version 2RC2 together with the OWL API on a 2.26 GHz standard laptop with 2GB RAM. The complete dataset as well as a more detailed presentation of the results is available at http://webrum.uni-mannheim.de/math/lski/matching/lod/. Due to the lack of space we can only present aggregated results in the following paragraphs.

Runtimes. Results related to runtime efficiency are presented in Table 1. In each row we aggregated the results of a specific matcher for one of the datasets explained above. For both Algorithm 2 and its brute-force counterpart Algorithm 1 the total of runtimes is displayed in milliseconds. Obviously, Algorithm 2 outperforms the brute force approach. Runtime performance increased by a coefficient of 1.8 to 9.3. To better understand under which circumstances Algorithm 2 performs better, we added columns presenting the size of the input alignment $\mathcal{A}$, the size of the debugged alignment $\mathcal{A}'$, and the size of the diagnosis $\Delta = \mathcal{A} \setminus \mathcal{A}'$. Furthermore, the column captioned with '$k \neq \text{NIL}$' refers to the number of correspondences that have additionally been detected due to complete reasoning techniques. In particular, it displays how often $k \neq \text{NIL}$ is evaluated as true in line 14 of Algorithm 2. Finally, we analyze the fraction of those correspondences that have been detected by efficient reasoning techniques.

Although we observe that absolute runtimes are affected by the alignment size (see for example the C_{07}-OLA row), the coefficient of runtimes seems not to be affected directly. The same holds for the size of the diagnosis Δ. Instead of that and in accordance to our theoretical considerations the runtime coefficient correlates with the fraction of conflicts that can be detected efficiently. While for the conference testcases results have to be considered inconclusive, this pattern clearly emerges for the benchmark testcases. The efficiency of Algorithm 2 is thus directly affected by the degree of completeness of POSSIBLYCOHERENT invoked as subprocedure.

Diagnostic Quality. In previous work we already argued that the coherence of an alignment is a quality of its own [8]. An incoherent alignment causes specific problems depending on the scenario in which the alignment is used. We now additionally investigate in how far the removal of the diagnosis increases the quality of the input alignment $\mathcal{A}$ by comparing it against reference alignment $\mathcal{R}$. In particular, we compute for both the input alignment $\mathcal{A}$ and the repaired alignment $\mathcal{A}' = \mathcal{A} \setminus \Delta$ the classical measures of precision and recall. The precision of an alignment describes its degree of correctness, while recall describes its degree of completeness. A definition of these measures with respect to alignment evaluation can be found in [2].

The results of our measurements are presented in Table 2. The first two columns identify datasets, followed by columns presenting the size of the input alignment $\mathcal{A}$, the size of the diagnosis $\Delta = \mathcal{A} \setminus \mathcal{A}'$, and the number of removed correspondences $\Delta \setminus \mathcal{R}$ that are actually incorrect i.e. those correspondences that have been removed correctly. The following three columns show how precision, recall and f-measure have been affected by the application of our algorithm. In the *Effect* column the results are aggregated as difference between the f-measure of the input alignment $\mathcal{A}$ and the f-measure of the repaired alignment $\mathcal{A}'$.

Table 1. Aggregated Runtime of EFFICIENTLOD-Algorithm (Alg.2) and BRUTEFORCELOD-Algorithm (Alg.1) and related characteristics

Testcase		Runtime Comparison			Alignment Size & Deleted Correspondences				
DS	Matcher	Alg.2	Alg.1	Coeff.	$|\mathcal{A}|$	$|\mathcal{A}'|$	$|\Delta|$	'$k \neq$ NIL'	Frac.
B_{08}^d	Aroma	8656	71846	8.3	202	194	8	2	75%
	ASMOV	6226	47714	7.7	222	218	4	1	75%
	CIDER	7530	70028	9.3	195	181	14	1	93%
	DSSim	3922	36343	9.3	184	179	5	0	100%
	Lily	15468	74352	4.8	218	210	8	4	50%
	RiMOM	15942	78219	4.9	235	221	14	5	64%
	SAMBO	3800	28655	7.5	197	196	1	0	100%
	SAMBOdtf	8586	59211	6.9	206	202	4	2	50%
C_{07}	Falcon	6847	12414	1.8	70	56	14	4	71%
	OLA	39830	73497	1.8	404	228	176	27	85%
C_{08}	ASMOV	13289	23425	1.8	153	128	25	10	60%
	Lily	4604	14609	3.2	78	63	15	3	80%

Based on the f-measure differences we conclude that in 13 of 16 testcases we increased the overall quality of the alignment. Notice again that these results are aggregated average values. Taking a closer look at the individual results for each generated alignment (not depicted in Table 2), we observe that in 15 cases our approach has negative effects on the f-measure, in 14 cases we observed no effects at all, and in 51 cases we measured an increased f-measure. Obviously, this effect is based on an increased precision and a stable or only slightly decreased recall. Nevertheless, there are some exceptions to this pattern.

On the one hand we have negative results for B_{08}^d-DSSim, C_{08}-ASMOV and C_{08}^d-ASMOV. Due to characteristics of a local optimal diagnosis an incorrect correspondence might cause the removal of all conflicting correspondences with lower confidence given that there exists no conflicting correspondence with higher confidence. An analysis of the individual results revealed that the negative effects are based on this pattern, i.e. an incorrect correspondence has been annotated with very high confidence and no 'antagonist' has been annotated with higher confidence.

On the other hand we measured strong positive effects for the OLA system on the conference dataset. These effects are associated with the large size of the alignments generated by OLA. It seems that, compared to the other submissions, the matching results of OLA have not been filtered or thresholded in an appropriate way. OLA generated a total of 404 correspondences with respect to our C datasets. For the original dataset C (no disjointness axioms added) 176 of these correspondences have been automatically removed by our approach and only 2 of these removals were incorrect, which raised the f-measure from 20.8% to 31.6% (from 20.8% to 38.1% for the C^d dataset). Notice that our algorithm expects no parameter which corresponds to a threshold or an estimated size of the reference alignment. Instead of that the algorithm automatically adapts to the quality of the input due to the fact that a highly incorrect alignment will be higly incoherent. Overall, the results indicate that our approach does not only ensure the quality of the input alignment but even more has significant positive effects.

Table 2. Alignment size, size of diagnosis and number of correctly removed correspondences; effects on precision, recall, and f-measure

DS	Matcher	$\|\mathcal{A}\|$	$\|\Delta\|$	$\|\Delta \setminus \mathcal{R}\|$	Prec. $\mathcal{A} \rightsquigarrow \mathcal{A}'$	Rec. $\mathcal{A} \rightsquigarrow \mathcal{A}'$	F-m. $\mathcal{A} \rightsquigarrow \mathcal{A}'$	Effect
B_{08}^d	Aroma	202	8	7	80.2 $\rightsquigarrow$ 83.0	70.1 $\rightsquigarrow$ 69.7	74.8 $\rightsquigarrow$ 75.8	+0.9
	ASMOV	222	4	3	78.4 $\rightsquigarrow$ 79.4	75.3 $\rightsquigarrow$ 74.9	76.8 $\rightsquigarrow$ 77.1	+0.2
	CIDER	195	14	5	87.2 $\rightsquigarrow$ 89.0	73.6 $\rightsquigarrow$ 69.7	79.8 $\rightsquigarrow$ 78.2	-1.7
	DSSim	184	5	5	87.5 $\rightsquigarrow$ 89.9	69.7 $\rightsquigarrow$ 69.7	77.6 $\rightsquigarrow$ 78.5	+0.9
	Lily	218	8	8	83.0 $\rightsquigarrow$ 86.2	78.4 $\rightsquigarrow$ 78.4	80.6 $\rightsquigarrow$ 82.1	+1.5
	RiMOM	235	14	14	78.3 $\rightsquigarrow$ 83.3	79.7 $\rightsquigarrow$ 79.7	79.0 $\rightsquigarrow$ 81.4	+2.4
	SAMBO	197	1	1	91.9 $\rightsquigarrow$ 92.3	78.4 $\rightsquigarrow$ 78.4	84.6 $\rightsquigarrow$ 84.8	+0.2
	SAMBOdtf	206	4	4	88.3 $\rightsquigarrow$ 90.1	78.8 $\rightsquigarrow$ 78.8	83.3 $\rightsquigarrow$ 84.1	+0.8
C_{07}	Falcon	70	14	11	65.7 $\rightsquigarrow$ 76.8	60.5 $\rightsquigarrow$ 56.6	63.0 $\rightsquigarrow$ 65.2	+2.1
	OLA	404	176	174	12.4 $\rightsquigarrow$ 21.1	65.8 $\rightsquigarrow$ 63.2	20.8 $\rightsquigarrow$ 31.6	+10.7
C_{08}	ASMOV	153	25	20	22.9 $\rightsquigarrow$ 23.4	46.1 $\rightsquigarrow$ 39.5	30.6 $\rightsquigarrow$ 29.4	-1.2
	Lily	78	15	13	44.9 $\rightsquigarrow$ 52.4	46.1 $\rightsquigarrow$ 43.4	45.5 $\rightsquigarrow$ 47.5	+2.0
C_{07}^d	Falcon	70	17	14	65.7 $\rightsquigarrow$ 81.1	60.5 $\rightsquigarrow$ 56.6	63.0 $\rightsquigarrow$ 66.7	+3.7
	OLA	404	228	226	12.4 $\rightsquigarrow$ 27.3	65.8 $\rightsquigarrow$ 63.2	20.8 $\rightsquigarrow$ 38.1	+17.3
C_{08}^d	ASMOV	153	33	27	22.9 $\rightsquigarrow$ 24.2	46.1 $\rightsquigarrow$ 38.2	30.6 $\rightsquigarrow$ 29.6	-1.0
	Lily	78	21	17	44.9 $\rightsquigarrow$ 54.4	46.1 $\rightsquigarrow$ 40.8	45.5 $\rightsquigarrow$ 46.6	+1.2

Related work. In [13] Qi et. al. propose a kernel revision operator for description logic-based ontologies. A revision deals with the problem of incorporating newly received information into accepted information consistently. Within their experiments the authors apply their approach amongst others to the revision of ontology alignments, where the matched ontologies are accepted information and the alignment between them is new and disputable information. Two of the algorithms proposed require to compute all $MIPS_{\mathcal{S}}(\mathcal{A}, \mathcal{O}_1, \mathcal{O}_2)$ in order to construct a minimal hitting set, while their third and most efficient algorithm cannot ensure the minimality of the constructed hitting set. We conducted additional experiments with the alignments used in [13]. We did not include these as part of the main experiments, because the datasets do not contain correspondences between properties and are not as comprehensive as the datasets used within our experiments. However, we observed runtimes between 50 and 250 milliseconds, while in [13] runtimes between 6 and 51 seconds have been reported for the fastest algorithm.

An approach, which aims to explain logical consequences of an alignment, has been proposed in [6]. Some of these consequences are unintended due to incorrect correspondences in $\mathcal{A}$ and cannot be accepted. An example of an unintended consequence is a concept becoming unsatisfiable due to $\mathcal{A}$. Such an alignment is referred to as incoherent within our framework. To generate plans for repairing a defect alignment, first, all justifications for the unintended consequences are computed. While in [13] all MIPS are used to compute a minimal hitting set, in [6] all justifications are used to compute minimal hitting sets referred to as repair plans. The authors point out, that the bottleneck of their approach is the computation of all justifications.

In summary, both approaches suffer from the incorrect assumption that a minimal hitting set can only be constructed given complete knowledge about all MIPS respectively all justifications. Contrary to this, we have shown that it is possible to compute a specific hitting set, namely a local optimal diagnosis, that is not only minimal but also takes into account confidence values in an appropriate manner.

6   Conclusion

We have presented a basic algorithm for computing a local optimal diagnosis as well as an efficient variant, which makes use of an intertwined combination of incomplete and complete reasoning techniques. These algorithms are based on precise logic-based semantics of an alignment. Although, we only focused on specific type of semantics, namely the natural semantics, there is some evidence that the principles of our approach can be applied to each reductionistic alignment semantics.

It turned out that the efficient variant of our algorithm outperformed the basic algorithm by a factor of ≈ 2 to 10. In particular, we observed that the runtime is first and foremost determined by the fraction of conflicts detectable by the incomplete reasoning procedures. In future work we will add additional reasoning patterns in order to detect more conflicts by efficient reasoning strategies.

Our algorithm improves in most cases an alignments f-measure due to an increased precision. However, we detected some outliers where a highly confident but incorrect correspondence had negative impact on the repairing process. An approach that removes a minimum number of correspondences would probably remove such a correspondence. Generally, it is not clear whether the *principle of minimal change* is a good guideline for repairing alignments. Experiments we conducted so far show inconclusive results and require additional analysis.

We already pointed to some problems of other approaches. We believe that these problems are based on not taking into account three specifics of the problem under discussion. First, correspondences are annotated with confidence values. Second, there are significantly less correspondences in an alignment than axioms in the matched ontologies. Third, given the monotonicity of S, everything that holds in $\mathcal{O}_1$ and $\mathcal{O}_2$ holds also in the merged ontology $\mathcal{O}_1 \cup_{\mathcal{A}}^S \mathcal{O}_2$. The first observation was taken into account in the definition of a local optimal diagnosis, the second observation points to the possibility of iterating over all correspondences (the main loop in both algorithms), and the third observation is exploited within the combination of pattern-based reasoning and reasoning in the merged ontology.

Acknowledgement. The work has been partially supported by the German Science Foundation (DFG) under contract STU 266/3-1 and STU 266/5-1.

References

1. Borgida, A., Serafini, L.: Distributed description logics: Assimilating information from peer sources. Journal on Data Semantics (2003)
2. Euzenat, J., Shvaiko, P.: Ontology Matching. Springer, Heidelberg (2007)
3. Garey, M.R., Johnson, D.S.: Computers and intractability: A guide to the theory of NP-completeness. W. H. Freeman and Company, New York (1979)
4. Giunchiglia, F., Yatskevich, M., Shvaiko, P.: Semantic matching: Algorithms and implementation. Journal on Data Semantics (2007)

5. Jean-Mary, Y.R., Kabuka, M.R.: Asmov. Results for OAEI 2008. In: Proc. of the ISWC 2008 workshop on ontology matching, Karlsruhe, Germany (2008)
6. Jimenez-Ruiz, E., Grau, B.C., Horrocks, I., Berlanga, R.: Ontology integration using mappings: Towards getting the right logical consequences. In: Proc. of the 6th Annual European Semantic Web Conference, Heraklion, Crete, Greece (2009)
7. Meilicke, C., Stuckenschmidt, H.: Applying logical constraints to ontology matching. In: Proc. of the 30th German Conference on Artificial Intelligence, Osnabrück, Germany (2007)
8. Meilicke, C., Stuckenschmidt, H.: Incoherence as a basis for measuring the quality of ontology mappings. In: Proc. of the ISWC 2008 Workshop on Ontology Matching, Karlsruhe, Germany (2008)
9. Meilicke, C., Stuckenschmidt, H.: An efficient method for computing a local optimal alignment diagnosis. Technical report, University Mannheim, Computer Science Institute (2009)
10. Meilicke, C., Stuckenschmidt, H., Svab-Zamazal, O.: A reasoning-based support tool for ontology mapping evaluation. In: Proc. of the European Semantic Web Conference, Heraklion, Greece (2009)
11. Meilicke, C., Tamilin, A., Stuckenschmidt, H.: Repairing ontology mappings. In: Proc. of the Twenty-Second Conference on Artificial Intelligence, Vancouver, Canada (2007)
12. Meilicke, C., Völker, J., Stuckenschmidt, H.: Learning disjointness for debugging mappings between lightweight ontologies. In: Proc. of the 16th International Conference on Knowledge Engineering and Knowledge Management, Acitrezza, Italy (2008)
13. Qi, G., Haase, P., Huang, Z., Ji, Q., Pan, J.Z., Völker, J.: A kernel revision operator for terminologies - algorithms and evaluation. In: Sheth, A.P., Staab, S., Dean, M., Paolucci, M., Maynard, D., Finin, T., Thirunarayan, K. (eds.) ISWC 2008. LNCS, vol. 5318, pp. 419–434. Springer, Heidelberg (2008)
14. Reiter, R.: A theory of diagnosis from first principles. Artificial Intelligence (1987)
15. Schlobach, S., Cornet, R.: Non-standard reasoning services for the debugging of description logic terminologies. In: Proc. of 18th International Joint Conference on Artificial Intelligence, Acapulco, Mexico (2003)
16. Serafini, L., Tamilin, A.: Local tableaux for reasoning in distributed description logics. In: Proc. of the Int. Workshop on Description Logics, Whistler, Canada (2004)
17. Sirin, E., Parsia, B., Grau, B.C., Kalyanpur, A., Katz, Y.: Pellet: A practical OWL-DL reasoner. Journal of Web Semantics (2007)

Paraconsistent Reasoning for OWL 2

Yue Ma[1] and Pascal Hitzler[2,*]

[1] Institute LIPN, Université Paris-Nord (LIPN - UMR 7030), France
[2] Institute AIFB, Universität Karlsruhe, Germany
yue.ma@lipn.univ-paris13.fr, pascal@pascal-hitzler.de

Abstract. A four-valued description logic has been proposed to reason with description logic based inconsistent knowledge bases. This approach has a distinct advantage that it can be implemented by invoking classical reasoners to keep the same complexity as under the classical semantics. However, this approach has so far only been studied for the basic description logic $\mathcal{ALC}$. In this paper, we further study how to extend the four-valued semantics to the more expressive description logic $\mathcal{SROIQ}$ which underlies the forthcoming revision of the Web Ontology Language, OWL 2, and also investigate how it fares when adapted to tractable description logics including $\mathcal{EL}++$, DL-Lite, and Horn-DLs. We define the four-valued semantics along the same lines as for $\mathcal{ALC}$ and show that we can retain most of the desired properties.

1 Introduction

Expressive and tractable description logics have been well-studied in the field of semantic web methods and applications, see e.g. [22,6]. In particular, description logics are the foundations of the Web Ontology Language OWL [7,17] and its forthcoming revision, OWL 2 [24]. However, real knowledge bases and data for Semantic Web applications will rarely be perfect. They will be distributed and multi-authored. They will be assembled from different sources and reused. It is unreasonable to expect such realistic knowledge bases to be always logically consistent, and it is therefore important to study ways of dealing with inconsistencies in both expressive and tractable description logic based ontologies, as classical description logics break down in the presence of inconsistent knowledge.

About inconsistency handling of ontologies based on description logics, two fundamentally different approaches can be distinguished. The first is based on the assumption that inconsistencies indicate erroneous data which is to be repaired in order to obtain a consistent knowledge base, e.g. by selecting consistent subsets for the reasoning process [21,8,5]. The other approach yields to the insight that inconsistencies are a natural phenomenon in realistic data which are to be handled by a logic which tolerates it [20,23,13]. Such logics are called paraconsistent, and the most prominent of them are based on the use of additional truth values standing for *underdefined* (i.e. neither true nor false) and *overdefined* (or *contradictory*, i.e. both true and false). Such logics are appropriately called *four-valued logics* [1]. We believe that either of the approaches

---

* The second author is now at Kno.e.sis Center, Wright State University, Dayton, OH, USA.

A. Polleres and T. Swift (Eds.): RR 2009, LNCS 5837, pp. 197–211, 2009.
© Springer-Verlag Berlin Heidelberg 2009

is useful, depending on the application scenario. Besides this, four-valued semantics proves useful for measuring inconsistency of ontologies [16], which can provide context information for facilitating inconsistency handling.

In this paper, we extend our study of paraconsistent semantics for $\mathcal{ALC}$ from [13]. This approach has the pleasing two properties that (1) reasoning under the paraconsistent semantics can be reduced to reasoning under classical semantics and (2) the transformations required for the reduction from the paraconsistent semantics to the classical semantics is linear in the size of the knowledge base. In this paper, we will carry these results over to $\mathcal{SROIQ}$, which underlies OWL 2,[1] and also study its impact for several tractable description logics around OWL 2. We also present a slight modification to the semantics presented in [13]. In more detail, the contributions of the paper are as follows.

- The extension of four-valued semantics to $\mathcal{SROIQ}$ is defined. Specially, we show that it still can be reduced to classical semantics regardless its high expressivity.
- The four-valued semantics is studied for the tractable description logics $\mathcal{EL}$++, Horn-DLs, and DL-Lite, for some of these adaptations of the semantics are made. We show that under certain restrictions our approach retains tractability.
- Compared with our existing work on four-valued semantics for $\mathcal{ALC}$, in this paper, we do not impose four-valued semantics on roles (with the exception of DL-Lite). The reasons are: (1) Negative roles are not used as concept constructors in $\mathcal{ALC}$, $\mathcal{SROIQ}$, $\mathcal{EL}$++, or Horn-DLs[2] such that contradiction caused directly by roles can safely be ignored. (2) This modified four-valued semantics is more similar to the classical semantics. (3) Four-valued semantics is semantically weaker than the classical semantics which means that there are undesired missing conclusions under the semantics from [13], which is not the case in our modified approach.

The paper is structured as follows. We first review briefly the four-valued semantics for $\mathcal{ALC}$ in Section 2. Then we study the four-valued semantics for expressive description logics in Section 3 and four-valued semantics for tractable description logics in Section 4, respectively. We conclude and discuss future work in Section 5.

This paper is an extension and revision of the workshop paper [14].

2 Preliminaries

2.1 The Four-Valued Semantics for $\mathcal{ALC}$ – with a Slight Modification

We describe the syntax and semantics of four-valued description logic $\mathcal{ALC}4$ from [13] with a slight modification. Syntactically, $\mathcal{ALC}4$ hardly differs from $\mathcal{ALC}$. Complex concepts and assertions are defined in exactly the same way. For general class inclusion

---

[1] We will actually not deal with aspects of OWL 2 which are not part of $\mathcal{SROIQ}$, such as datatypes, keys, etc. We also ignore disjoint roles.

[2] Note that OWL 2 allows negative property assertions like $\neg R(a, b)$ in the ABox; however they can be considered syntactic sugar on top of $\mathcal{SROIQ}$ since they can be written as $\{a\} \sqsubseteq \forall R.\neg\{b\}$. In this paper, we assume that all negative property assertions have been rewritten in this way.

(GCI) axioms, however, significant effort has been devoted on the intuitions behind these different implications in [13]. For the four-valued semantics, three different kinds of inclusions can be used, as follows. They serve different underlying intuitions and differ in inferential strength. A detailed discussion of this has been presented in [13] which we will not repeat here.

$$C \mapsto D \text{ (material inclusion axiom)},$$
$$C \sqsubset D \text{ (internal inclusion axiom)},$$
$$C \rightarrow D \text{ (strong inclusion axiom)}.$$

Semantically, interpretations map individuals to elements of the domain of the interpretation, as usual. For concepts, however, modifications are made to the notion of interpretation in order to allow for reasoning with inconsistencies.

Intuitively, in four-valued logic we need to consider four situations which can occur in terms of containment of an individual in a concept: (1) we know it is contained, (2) we know it is not contained, (3) we have no knowledge whether or not the individual is contained, (4) we have contradictory information, namely that the individual is both contained in the concept and not contained in the concept. There are several equivalent ways how this intuition can be formalized, one of which is described in the following.

For a given domain Δ^I and a concept C, an interpretation over Δ^I assigns to C a pair $\langle P, N \rangle$ of (not necessarily disjoint) subsets of Δ^I. Intuitively, P is the set of elements known to belong to the extension of C, while N is the set of elements known to be not contained in the extension of C. For simplicity of notation, we define functions $proj^+(\cdot)$ and $proj^-(\cdot)$ by $proj^+\langle P, N \rangle = P$ and $proj^-\langle P, N \rangle = N$. If for some $x \in \Delta^I$ and some concept C, $x \in proj^+(C^I) \cap proj^-(C^I)$, then we write $C^I(x) = B$, where B is a truth value under four-valued semantics representing contradiction state.

Formally, a four-valued interpretation is a pair $I = (\Delta^I, \cdot^I)$ with Δ^I as domain, where $\cdot^I$ is a function assigning elements of Δ^I to individuals, and subsets of $(\Delta^I)^2$ to concepts, such that the conditions in Table 1 are satisfied. Note that the semantics of roles here remains unchanged from the classical two-valued case, and in this point the semantics presented here differs from that in [13]. Intuitively, inconsistencies always arise on concepts, and not on roles, at least in the absence of role negation, which is often assumed when studying DLs. We will see in this paper that this approach can be used to tolerate inconsistency, not only for $\mathcal{ALC}$ but also for more expressive description logics. This is an improvement over [13] in the sense that we would like to make as few changes as possible when extending the classical semantics to a four-valued semantics for handling inconsistency.

The semantics of the three different types of inclusion axioms is formally defined in Table 2 (together with the semantics of concept assertions). Again we refer to the discussions in [13] for details.

We say that a four-valued interpretation (a *4-interpretation*) I satisfies a four-valued knowledge base O (i.e. is a model, or *4-model*, of it) iff it satisfies each assertion and each inclusion axiom in O. A knowledge base O is satisfiable (unsatisfiable) iff there exists (does not exist) such a model. We write $O \models_4 \alpha$ for O and an axiom α if and only if each 4-valued model of O is a model of α. Moreover, if O and O' have a same set of 4-valued models, we denote $O =_4 O'$.

Table 1. Semantics of $\mathcal{ALC}4$ Concepts

Constructor Syntax	Semantics
A	$A^I = \langle P, N \rangle$, where $P, N \subseteq \Delta^I$
R	$R^I \subseteq \Delta^I \times \Delta^I$
o	$o^I \in \Delta^I$
$\top$	$\langle \Delta^I, \emptyset \rangle$
$\bot$	$\langle \emptyset, \Delta^I \rangle$
$C_1 \sqcap C_2$	$\langle P_1 \cap P_2, N_1 \cup N_2 \rangle$, if $C_i^I = \langle P_i, N_i \rangle$ for $i = 1, 2$
$C_1 \sqcup C_2$	$\langle P_1 \cup P_2, N_1 \cap N_2 \rangle$, if $C_i^I = \langle P_i, N_i \rangle$ for $i = 1, 2$
$\neg C$	$(\neg C)^I = \langle N, P \rangle$, if $C^I = \langle P, N \rangle$
$\exists R.C$	$\langle \{x \mid \exists y, (x,y) \in R^I \text{ and } y \in proj^+(C^I)\},$ $\{x \mid \forall y, (x,y) \in R^I \text{ implies } y \in proj^-(C^I)\} \rangle$
$\forall R.C$	$\langle \{x \mid \forall y, (x,y) \in R^I \text{ implies } y \in proj^+(C^I)\},$ $\{x \mid \exists y, (x,y) \in R^I \text{ and } y \in proj^-(C^I)\} \rangle$

Table 2. Semantics of inclusion axioms in $\mathcal{ALC}4$

Axiom Name	Syntax	Semantics
material inclusion	$C_1 \mapsto C_2$	$\Delta^I \setminus proj^-(C_1^I) \subseteq proj^+(C_2^I)$
internal inclusion	$C_1 \sqsubset C_2$	$proj^+(C_1^I) \subseteq proj^+(C_2^I)$
strong inclusion	$C_1 \rightarrow C_2$	$proj^+(C_1^I) \subseteq proj^+(C_2^I)$ and $proj^-(C_2^I) \subseteq proj^-(C_1^I)$
individual assertions	$C(a)$	$a^I \in proj^+(C^I)$
	$R(a,b)$	$(a^I, b^I) \in R^I$

2.2 Reduction from Four-Valued Semantics of $\mathcal{ALC}$ to Classical Semantics

It is a pleasing property of $\mathcal{ALC}4$, that it can be translated easily into classical $\mathcal{ALC}$, such that paraconsistent reasoning can be simulated by using standard $\mathcal{ALC}$ reasoning algorithms.

Definition 1. *(Concept transformation) For any given concept C, its transformation $\pi(C)$ is the concept obtained from C by the following inductively defined transformation.*

- *If $C = A$ for A an atomic concept, then $\pi(C) = A^+$, where A^+ is a new concept;*
- *If $C = \neg A$ for A an atomic concept, then $\pi(C) = A'$, where A' is a new concept;*
- *If $C = \top$, then $\pi(C) = \top$;*
- *If $C = \bot$, then $\pi(C) = \bot$;*
- *If $C = E \sqcap D$ for concepts D, E, then $\pi(C) = \pi(E) \sqcap \pi(D)$;*
- *If $C = E \sqcup D$ for concepts D, E, then $\pi(C) = \pi(E) \sqcup \pi(D)$;*
- *If $C = \exists R.D$ for D a concept and R is a role, then $\pi(C) = \exists R.\pi(D)$;*
- *If $C = \forall R.D$ for D a concept and R is a role, then $\pi(C) = \forall R.\pi(D)$;*
- *If $C = \neg\neg D$ for a concept D, then $\pi(C) = \pi(D)$;*

- If $C = \neg(E \sqcap D)$ for concepts D, E, then $\pi(C) = \pi(\neg E) \sqcup \pi(\neg D)$;
- If $C = \neg(E \sqcup D)$ for concepts D, E, then $\pi(C) = \pi(\neg E) \sqcap \pi(\neg D)$;
- If $C = \neg(\exists R.D)$ for D a concept and R is a role, then $\pi(C) = \forall R.\pi(\neg D)$;
- If $C = \neg(\forall R.D)$ for D a concept and R is a role, then $\pi(C) = \exists R.\pi(\neg D)$;

Based on this, axioms are transformed as follows.

Definition 2. *(Axiom Transformations) For any ontology O, $\pi(O)$ is defined as the set $\{\pi(\alpha) \mid \alpha$ is an axiom of $O\}$, where $\pi(\alpha)$ is the transformation performed on each axiom defined as follows:*

- $\pi(\alpha) = \neg\pi(\neg C_1) \sqsubseteq \pi(C_2)$, if $\alpha = C_1 \mapsto C_2$;
- $\pi(\alpha) = \pi(C_1) \sqsubseteq \pi(C_2)$, if $\alpha = C_1 \sqsubseteq C_2$;
- $\pi(\alpha) = \{\pi(C_1) \sqsubseteq \pi(C_2), \pi(\neg C_2) \sqsubseteq \pi(\neg C_1)\}$, if $\alpha = C_1 \rightarrow C_2$;.
- $\pi(C(a)) = \pi(C)(a), \pi(R)(a, b) = R(a, b)$,

where a, b are individuals, C_1, C_2, C are concepts, R is a role.

We note two issues. First, the transformation algorithm is linear in the size of the ontology. Second, for any $\mathcal{ALC}$ ontology O, $\pi(O)$ is still an $\mathcal{ALC}$ ontology. Based on these two observations as well as the following theorem, we can see that paraconsistent reasoning of $\mathcal{ALC}$ can indeed be simulated on standard reasoners by means of the transformation just given.

Theorem 1. *For any ontology O in $\mathcal{ALC}$ we have $O \models_4 \alpha$ if and only if $\pi(O) \models_2 \pi(\alpha)$, where $\models_2$ is the classical $\mathcal{ALC}$ entailment.*

The following definition, also employed in [13], will be required to ensure that knowledge bases which are inconsistent under the classical semantics become consistent, after transformation, under the four-valued semantics – see Proposition 4.

Definition 3. *Given a knowledge base O, the satisfiable form of O, written $SF(O)$, is a knowledge base obtained by replacing each occurrence of $\bot$ in O with $A_{new} \sqcap \neg A_{new}$, and replacing each occurrence of $\top$ in (O) with $A_{new} \sqcup \neg A_{new}$, where A_{new} is a new atomic concept.*

3 Paraconsistent Semantics for Expressive DLs

In this section, we study how to extend four-valued semantics to $\mathcal{SROIQ}$. For the conflicting assertion set $\{\geq (n + 1)R.C(a), \leq nR.C(a)\}$, intuitively, it is caused by the contradiction that there should be less than n different individuals related to a via the R relation, and also there should be more than $n + 1$ different individuals related to a via R. That is, the contradiction is from the set of individuals of concept C which relate a via R. By this idea, we extend the four-valued semantics to the constructors for number restrictions (with four-valued semantics to nominal) in Table 3, where $\#(S)$ stands for the cardinality of a set S. We remark that the semantics of roles is just the classical semantics. So the semantics for role axioms are still classical.

We give the following example to illustrate the intuition of our four-valued semantics for number restrictions $\geq nR.C$ given above.

Table 3. Four-valued Semantics Extension to Number Restrictions and Nominals

Constructor	Semantics
$\geq nR.C$	$\langle\{x \mid \#(y.(x,y) \in R^I \wedge y \in proj^+(C^I)) \geq n\},$ $\{x \mid \#(y.(x,y) \in R^I \wedge y \notin proj^-(C^I)) < n\}\rangle$
$\leq nR.C$	$\langle\{x \mid \#(y.(x,y) \in R^I \wedge y \notin proj^-(C^I)) \leq n\},$ $\{x \mid \#(y.(x,y) \in R^I \wedge y \in proj^+(C^I)) > n\}\rangle$
$\{o_1,...o_n\}$	$\langle\{o_1^I, ..., o_n^I\}, N\rangle$, where $N \subseteq \Delta^I$

Example 1. *Consider the knowledge base*

$$\{\geq 2hasStu.PhD(Green), \leq 1hasStu.PhD(Green)\}$$

which states the conflicting facts that Green has at least two and at most one PhD student. Consider a 4-interpretation $I = (\Delta^I, \cdot^I)$ *where* $\Delta^I = \{a_1, a_2, b_1, b_2, Green\}$, $PhD^I = \langle\{a_1, b_1\}, \{b_1, b_2, a_2\}\rangle$, $hasStu^I = \{(Green, a_1), (Green, a_2), (Green, b_1),$ $(Green, b_2)\}$. *According to Table 3, I is a 4-model because* $(\geq 2hasStu.PhD(Green))^I$ $= (\leq 1hasStu.PhD(Green))^I = B$ *and by checking*

$$Green \in \{x \mid \#(y.(x,y) \in hasStu^I \wedge y \in proj^+(PhD^I)) \geq 2\},$$
$$Green \in \{x \mid \#(y.(x,y) \in hasStu^I \wedge y \notin proj^-(PhD^I)) < 2\}.$$

This example shows that the contradictions on the constructor of number restriction $\geq nR.C$ is reflected by the contradiction on C, which is our underlying idea of Table 3. Generalizing this example, we have the following property which shows that if we have contradictions of the form of $\{\geq nR.C(a), \leq mR.C(a)\} \subseteq O$ with $(m < n)$, then there will be at least $n - m$ individuals relating a via R and contradictorily belonging to concept C under its four-valued model:

Proposition 2. *Given an ontology* O, *if* $\{\geq nR.C(a), \leq mR.C(a)\} \subseteq O$ *and* $m < n$, *then for any four-valued interpretation* I *of* O, *we have*

$$\#\{b \mid (a, b) \in R^I \text{ and } C^I(b) = B\} \geq n - m.$$

Proof. Suppose $C^I = \langle P, N\rangle$, denote $T = \{y \mid (a, y) \in R^I\}$, $T_1 = \{y \mid (a, y) \in R^I \wedge y \in P\}$, $T_1 = \{y \mid (a, y) \in R^I \wedge y \in N\}$. It is equal to prove that $|T_1 \cap T_2| \geq n - m$.

From the assumption and Table 3, we have $a \in \{x \mid \#(y.(x,y) \in R^I \wedge y \in P) \geq n\}$ and $a \in \{x \mid \#(y.(x,y) \in R^I \wedge y \notin N) \leq m\}$. That is, $\#(y.(a,y) \in R^I \wedge y \in P) \geq n$ and $\#(y.(a,y) \in R^I \wedge y \notin N) \leq m$, which means $|T_1| \geq n$ and $|T| - |T_2| \leq m$, Then, it is easy to see that $|T_1| + |T_2| \geq |T| + n - m$. Because $T_1 \subseteq T, T_2 \subseteq T$, we have $|T_1 \cap T_2| \geq n - m$. $\qquad\square$

The underlying idea of the four-valued semantics for nominals is that if the contradiction occurs on a nominal concept, then we explicitly collect the contradictory individuals into the *proj⁻* part of the four-valued semantics of the nominal such that a four-valued model exists. We explain this by the following example.

Let $O = \{EuropeanState \sqsubseteq \exists currency.\{euro\}, (\forall currency.\neg\{euro\})(UK),$ $EuropeanState(UK)\}$ which says that European countries have euro as their currency and UK is a European country whose currency is not euro. We can find a 4-model $I = \langle \Delta^I, \cdot^I \rangle$ for O, with $\Delta^I = \{UnitedKingdom, curreuro\}$ and $UK^I = UnitedKingdom$, $euro^I = curreuro$, $EuropeanState^I = \langle\{UnitedKingdom\}, \emptyset\rangle$, $currency^I = \{(UnitedKingdom, curreuro)\}$ and the contradictory $(\{euro\})^I = \langle\{curreuro\}, \{curreuro\}\rangle$. This model says that the currency $curreuro$ belongs to the concept $\{euro\}$ contradictorily. That is, we have conflicting information about whether the currency of Uk, which reflects the contradictory situation described in O.

For the extended four-valued semantics defined in Table 3, we have that the following properties hold as under the classical semantics; proofs can be obtained by carefully checking the definition of four-valued semantics.

Proposition 3. *Let C be a concept and R be an object role name. For any four-valued interpretation I defined satisfying Table 3, we have*

$$(\neg(\leq nR.C))^I =_4 (> nR.C)^I \quad and \quad (\neg(\geq nR.C))^I =_4 (< nR.C)^I.$$
$$(\exists R.C)^I =_4 (\geq 1R.C)^I \quad and \quad (\forall R.C)^I =_4 (< 1R.\neg C)^I.$$

Propositions 3 shows that many intuitive relations between different concept constructors still hold under the four-valued semantics, which is one of the nice properties of our four-valued semantics for handling inconsistency.

The next proposition shows that our definition of four-valued semantics for $\mathcal{SROIQ}$ is enough to handle inconsistencies in a $\mathcal{SROIQ}$ knowledge base.

Proposition 4. *For any $\mathcal{SROIQ}$ knowledge base O, $SF(O)$ always has at least one 4-valued model, where $SF(\cdot)$ operator is defined in Definition 3.*

Proof. We can prove that $SF(O)$ has the following 4-valued model I: $A^I = \langle \Delta^I, \cdot^I \rangle$ for each concept name $A \in SF(O)$, $\{o_1, ..., o_n\}^I = \rangle\{o_1^I, ..., o_n^I\}$, $\Delta^I \langle$ and $proj^+(R^I) = \Delta^I \times \Delta^I$ for each role name R, where $\#(\Delta^I) \geq n$. We prove this in two steps. First, it is not difficult to see that for any instance $a \in \Delta^I$,

$$a \in proj^+((\geq nR.C)^I) = \{x \mid \#(y.(x, y) \in proj^+(R^I) \wedge y \in proj^+(C^I)) \geq n\} \text{ and}$$
$$a \in proj^-((\geq nR.C)^I) = \{x \mid \#(y.(x, y) \in proj^+(R^I) \wedge y \notin proj^-(C^I)) < n\}.$$

So $(\geq nR.C)^I = \langle \Delta^I, \Delta^I \rangle$. Similarly, $(\leq nR.C(a))^I = \langle \Delta^I, \Delta^I \rangle$ can be proved.

Secondly, we can easy see that every GCI axiom, in the form of $C(a), R(a, b), C \mapsto D, C \sqsubseteq D$ or $C \to D$, is satisfied in I according to Table 2. For every role inclusion axiom $R \sqsubseteq S$ and role transitivity axiom $Trans(R)$, they hold because all roles are interpreted on $\Delta^I \times \Delta^I$. $\qquad\square$

Note that unqualified number restrictions, $\geq n.R$ and $\leq n.R$ are special forms of number restrictions because $\leq n.R$ equals to $\leq nR.\top$ and $\geq n.R$ equals to $\geq nR.\top$. However, if we defined the four-valued semantics of $\leq n.R (\geq n.R)$ by the four-valued semantics of $\leq nR.\top (\geq nR.\top)$ defined in Table 3 and Table 1, we would find that $\{\leq n.R(a), \geq n+1.R(a)\}$ is still an unsatisfiable set. This is because the following two inequations cannot hold simultaneously since $\top^I = \langle \Delta^I, \emptyset \rangle$:

$$\#(y.(a,y) \in proj(R^I) \wedge y \in proj^+(\top^I)) \geq n + 1$$
$$\#(y.(a,y) \in proj(R^I) \wedge y \notin proj^-(\top^I)) \leq n$$

To address this problem, we also adopt the *substitution* defined by Definition 3. By substituting $\top$ by $A_{new} \sqcup \neg A_{new}$ in $\geq (n+1)R.\top$ and $\leq nR.\top$, we can see that $\{\leq n.R(a), \geq n + 1.R(a)\}$ has a four-valued model with $\Delta^I = \{a, b_1, ..., b_{n+1}\}$, $(a, b_i) \in R^I$ for $1 \leq i \leq n + 1$, and $A_{new}^I = \langle \Delta^I, \Delta^I \rangle$. By doing this, we get a four-valued model I which pushes the contradiction onto the new atomic concept A_{new}.

Next we study how to extend the reduction algorithm to the case of four-valued semantics of $\mathcal{SROIQ}$.

Definition 4. *(Definition 1 extended) For any concept C, its transformation $\pi(C)$ is the concept obtained from C by the following inductively defined transformation.*

- *If $C = \geq nR.D$ for D a concept and R a role, then $\pi(C) = \geq nR.\pi(D)$;*
- *If $C = \leq nR.D$ for D a concept and R a role, then $\pi(C) = \leq nR.\neg\pi(\neg D)$;*
- *If $C = \neg(\geq nR.D)$ for D a concept and R a role, then $\pi(C) = < nR.\neg\pi(\neg D)$;*
- *If $C = \neg(\leq nR.D)$ for D a concept and R a role, then $\pi(C) = > nR.\pi(D)$;*
- *For nominal $\{o_1, ..., o_n\}$, $\pi(\{o_1, ..., o_n\}) = \{o_1, ..., o_n\}$.*
- *For negated nominal $\neg\{o_1, ..., o_n\}$, $\pi(\neg\{o_1, ..., o_n\}) = \{o_1, ..., o_n\}'$ which is a new nominal.*

Regarding both the extension of number restrictions and of nominals, the following theorem holds, which lays the theoretical foundation for the algorithm of four-valued semantics for expressive DLs.

Theorem 5. *(Theorem 1 extended) For any ontology O in $\mathcal{SROIQ}$, we have $O \models_4 \alpha$ if and only if $\pi(O) \models_2 \pi(\alpha)$, where $\models_2$ is the entailment in classical $\mathcal{SROIQ}$.*

Proof. By carefully checking the proof of Theorem 1 [13], we find that the decomposability of four-valued semantics to two-valued semantics [15] is key to the claim. It is not difficult to check that the number restriction and nominal constructors satisfy the decomposability. □

4 Tractable DLs

The forthcoming revision of the Web Ontology Language features so-called *profiles* which are sublanguages of OWL 2 that have desirable properties like polynomial time complexities [19]. In the following, we examine such tractable languages, more precisely $\mathcal{EL}$++, which corresponds to OWL 2 EL, DL-Lite, which corresponds to OWL 2 QL, and Horn-$\mathcal{SHOIQ}$, which is an extension of OWL 2 RL.

We will see that inconsistencies are also unavoidable in these tractable DLs, therefore we consider how to deal with inconsistencies by our approach. We focus on discussing whether the four-valued semantics can preserve the tractability of these tractable DLs. That is, whether the reduction for computing the four-valued semantics transfers tractable DLs still into tractable DLs. If it does, then we can use the four-valued semantics to deal with inconsistency without having to worry about intractability.

4.1 $\mathcal{EL}$++

We do not consider concrete domains. The syntax definition of $\mathcal{EL}$++ knowledge bases is shown in Table 4. $\mathcal{EL}$++ ontologies may also contain role inclusions (RI) of the form $r_1 \circ \cdots \circ r_k \sqsubseteq r$, where $\circ$ denotes role composition.

Table 4. $\mathcal{EL}$++ and Horn-$\mathcal{SHOIQ}$$\circ$. The Horn-$\mathcal{SHOIQ}$$\circ$ normal form used is due to [11].

Language	GCIs	Tractability-preserving Inclusions
$\mathcal{EL}++$	$C \sqsubseteq D$, where $C, D = \top \mid \bot \mid \{a\} \mid$ $C_1 \sqcap C_2 \mid \exists r.C$	internal inclusion (only)
Horn-$\mathcal{SHOIQ}$$\circ$	$\top \sqsubseteq A$, $A \sqsubseteq \bot$, $A \sqcap A' \sqsubseteq B$, $\exists R.A \sqsubseteq B$, $A \sqsubseteq \exists R.B$, $A \sqsubseteq \forall S.B$, $A \sqsubseteq \geq nR.B$, $A \sqsubseteq \leq 1R.B$.	internal inclusion (only)

It is easy to see that an $\mathcal{EL}$++ knowledge base may be inconsistent if we consider the knowledge base $\{A \sqsubseteq \bot, A(a)\}^3$. So we still hope that the 4-valued semantics can help us to handle inconsistency in $\mathcal{EL}$++ knowledge bases. However, we will see that we don't have as many choices of class inclusion as in $\mathcal{ALC}$ and $\mathcal{SROIQ}$ if we want to maintain the tractability of the 4-valued entailment relationship of $\mathcal{EL}$++. The analysis is as follows.

Obviously, the concept transformation of Definition 1 performing on an $\mathcal{EL}$++ concept produces an $\mathcal{EL}$++ concept. For the transformation of internal inclusion, each $\mathcal{EL}$++ axiom $C \sqsubseteq D$ is transformed into $\pi(C) \sqsubseteq \pi(D)$ where $\pi(C)$ and $\pi(D)$ are still $\mathcal{EL}$++ concepts, so that $\pi(C) \sqsubseteq \pi(D)$ is still an $\mathcal{EL}$++ axiom. So internal class inclusion does not destroy the tractability of $\mathcal{EL}$++. This property does not hold for material and strong class inclusions as shown by the following counterexamples: $A \sqcap AA \mapsto B$ and $A \sqcap AA \rightarrow B$. They will be transformed into $\neg(A' \sqcup AA') \sqsubseteq B^+$ and $\{A^+ \sqcap AA^+ \sqsubseteq B^+, B' \sqsubseteq (A' \sqcup AA')\}$ by Definition 2, which are not within the expressivity of $\mathcal{EL}$++. This is mainly because of no negative constructor in $\mathcal{EL}$++.

For role inclusions in $\mathcal{EL}$++, since there is no negative role constructor which can cause inconsistency, we only need to use the classical interpretation for roles as what we do for $\mathcal{ALC}$. So adaptation of 4-valued semantics does not affect the role inclusions axioms.

Theorem 6. *For any give $\mathcal{EL}$++ ontology O and axiom α, $O \models_4 \alpha$ if and only if $\pi(O) \models_2 \pi(\alpha)$. Moreover, if all of the inclusion axioms in O are interpreted as internal inclusion under its four-valued semantics, then $\pi(O)$ is an $\mathcal{EL}$++ ontology.*

Proof. By Theorem 1, it is obvious that O is 4-satisfiable if and only if $\pi(O)$ is two-valued satisfiable. For any inclusion axiom of O, by definition 1 and definition 4, by induction on the construe of concepts of $\mathcal{EL}$++, we can easily get that for any $\mathcal{EL}$++

---

^{3} Note that to enable four-valued models on this ontology, we still need to first perform the substitution defined in Definition 3.

concept C, $\pi(C)$ is still an $\mathcal{EL}$++ concept. Because for any internal inclusion axiom $C \sqsubseteq D$, $\pi(C \sqsubseteq D) = \pi(C) \sqsubseteq \pi(D)$ is an $\mathcal{EL}$++ axiom since $\pi(C)$ and $\pi(D)$ are $\mathcal{EL}$++ concepts. Therefore, $\pi(O) = \{\pi(C) \sqsubseteq \pi(D) \mid C \sqsubseteq D \in O\}$ is a classical $\mathcal{EL}$++ ontology.   □

4.2   Horn-DLs

We ground our discussion on Horn-$\mathcal{SHOIQ}$∘ as defined in [11]. Then we will point out that the same conclusion holds for other Horn-DLs, like Horn-$\mathcal{SHOIQ}$ [18], which has tractable data complexiy [9]. We define Horn-$\mathcal{SHOIQ}$∘ by means of a normal form given in [11], which can be found in Table 4 where A, A', B are concept names.

We can see that all of the Horn-$\mathcal{SHOIQ}$∘ concept constructors preserve its form under $\pi(\cdot)$ operator except $\leq 1R.B$, because $\pi(\leq 1R.B) = \leq 1R.\neg B'$ according to Definition 4. To still maintain the concept structure of $\leq 1R.B$ within Horn-$\mathcal{SHOIQ}$∘, we redefine the $\pi(\cdot)$ as follows

Definition 5. *For any Horn-$\mathcal{SHOIQ}$∘ concept C, $\pi_{Horn}(C)$ is inductively defined as follows:*

- *$\pi_{Horn}(C) = \pi(C)$, if $C = \top, A, \bot, A \sqcap B, \exists R.A, \forall S.B, \geq nR.B$;*
- *$\pi_{Horn}(\leq 1R.B) = \leq 1R.B^=$, where $B^= \sqcap B' \sqsubseteq \bot$, $B^=$ is a new concept name,*
- *$\pi_{Horn}(\neg(\leq 1R.B)) = \geq 2R.B$.*

By Definitions 5 and 2, the transformations for material inclusion axiom $A \mapsto \leq 1R.B$, internal inclusion axiom $A \sqsubseteq \leq 1R.B$ and strong inclusion axiom $A \to \leq 1R.B$ are as follows:

$$\pi_{\text{Horn}}(A \mapsto \leq 1R.B) = \{\neg A' \sqsubseteq \leq 1R.B^=, B^= \sqcap B' \sqsubseteq \bot\}.$$
$$\pi_{\text{Horn}}(A \sqsubseteq \leq 1R.B) = \{A \sqsubseteq \leq 1R.B^=, B^= \sqcap B' \sqsubseteq \bot\}.$$
$$\pi_{\text{Horn}}(A \to \leq 1R.B) = \{A \sqsubseteq \leq 1R.B^=, B^= \sqcap B' \sqsubseteq \bot, \geq 2R.B \sqsubseteq A'\}.$$

Obviously, $\pi_{\text{Horn}}(A \sqsubseteq \leq 1R.B)$ is Horn-$\mathcal{SHOIQ}$∘ ontology, but others are not. Another counterexample for material inclusion and strong inclusion that their transformation cannot guarantee within Horn-$\mathcal{SHOIQ}$∘ expressivity is the one used in the $\mathcal{EL}$++ case. The transformed forms $\neg(A' \sqcup AA') \sqsubseteq B^+$ and $\{A^+ \sqcap AA^+ \sqsubseteq B^+, B' \sqsubseteq (A' \sqcup AA')\}$ are not within the expressivity of Horn-$\mathcal{SHOIQ}$∘. Since $A \sqcap AA \sqsubseteq B$ is allowed in other Horn-DLs, the same conclusion as for Horn-$\mathcal{SHOIQ}$∘ holds. This means that when we want ro preserve the structure of tractable Horn-DLs, we have to choose internal inclusion as the only inclusion form to perform paraconsistent reasoning.

Similarly to the case of $\mathcal{EL}$++, we have the following theorem, which guarantees that internal inclusion axiom can preserve the expressivity of Horn-$\mathcal{SHOIQ}$∘:

Theorem 7. *For any Horn-$\mathcal{SHOIQ}$∘ ontology O, suppose any class inclusion axiom in O is interpreted as internal inclusion. Then (1) $\pi_{Horn}(O)$ is a Horn-$\mathcal{SHOIQ}$∘ ontology; (2) $O \models_4 \alpha$ if and only if $\pi_{Horn}(O) \models_2 \pi_{Horn}(\alpha)$.*

Proof. By Definitions 5 and 2, the first conclusion holds obviously by observing that $\pi_{\mathrm{Horn}}(\leq 1R.A)$ contains a Horn-$\mathcal{SHOIQ}\circ$ concept and an additional Horn-$\mathcal{SHOIQ}\circ$ inclusion axiom under classical semantics. The second claim holds because an Horn-$\mathcal{SHOIQ}\circ$ ontology is a subset of $\mathcal{SROIQ}$ ontology, which makes Theorem 1 guarantee the validity. $\qquad\square$

Note that the case just treated covers DLP [4], which corresponds to OWL 2 RL. In particular, the above transformation, properly restricted, shows that DLP transforms into DLP under internal inclusion, so tractability is preserved when the four-valued semantics is applied.

4.3 DL-Lite

The DL-Lite family includes DL-Lite$_{core}$, DL-Lite$_{\mathcal{F}}$, and DL-Lite$_{\mathcal{R}}$; the latter corresponds to OWL 2 QL. The logics of the DL-Lite family are the maximal DLs supporting efficient query answering over large amounts of instances. In [3], the usual DL reasoning tasks on DL-Lite family are shown to be polynomial in the size of the TBox, and query answering is LogSpace in the size of the ABox. Moreover, the DL-Lite family allows for separation between TBox and ABox reasoning during query evaluation: the part of the process requiring TBox reasoning is independent of the ABox, and the part of the process requiring the ABox can be carried out by a SQL engine [3].

Concepts and roles of DL-Lite family are formed by the following syntax [3]:

$$B ::= A \mid \exists R \qquad R ::= P \mid P^-$$
$$C ::= B \mid \neg B \qquad E ::= R \mid \neg R$$

where A denotes an atomic concept, P an atomic role, and P^- the inverse of the atomic role P. See to Table 5 for the syntax definitions of GCIs and Role Inclusions.

Table 5. DL-Lite Family

Language	GCIs	Other Axioms	Tractability-preserving Inclusions
DL-Lite$_{core}$	$B \sqsubseteq C$	$\emptyset$	internal inclusion (only)
DL-Lite$_{\mathcal{R}}$	$B \sqsubseteq C$	$R \sqsubseteq E$	internal inclusion (only)
DL-Lite$_{\mathcal{F}}$	$B \sqsubseteq C$	(funct R)	internal inclusion (only)

It is also easy to construct an inconsistent knowledge base even for DL-Lite$_{core}$. For instance, $KB = \{B \sqsubseteq \neg A, B(a), A(a)\}$. Moreover, conflictions about roles possibly occur on DL-Lite$_{\mathcal{R}}$, such as $\{P \sqsubseteq S, P \sqsubseteq \neg S, P(a,b)\}$.

In order to still adopt 4-valued semantics for the DL-Lite family, we define the four-valued semantics extension for roles. Just as the four-valued semantics for concepts, a pair $\langle R_P, R_N \rangle$ $(R_P, R_N \subseteq (\Delta^I)^2)$ denotes the four-valued semantics of a role R under interpretation I, where R_P stands for the set of pairs of individuals which are related via R and R_N explicitly represents the set of pairs of individuals which are not related via R. Table 6 gives the formal definition.

Table 6. Four-valued Semantics of DL-Lite

Syntax about Roles	Semantics
R	$R^I = \langle R_P, R_N \rangle$, where $R_P, R_N \subseteq \Delta^I \times \Delta^I$
R^-	$(R^-)^I = \langle R_P^-, R_N^- \rangle$, where R_P^-, R_N^- represent the inverse relations on R_P^- and R_N^-, respectively.
$\neg R$	$(\neg R)^I = \langle R_N, R_P \rangle$
$\exists R$	$\langle \{x \mid \exists y, (x,y) \in R_P^I\}, \{x \mid \neg \exists y, (x,y) \notin R_N^I\} \rangle$
$\neg \exists R$	$\langle \{x \mid \neg \exists y, (x,y) \notin R_N^I\}, \{x \mid \exists y, (x,y) \in R_P^I\} \rangle$
$=_4$	$(=_4)^I = \langle =_P, =_N \rangle$, where $=_P, =_N \subseteq (\Delta^I)^2$
$(Func\ R)$	for any $x, y, z \in \Delta^I$, if $(x,y) \in R_P$ and $(x,z) \in R_P$, then $(y,z) \in =_P$

For simplifying notation, we say that x and y are *positively related* via R under interpretation I if $(x,y) \in R_P^I$, and that x and y are *negatively related* via R under interpretation I if $(x,y) \in R_N^I$.

Intuitively, the first part of the four-valued semantics $\exists R$ denotes the set of individuals x which have an individual y positively related to x via R. While the second part of the four-valued semantics $\exists R$ in Table 6 denotes the set of individuals x which negatively relates to any individual y via R. Note that x is not negatively related to y does not mean x and y are positively related under the four-valued semantics, since $R_P^I \cup R_N^I = \Delta^I \times \Delta^I$ and $R_P^I \cap R_N^I = \emptyset$ are not necessary to hold under the four-valued semantics. This is also the key point why our four-valued semantics can tolerate conflicts caused by role assertions, by allowing a, b both positively related and negatively related via R under a four-valued interpretation I. Similarly as the four-valued semantics for concepts, by imposing $R_P^I \cup R_N^I = \Delta^I \times \Delta^I$ and $R_P^I \cap R_N^I = \emptyset$ on a four-valued interpretation I, I degenerates into a two-valued interpretation.

By the following example, we can see more clearly the intuition underlying the four-valued semantics of $\exists R$:

Example 2. *Given ontology* $O = \{A \sqsubseteq \exists hasStud, A \sqsubseteq \neg \exists hasStud, A(Green)\}$ *which is inconsistent, consider the following four-valued interpretation* $I = (\Delta^I, \cdot^I)$ *with* $\Delta^I = \{a, b, Green\}$ *and* $hasStud^I = \langle \{(Green, a)\}, \{(Green, a), (Green, b), (Green, Green)\} \rangle$. *Under this interpretation, it means that there is information that supports Green having a student a, and there is also information which shows that Green does not relate to any individual via role hasStudent. By checking the following formula and by Table 6, we know that I is a 4-model of O:*

$$Green \in \{three\ exists\ y \in \Delta^I,\ such\ that\ (Green, y) \in hasStud_P^I\}$$
$$Green \in \{for\ all\ y \in \Delta^I, (Green, y) \in hasStud_N^I\}.$$

Intuitively, this 4-model reflects the contradiction about Green having a student.

To define a four-valued semantics for DL-Lite$_\mathcal{F}$ which can tolerate inconsistency, we need to give a four-valued semantics for equality as shown in Table 6, where we use $=_4$ to emphasize the four-valued semantics version of equality and to distinguish from classical equality $=$, and $=_P$ stands for the set of pairs of equal individuals and $=_N$ for the pairs of inequal individuals. To allow expressing inconsistency, the unique name assumption (UNA) is interpreted as: for any $a, b \in ABox, (a^I, b^I) \in =_N$ for any 4-interpretation I. By $=_4$, we can say that two individuals have the information to be same. Based on this, we can define the four-valued semantics for functionality axioms as shown in Table 6. Then if we have an ontology which contains $\{(Func\ R), R(a, b), R(a, c)\}$ and which is inconsistent under the UNA, it can have a 4-model I by assigning $(a^I, b^I) \in =_P \cap =_N$. Now we turn to define the concept transformations for DL-Lite.

Definition 6. *The concept and role transformations for DL-Lite concepts are defined by structural induction as follows.*

- *For $E = R$, $\pi_{Lite}(E) = R$;*
- *For $E = \neg R$, $\pi_{Lite}(E) = R'$, where R' is a new role;*
- *For $C = \exists R$, $\pi_{Lite}(C) = \exists R$;*
- *For $C = \neg \exists R$, $\pi_{Lite}(C) = \neg \exists R^=$, where $R^=$ is a new role name and $R^= \sqsubseteq \neg R'$.*

Considering the internal inclusion transformation, we have that all the GCIs $B \sqsubseteq C$ of DL-Lite will be transferred into the form $B \sqsubseteq C$ with at most an additional role inclusion because $\pi_{Lite}(B \sqsubseteq \neg \exists R) = \{B \sqsubseteq \neg \exists R^=, R^= \sqsubseteq \neg R'\}$. For material inclusion and strong inclusion, because the negative concept is not allowed to occur on the left of a GCI, they do not preserve the DL-Lite structure. So only internal inclusion works under the reduction from four-valued semantics to classical semantics of the DL-Lite family to keep tractability.

Theorem 8. *For any ontology DL-Lite O, $O \models_4 \alpha$ if and only if $\pi_{Lite}(O) \models_{DL-Lite} \pi_{Lite}(\alpha)$, where $\models_{DL-Lite}$ means classical DL-Lite$_\mathcal{R}$ entailment if O is DL-Lite$_{core}$ or DL-Lite$_\mathcal{R}$; and means classical DL-Lite$_\mathcal{A}$ [2] entailment if O is DL-Lite$_\mathcal{F}$.*

Proof. Similarly to the cases of $\mathcal{ALC}$ and $\mathcal{SROIQ}$, by checking the decomposability of four-valued DL-Lite semantics to classical semantics, we can see the theorem holds by further noting the following two facts: 1) The operator $\pi_{Lite}(\cdot)$ performing on internal axioms which contain $\neg \exists R$ may produce a new role axiom in the form of $R^= \sqsubseteq \neg R'$. 2) The produced new role axioms $R^= \sqsubseteq \neg R'$ combining with function assertions in DL-Lite$_\mathcal{F}$ fall into DL-Lite$_\mathcal{A}$ because every right-hand side of the new produced role axiom is like $\neg R'$ which won't occur in function assertions. $\square$

5 Conclusions

In this paper, we extended on our previous study of the four-valued semantics for description logics, and especially adapted it for OWL2 and its tractable DLs. We formally defined their four-valued semantics and proper reductions to the classical semantics, such that all the benefits from existing reasoners on these DLs can be taken advantage of by invoking classical reasoners after employing the presented reduction algorithms in a preprocessing manner. Furthermore, the preprocessing transformations are linear in the size of the knowledge bases. Unlike the four-valued semantics for $\mathcal{ALC}$ and

$\mathcal{SROIQ}$, we showed that in order to preserve the tractability of tractable DLs, only internal class inclusion among the three class inclusion forms is suitable.

Our approach has already been implemented as part of the NeOn Toolkit[4] plugin RaDON. The plugin, which is described in [10], encompasses several methods for inconsistency handling in OWL ontologies. The paraconsistent reasoning algorithm of RaDON, which is based on the work presented in this paper, leaves it to the user to decide how class inclusion axioms are transformed.

Future work on this topic can go into several directions. Adaptation of our approach to obtain tractable paraconsistent reasoning support for larger tractable languages than those presented here, e.g. for ELP [12], will enhance its potential applicability. We also consider it important to investigate on which grounds reasonable choices for transforming inclusion axioms can be made; indeed there may be alternative choices to the three inclusion axioms presented here, which may be useful in certain contexts.

Acknowledgement. We acknowledge support by the Deutsche Forschungsgemeinschaft (DFG) in the ReaSem project; and by OSEO, agence nationale de valorisation de la recherche in the Quaero project. We appreciate the anonymous reviewers for their valuable comments.

References

1. Belnap, N.D.: A useful four-valued logic. Modern uses of multiple-valued logics, 7–73 (1977)
2. Calvanese, D., De Giacomo, G., Lembo, D., Lenzerini, M., Poggi, A., Rosati, R.: Linking data to ontologies: The description logic DL-Lite-A. In: Proc. of the 2nd Int. Workshop on OWL: Experiences and Directions (OWLED 2006). CEUR Electronic Workshop Proceedings, vol. 216 (2006), http://ceur-ws.org/
3. Calvanese, D., De Giacomo, G., Lembo, D., Lenzerini, M., Rosati, R.: Tractable reasoning and efficient query answering in description logics: The DL-Lite family. J. Autom. Reasoning 39(3), 385–429 (2007)
4. Grosof, B.N., Horrocks, I., Volz, R., Decker, S.: Description logic programs: combining logic programs with description logic. In: Proceedings of the 12th International World Wide Web Conference (WWW 2003), Budapest, Hungary, pp. 48–57. ACM, New York (2003)
5. Haase, P., van Harmelen, F., Huang, Z., Stuckenschmidt, H., Sure, Y.: A Framework for Handling Inconsistency in Changing Ontologies. In: Gil, Y., Motta, E., Benjamins, V.R., Musen, M.A. (eds.) ISWC 2005. LNCS, vol. 3729, pp. 353–367. Springer, Heidelberg (2005)
6. Hitzler, P., Krötzsch, M., Rudolph, S.: Foundations of Semantic Web Technologies. Chapman & Hall/CRC (2009)
7. Horrocks, I., Patel-Schneider, P.F., van Harmelen, F.: From $\mathcal{SHIQ}$ and RDF to OWL: The making of a web ontology language. Journal of Web Semantics 1(1), 7–26 (2003)
8. Huang, Z., van Harmelen, F., ten Teije, A.: Reasoning with inconsistent ontologies. In: Kaelbling, L.P., Saffiotti, A. (eds.) IJCAI, pp. 454–459. Professional Book Center (2005)
9. Hustadt, U., Motik, B., Sattler, U.: Data Complexity of Reasoning in Very Expressive Description Logics. In: Proc. of the 19th Int. Joint Conference on Artificial Intelligence (IJCAI 2005), Edinburgh, UK, July 30 – August 5, pp. 466–471. Morgan Kaufmann Publishers, San Francisco (2005)

---

[4] http://www.neon-toolkit.org/

10. Ji, Q., Haase, P., Qi, G., Hitzler, P., Stadtmüller, S.: Radon - repair and diagnosis in ontology networks. In: Aroyo, L., Traverso, P., Ciravegna, F., Cimiano, P., Heath, T., Hyvönen, E., Mizoguchi, R., Oren, E., Sabou, M., Simperl, E.P.B. (eds.) ESWC 2009. LNCS, vol. 5554, pp. 863–867. Springer, Heidelberg (2009)

11. Krötzsch, M., Rudolph, S., Hitzler, P.: Complexity boundaries for Horn description logics. In: AAAI, pp. 452–457. AAAI Press, Menlo Park (2007)

12. Krötzsch, M., Rudolph, S., Hitzler, P.: ELP: Tractable rules for OWL 2. In: Sheth, A.P., Staab, S., Dean, M., Paolucci, M., Maynard, D., Finin, T., Thirunarayan, K. (eds.) ISWC 2008. LNCS, vol. 5318, pp. 649–664. Springer, Heidelberg (2008)

13. Ma, Y., Hitzler, P., Lin, Z.: Algorithms for paraconsistent reasoning with OWL. In: Franconi, E., Kifer, M., May, W. (eds.) ESWC 2007. LNCS, vol. 4519, pp. 399–413. Springer, Heidelberg (2007)

14. Ma, Y., Hitzler, P., Lin, Z.: Paraconsistent reasoning for expressive and tractable description logics. In: Baader, F., Lutz, C., Motik, B. (eds.) Proceedings of the 21st International Workshop on Description Logics, DL 2008, Dresden, Germany, May 2008. CEUR Workshop Proceedings, vol. 353 (2008)

15. Ma, Y., Lin, Z., Lin, Z.: Inferring with inconsistent OWL DL ontology: A multi-valued logic approach. In: Grust, T., Höpfner, H., Illarramendi, A., Jablonski, S., Mesiti, M., Müller, S., Patranjan, P.-L., Sattler, K.-U., Spiliopoulou, M., Wijsen, J. (eds.) EDBT 2006. LNCS, vol. 4254, pp. 535–553. Springer, Heidelberg (2006)

16. Ma, Y., Qi, G., Hitzler, P., Lin, Z.: Measuring inconsistency for description logics based on paraconsistent semantics. In: Mellouli, K. (ed.) ECSQARU 2007. LNCS (LNAI), vol. 4724, pp. 30–41. Springer, Heidelberg (2007)

17. McGuinness, D.L., van Harmelen, F. (eds.): OWL Web Ontology Language Overview. W3C Recommendation, February 10 (2004), http://www.w3.org/TR/owl-features/.

18. Motik, B.: Reasoning in description logics using resolution and deductive databases. PhD thesis, University of Karlsruhe, Germany (2006)

19. Motik, B., Grau, B.C., Horrocks, I., Wu, Z., Fokoue, A., Lutz, C. (eds.): OWL 2 Web Ontology Language: Profiles. W3C Candidate Recommendation, June 11 (2009), http://www.w3.org/TR/2009/CR-owl2-profiles-20090611/

20. Patel-Schneider, P.F.: A four-valued semantics for terminological logics. Artificial Intelligence 38, 319–351 (1989)

21. Schlobach, S., Cornet, R.: Non-standard reasoning services for the debugging of description logic terminologies. In: Gottlob, G., Walsh, T. (eds.) IJCAI, pp. 355–362. Morgan Kaufmann, San Francisco (2003)

22. Staab, S., Studer, R.: Handbook on Ontologies, 2nd edn. International Handbooks on Information Systems. Springer, Heidelberg (2009)

23. Straccia, U.: A sequent calculus for reasoning in four-valued description logics. In: Galmiche, D. (ed.) TABLEAUX 1997. LNCS, vol. 1227, pp. 343–357. Springer, Heidelberg (1997)

24. W3C OWL Working Group. OWL 2 Web Ontology Language: Document Overview (2009), http://www.w3.org/TR/owl2-overview/

A Formal Theory for Modular ERDF Ontologies

Anastasia Analyti[1], Grigoris Antoniou[1,2], and Carlos Viegas Damásio[3]

[1] Institute of Computer Science, FORTH-ICS, Greece
[2] Department of Computer Science, University of Crete, Greece
[3] CENTRIA, Departamento de Informatica, Faculdade de Ciencias e Tecnologia,
Universidade Nova de Lisboa, 2829-516 Caparica, Portugal
{analyti,antoniou}@ics.forth.gr, cd@di.fct.unl.pt

Abstract. The success of the Semantic Web is impossible without any form of modularity, encapsulation, and access control. In an earlier paper, we extended RDF graphs with weak and strong negation, as well as derivation rules. The ERDF #*n*-stable model semantics of the extended RDF framework (*ERDF*) is defined, extending RDF(S) semantics. In this paper, we propose a framework for modular ERDF ontologies, called *modular ERDF framework*, which enables collaborative reasoning over a set of ERDF ontologies, while support for hidden knowledge is also provided. In particular, the *modular ERDF stable model semantics* of modular ERDF ontologies is defined, extending the ERDF #*n*-stable model semantics. Our proposed framework supports local semantics and different points of view, local closed-world and open-world assumptions, and scoped negation-as-failure. Several complexity results are provided.

1 Introduction

Ontologies and automated reasoning are the building blocks of the Semantic Web initiative. Derivation rules can be included in an ontology to define derived concepts based on base concepts. For example, rules allow to define the extension of a class or property based on a complex relation between the extensions of the same or other classes and properties. On the other hand, the inclusion of negative information both in the form of negation-as-failure and explicit negative information is also needed to enable various forms of reasoning. In [1], the Semantic Web language RDFS [8,6] is extended to accommodate the two negations of Partial Logic [7], namely *weak negation* ∼ (expressing negation-as-failure or non-truth) and *strong negation* ¬ (expressing explicit negative information or falsity), as well as derivation rules. The new language is called *Extended RDF* (*ERDF*). Specifically, in [1], the *stable model semantics* of ERDF ontologies is developed, based on Partial Logic, extending the model-theoretic semantics of RDFS [6]. The concrete syntax of ERDF is presented in [14].

ERDF enables the combination of closed-world (non-monotonic) and open-world (monotonic) reasoning, in the same framework, through the presence of weak negation (in the body of the program rules) and the new metaclasses *erdf:TotalProperty* and *erdf:TotalClass*, respectively. In particular, relating

A. Polleres and T. Swift (Eds.): RR 2009, LNCS 5837, pp. 212–226, 2009.

strong and weak negation at the interpretation level, ERDF distinguishes two categories of properties and classes [1]. *Partial properties* are properties p that may have truth-value gaps, that is $p(x, y)$ is possibly neither true nor false. *Total properties* are properties p that satisfy *totalness*, that is $p(x, y)$ is either true or false. Partial and total classes c are defined similarly, by replacing $p(x, y)$ by $rdf{:}type(x, c)$. In [1], it is shown that on total properties and total classes, the *Open-World Assumption (OWA)* applies.

ERDF also distinguishes properties and classes that are completely represented in a knowledge base with respect to an (optional) ERDF formula F, corresponding to the *context* where the completion takes place. Such a completeness assumption for *closing* a partial property p by default may be expressed in ERDF by means of the rule $\neg p(?x, ?y) \leftarrow F \wedge \sim p(?x, ?y)$ and for a partial class c, by means of the rule $\neg rdf{:}type(?x, c) \leftarrow F \wedge \sim rdf{:}type(?x, c)$, where F is an ERDF formula.

Intuitively, an ERDF ontology is the combination of (i) an ERDF graph G containing (implicitly existentially quantified) positive and negative information, and (ii) an ERDF program P containing derivation rules, with possibly all connectives $\sim, \neg, \supset, \wedge, \vee, \forall, \exists$ in the body of a rule, and strong negation $\neg$ in the head of a rule.

In [1], it is shown that stable model entailment conservatively extends RDFS entailment from RDF graphs to ERDF ontologies. Unfortunately, satisfiability and entailment under the ERDF stable model semantics are in general undecidable. This is due to the fact that the RDF vocabulary is infinite. Therefore, to achieve decidability of reasoning in the general case, in [2], we propose a modified semantics, called *ERDF #n-stable model semantics* (for $n \in I\!N$), in which from the RDF vocabulary, we remove the infinite set of terms $\{rdf{:}\_i \mid i > n\}$. The new semantics also extends RDFS entailment from RDF graphs to ERDF ontologies. Additionally, in [2], we provide an equivalence statement between ERDF stable model entailment and ERDF #n-stable model entailment on an ERDF ontology O, in the case that the bodies of the rules in O contain only the connectives $\neg$ and $\wedge$.

The success of the Semantic Web is impossible without any form of modularity, encapsulation, and access control. In this paper, we propose a framework for modular ERDF ontologies, called *modular ERDF framework*, in which a modular ERDF ontology $\mathcal{R}$ is a set of **r**-ERDF ontologies. Intuitively, an **r**-ERDF ontology $O \in \mathcal{R}$ is an ERDF ontology that can import or just reference knowledge about a property or class x from other **r**-ERDF ontologies in $\mathcal{R}$ that define x and are willing to export this knowledge to O. Thus, our modular ERDF framework enables collaborative reasoning over a set of **r**-ERDF ontologies, while support for hidden knowledge is also provided. Additionally, it supports local semantics and different points of view, local closed-world and open-world assumptions, and scoped negation-as-failure.

Specifically, in this paper, we define the *modular (ERDF) stable models* of an **r**-ERDF ontology w.r.t. a modular ERDF ontology. Several properties of the modular stable model semantics are provided, including that modular stable model entailment extends #n-stable model entailment on ERDF ontologies, and thus also, RDFS entailment on RDF graphs. We show that if $\mathcal{R}$ is a simple modular ERDF ontology (i.e., the bodies of the rules of the r-ERDF ontologies in $\mathcal{R}$ contain only the connectives $\sim$, $\neg$, $\wedge$) then query answering under the modular ERDF stable model semantics reduces to query answering under the answer set semantics [5]. Moreover, we provide complexity results for the modular ERDF stable model semantics on (i) simple modular ERDF ontologies, (ii) modular ERDF ontologies without quantifiers, and (ii) general modular ERDF ontologies.

We would like to mention that the goal of our modular ERDF framework is on interconnecting independently developed r-ERDF ontologies over the web and *not* on querying a large ontology by decomposing it into smaller sub-ontologies. The latter problem has been considered for answer set semantics in [10], but [10] prohibits the existence of positive recursion among modules, a serious limitation for the Semantic Web setting. In contrast, in our framework, considered r-ERDF ontologies may be interconnected via cyclic references. For example, an r-ERDF ontology O may be created any time after the independent creation of the r-ERDF ontologies on which it depends (which later may be updated, possibly referring to O).

The rest of the paper is organized as follows: In Section 2, we review ERDF graphs, which we extend to r-ERDF formulas. Then, we define r-ERDF ontologies and valid modular ERDF ontologies. Section 3 defines the modular ERDF interpretations of an r-ERDF ontology w.r.t. a modular ERDF ontology. Then, it defines satisfiability of an r-ERDF formula by such a modular ERDF interpretation and an r-ERDF ontology. In Section 4, we define the modular stable semantics of an r-ERDF ontology w.r.t. a modular ERDF ontology, and provide its properties. Further, we provide several complexity results for the modular ERDF stable model semantics. Section 5 reviews related work and concludes the paper.

2   Modular ERDF Ontologies

In this Section, we define r-ERDF formulas, valid r-ERDF ontologies, and valid modular ERDF ontologies. Additionally, we provide a comprehensive example of a modular ERDF ontology.

A (Web) *vocabulary* V is a set of URI references and/or literals (plain or typed) [6]. We denote the set of all URI references by URI, the set of all plain literals by $\mathcal{PL}$, the set of all typed literals by $\mathcal{TL}$, and the set of all literals by $\mathcal{LIT}$. We consider a set Var of variable symbols, such that the sets Var, URI, $\mathcal{LIT}$ are pairwise disjoint. In our examples, variable symbols are prefixed by "?".

Below, we review the definition of an ERDF triple from [1]. Let V be a vocabulary. A (*normal*) *ERDF triple* over V is an expression of the form $p(s,o)$ or $\neg p(s,o)$, where $s, o \in V \cup Var$ are called *subject* and *object*, respectively, and $p \in V \cap URI$ is called *property*.

Below we extend the definition of an ERDF formula, provided in [1], to r-ERDF formulas. We consider the connectives $\{\sim, \neg, \wedge, \vee, \supset, \exists, \forall\}$, where $\neg$, $\sim$, and $\supset$ are called *strong negation*, *weak negation*, and *material implication* respectively. Let V be a vocabulary and let $O_{\mathrm{nam}} \subseteq URI$ be a set of r-ERDF ontology names. We define $L(V)$ to be the smallest set that contains the ERDF triples over V and is closed with respect to the following conditions: if $F, G \in L(V)$ then $\{\sim F,\ F \wedge G,\ F \vee G,\ F \supset G,\ \exists x F,\ \forall x F\} \subseteq L(V)$, where $x \in Var$. A *(normal) ERDF formula* over V is an element of $L(V)$. A *qualified ERDF formula* over V and O_{nam} has the form $F@oname$, where $F \in L(V)$ and $oname \in O_{\mathrm{nam}}$ (i.e., F will be evaluated at the r-ERDF ontology identified by $oname$).

Definition 1 (r-ERDF formula). Let V be a vocabulary and let $O_{\mathrm{nam}} \subseteq URI$. We define $L(V, O_{\mathrm{nam}})$ to be the smallest set that (i) contains the ERDF formulas over V and the qualified ERDF formulas over V and O_{nam}, and (ii) is closed with respect to the following conditions: if $F, G \in L(V, O_{\mathrm{nam}})$ then $\{\sim F,\ F \wedge G,\ F \vee G,\ F \supset G,\ \exists x F,\ \forall x F\} \subseteq L(V, O_{\mathrm{nam}})$, where $x \in Var$. An *r-ERDF formula* F over V and O_{nam} is an element of $L(V, O_{\mathrm{nam}})$. We denote the set of variables appearing in F by $Var(F)$, and the set of free variables[1] appearing in F by $FVar(F)$. $\square$

Next, we review the definition of an ERDF graph G and the skolemization of G from [1]. An *ERDF graph* G over a vocabulary V is a set of ERDF triples over V. We denote the variables appearing in G by $Var(G)$, and the set of URI references and literals appearing in G by V_G. Intuitively, an ERDF graph G represents an existentially quantified conjunction of ERDF triples. Specifically, let $G = \{t_1, ..., t_m\}$ be an ERDF graph, and let $Var(G) = \{x_1, ..., x_k\}$. Then, G represents the ERDF formula $formula(G) = \exists?x_1, ..., \exists?x_k\ t_1 \wedge ... \wedge t_m$. Existentially quantified variables in ERDF graphs are handled by *skolemization*. Let G be an ERDF graph. The *skolemization function* of G is an 1:1 mapping $sk_G : Var(G) \rightarrow URI$, where for each $x \in Var(G)$, $sk_G(x)$ is an artificial URI, denoted by $G{:}x$. The *skolemization* of G, denoted by $sk(G)$, is the ground ERDF graph derived from G after replacing each $x \in Var(G)$ by $sk_G(x)$.

Below, we extend the definitions of ERDF rule and ERDF program, provided in [1], to r-ERDF rule and r-ERDF program, respectively.

Definition 2 (r-ERDF rule, r-ERDF program). An *r-ERDF rule* r over a vocabulary V and $O_{\mathrm{nam}} \subseteq URI$ is an expression of the form: $G \leftarrow F$, where $F \in L(V, O_{\mathrm{nam}}) \cup \{\mathtt{true}\}$ (called *condition*) and G (called *conclusion*) is either an ERDF triple over V or $\mathtt{false}$. We assume that no bound variable in F appears free in G. We denote the set of variables and the set of free variables of r by $Var(r)$ and $FVar(r)$[2], respectively. Additionally, we write $Cond(r) = F$ and $Concl(r) = G$.

An *r-ERDF program* P over a vocabulary V and $O_{\mathrm{nam}} \subseteq URI$ is a finite set of r-ERDF rules over V and O_{nam}. We denote the set of URI references and literals appearing in P by V_P. $\square$

---

[1] Without loss of generality, we assume that a variable cannot have both free and bound occurrences in F, and more than one bound occurrence.

[2] $FVar(r) = FVar(F) \cup FVar(G)$.

Below, we extend the definition of an ERDF ontology, provided in [1], to an r-ERDF ontology.

Definition 3 (r-ERDF ontology). An r-*ERDF ontology* O over a vocabulary V and $O_{nam} \subseteq URI$ is a triple $O = \langle Nam_O, L_O, Int_O \rangle$, where: (i) $Nam_O \in O_{nam}$ is the *name* of O, (ii) $L_O = \langle G_O, P_O, \rangle$, is the *logic* of O, where G_O is an ERDF graph over V and P_O is an r-ERDF program over V and O_{nam}, and (iii) $Int_O = \langle Exp_O^{pr}, Exp_O^{cl}, Imp_O^{pr}, Imp_O^{cl} \rangle$ is the *interface* of O, where: For $\mathtt{t} \in \{\mathtt{pr}, \mathtt{cl}\}$, it holds that:

- $Exp_O^{\mathtt{t}}$ is a set of pairs $\langle x, Exp \rangle$, where $x \in V$ and $Exp \subseteq O_{nam} - \{Nam_O\}$ or $Exp = \{*\}$. It holds that if $\langle x, Exp \rangle$ and $\langle x, Exp' \rangle \in Exp_O^{\mathtt{t}}$ then $Exp = Exp'$.
  We define: $Exported_O^{\mathtt{t}} = \{x \mid \exists \langle x, Exp \rangle \in Exp_O^{\mathtt{t}}\}$ and $Export_O^{\mathtt{t}}(x) = Exp$.
- $Imp_O^{\mathtt{t}}$ is a set of pairs $\langle x, Imp \rangle$, where $x \in V$, and $Imp \subseteq O_{nam} - \{Nam_O\}$ or $Imp = \{*\}$. It holds that if $\langle x, Imp \rangle$ and $\langle x, Imp' \rangle \in Imp_O^{\mathtt{t}}$ then $Imp = Imp'$.
  We define: $Imported_O^{\mathtt{t}} = \{x \mid \exists \langle x, Imp \rangle \in Imp_O^{\mathtt{t}}\}$ and $Import_O^{\mathtt{t}}(x) = Imp$.

Let O be an r-ERDF ontology. Intuitively, each pair $\langle x, Exp \rangle \in Exp_O^{pr}$ (resp. $\langle x, Exp \rangle \in Exp_O^{cl}$) corresponds to an $\mathtt{export}$ declaration of O, where x is a property (resp. class) exported by O and Exp is the list of r-ERDF ontologies to which O is willing to export x. If O is willing to export x to any requesting r-ERDF ontology then $Exp = \{*\}$.

Similarly, each pair $\langle x, Imp \rangle \in Imp_O^{pr}$ (resp. $\langle x, Imp \rangle \in Imp_O^{cl}$) corresponds to an $\mathtt{import}$ declaration of O, where x is a property (resp. class) requested by O, and Imp is the list of r-ERDF ontologies from which x is requested. If O requests x from any providing r-ERDF ontology then $Imp = \{*\}$. Obviously, we do not allow duplicate $\mathtt{export}$ and $\mathtt{import}$ declarations for classes and properties in O.

Definition 4 (Modular ERDF ontology). A *modular ERDF ontology* (*MEO*) $\mathcal{R}$ is a set of r-ERDF ontologies.   □

Example 1. Consider the modular ERDF ontology $\mathcal{R} = \{O_1, O_2, O_3, O_4, O_5\}$, shown in Figure 1[3]. Ontology O_1, with Nam_{O_1} =<http://geography.int>, provides geographical information, stating that the list of European countries is positively closed (w.r.t. the list of countries). This local CWA is expressed by the single rule in P_{O_1}. Ontology O_2, with Nam_{O_2} =<http://europa.eu>, defines the list of European Union countries (which does not include $\mathtt{Croatia}$) and states that this list is open (w.r.t. the resources of O_2)[4] by declaring the class $\mathtt{eu:CountryEU}$ as total. This local OWA is expressed by the first ERDF triple in G_{O_2}. Ontology O_3, with Nam_{O_3} =<http://www.pyramis.gr>, provides information regarding the package tours of the greek travel agency *Pyramis*. Similarly, ontology O_4, with Nam_{O_4} =<http://www.travel_plan.gr>, provides information regarding the package tours of the greek travel agency *Travel Plan*.

Finally, ontology O_5, with Nam_{O_5} =<http://www.anne_travel_pref.gr>, presents the travel preferences of Anne. Specifically, Anne prefers either (i) a

---

[3] Following usual convention, we have replaced ∧ by "," in the program rules.
[4] Note that ontology O_2 imports class $\mathtt{geo:Country}$ from ontology O_1.

Ontology O_1

$\langle http://geography.int\rangle$

exports class geo:Country to *.
exports class geo:Europ_Country to *.
exports property geo:capital to *.

$G_{O_1} =$
rdfs:subclass(geo:Europ_Country,
                    geo:Country).
rdf:type(geo:Egypt,geo:Country).
rdf:type(geo:Italy,geo:Europ_Country).
rdf:type(geo:Croatia,geo:Europ_Country).
geo:capital(geo:Cairo,geo:Egypt).
geo:capital(geo:Zagreb,geo:Croatia).
...

$P_{O_1} =$
$\neg$ rdf:type(?x,geo:Europ_Country) $\leftarrow$
        rdf:type(?x,geo:Country),
        $\sim$ rdf:type(?x,geo:Europ_Country).

Ontology O_2

$\langle http://europa.eu\rangle$

imports class geo:Country from
            $\langle http://geography.int\rangle$.
exports class eu:CountryEU to *.

$G_{O_2} =$
rdf:type(eu:CountryEU, erdf:TotalClass).
rdf:type(geo:Italy,eu:CountryEU).
rdf:type(geo:Greece,eu:CountryEU).
...

Ontology O_3

$\langle http://www.pyramis.gr\rangle$

exports property vac:travel to *.
exports property vac:visit to *.

$G_{O_3} =$
vac:travel(pyr:package1,geo:Egypt).
vac:visit(pyr:package1,geo:Cairo).
vac:travel(pyr:package2,geo:Egypt).
vac:visit(pyr:package2,geo:Cairo).
vac:visit(pyr:package2,geo:Luxor).

Ontology O_4

$\langle http://www.travel\_plan.gr\rangle$

exports property vac:travel to *.
exports property vac:visit to *.

$G_{O_4} =$
vac:travel(trav:package1,geo:Italy).
vac:visit(trav:package1,geo:Rome).
vac:travel(trav:package2,geo:Croatia).
vac:visit(trav:package2,geo:Zagreb).
vac:visit(trav:package2,geo:Trogir).

Ontology O_5

$\langle http://www.anne\_travel\_pref.gr\rangle$

imports class geo:Europ_Country from $\langle http://geography.int\rangle$.
imports property geo:capital from $\langle http://geography.int\rangle$.
imports class eu:CountryEU from $\langle http://europa.eu\rangle$.
imports property vac:travel from *.
imports property vac:visit from *.
exports property ann:choose_trav_package to $\langle http://www.peter\_travel\_pref.gr\rangle$.

$P_{O_5} =$
eq:id(?x,?x) $\leftarrow$ true.

ann:choose_trav_package(?package,?country) $\leftarrow$ $\neg$ rdf:type(?country,geo:Europ_Country),
  (vac:travel(?package,?country), vac:visit(?package,?city))@$\langle http://www.pyramis.gr\rangle$,
  $\forall$?city$'$ vac:visit(?package,?city$'$)@$\langle http://www.pyramis.gr\rangle$ $\supset$ eq:id(?city,?city$'$).

ann:choose_trav_package(?package,?country) $\leftarrow$ rdf:type(?country,geo:CountryEU),
  (vac:travel(?package,?country), vac:visit(?package,?city),
   vac:visit(?package,?city$'$))@$\langle http://www.travel\_plan.gr\rangle$, $\sim$ eq:id(?city,?city$'$).

ann:choose_trav_package(?package,?country) $\leftarrow$ rdf:type(?country,geo:Europ_Country),
  $\neg$ rdf:type(?country,geo:CountryEU),
   (vac:travel(?package,?country), vac:visit(?package,?city))@$\langle http://www.travel\_plan.gr\rangle$,
   geo:capital(?city,?country).

Fig. 1. A modular ERDF ontology

trip to a non-European country by *Pyramis* that visits only one city, or (ii) a trip to an EU country by *Travel Plan* that visits at least one city, or (iii) a trip to a European but not EU country by *Travel Plan* that visits the capital of the country. Note that O_5 imports the properties vac:travel and vac:visit from any providing r-ERDF ontology in $\mathcal{R}$ (that is, O_3 and O_4). Additionally, note that r-ERDF ontology O_5 exports property ann:choose_trav_package to an r-ERDF ontology, named <http://www.peter_travel_pref.gr>, not in $\mathcal{R}$. □

Let $\mathcal{R}$ be a modular ERDF ontology, let $O \in \mathcal{R}$, and let $x \in \mathit{Exported}_O^t$, for $t \in \{\mathtt{pr}, \mathtt{cl}\}$. We define:

$$\mathit{Export}_{O,\mathcal{R}}^t(x) = \begin{cases} \{\mathit{Nam}_{O'} \mid O' \in \mathcal{R} - \{O\}\} & \text{if } \mathit{Export}_O^t(x) = \{*\} \\ \mathit{Export}_O^t(x) \cap \{\mathit{Nam}_{O'} \mid O' \in \mathcal{R}\} & \text{otherwise} \end{cases}$$

Intuitively, $\mathit{Export}_{O,\mathcal{R}}^{\mathtt{pr}}(x)$ (resp. $\mathit{Export}_{O,\mathcal{R}}^{\mathtt{cl}}(x)$) denotes the r-ERDF ontologies in $\mathcal{R}$ to which O is willing to export property (resp. class) x.

Example 2. Consider the modular ERDF ontology $\mathcal{R}$ of Example 1. Then, it holds that: $\mathit{Export}_{O_2,\mathcal{R}}^{\mathtt{cl}}(\mathtt{eu:CountryEU}) = \{O_1, O_3, O_4, O_5\}$, because $\mathit{Export}_{O_2}^{\mathtt{cl}}$ (eu:CountryEU) $= \{*\}$. Additionally, it holds that $\mathit{Export}_{O_5,\mathcal{R}}^{\mathtt{pr}}(\mathtt{ann:choose\_trav\_package})=\{\}$. □

Let $\mathcal{R}$ be a modular ERDF ontology, let $O \in \mathcal{R}$, and let $x \in \mathit{Imported}_O^t$, for $t \in \{\mathtt{pr}, \mathtt{cl}\}$. We define:

$$\mathit{Import}_{O,\mathcal{R}}^t(x) = \begin{cases} \mathit{ExportingTo}_\mathcal{R}^t(x, O) & \text{if } \mathit{Import}_O^t(x) = \{*\} \\ \mathit{Import}_O^t(x) \cap \mathit{ExportingTo}_\mathcal{R}^t(x, O) & \text{otherwise,} \end{cases}$$

where $\mathit{ExportingTo}_\mathcal{R}^t(x, O) = \{\mathit{Nam}_{O'} \mid O' \in \mathcal{R}, \mathit{Nam}_O \in \mathit{Export}_{O',\mathcal{R}}^t(x)\}$. Intuitively, $\mathit{ExportingTo}_\mathcal{R}^{\mathtt{pr}}(x, O)$ (resp. $\mathit{ExportingTo}_\mathcal{R}^{\mathtt{cl}}(x, O)$) denotes the r-ERDF ontologies in $\mathcal{R}$ that are willing to export property (resp. class) x to O. Additionally, $\mathit{Import}_{O,\mathcal{R}}^{\mathtt{pr}}(x)$ (resp. $\mathit{Import}_{O,\mathcal{R}}^{\mathtt{cl}}(x)$) denotes the r-ERDF ontologies in $\mathcal{R}$ from which O imports property (resp. class) x.

Example 3. For the modular ERDF ontology $\mathcal{R}$ of Example 1, $\mathit{ExportingTo}_\mathcal{R}^{\mathtt{pr}}$ (vac:travel, O_5) $= \{O_3, O_4\}$. Additionally, $\mathit{Import}_{O_5,\mathcal{R}}^{\mathtt{pr}}(\mathtt{vac:travel}) = \{O_3, O_4\}$. □

In order for a modular rule base to be *valid*, it has to satisfy a number of validity constraints.

Definition 5 (Valid modular ERDF ontology). A modular ERDF ontology $\mathcal{R}$ is *valid* iff:

1. If $O, O' \in \mathcal{R}$ and $O \neq O'$ then $\mathit{Nam}_O \neq \mathit{Nam}_{O'}$.
2. If $O \in \mathcal{R}$ and $x \in \mathit{Imported}_O^t$, for $t \in \{\mathtt{pr}, \mathtt{cl}\}$, then $\mathit{Import}_O^t(x) = \{*\}$ or $\mathit{Import}_O^t(x) \subseteq \mathit{ExportingTo}_\mathcal{R}^t(x, O)$.
3. If $O \in \mathcal{R}$ and $r \in P_O$ such that a qualified ERDF formula $F@\mathit{Nam}_{O'}$ appears in $\mathit{Cond}(r)$ then: (i) $O' \in \mathcal{R}$, (ii) for each $p(s, o)$, where $p \neq \mathit{rdf:type}$, appearing in F, it holds that $O \in \mathit{Export}_{O',\mathcal{R}}^{\mathtt{pr}}(p)$, and (iii) for each $\mathit{rdf:type}(x, c)$, appearing in F, it holds that (a) $O \in \mathit{Export}_{O',\mathcal{R}}^{\mathtt{pr}}(\mathit{rdf:type})$ or (b) $c \in URI$ and $O \in \mathit{Export}_{O',\mathcal{R}}^{\mathtt{cl}}(c)$. □

Let $\mathcal{R}$ be a valid modular ERDF ontology. Constraint (1) of Definition 5 expresses that different r-ERDF ontologies in $\mathcal{R}$ should have different names in order to be uniquely identified. Let $O \in \mathcal{R}$. Constraint (2) expresses that if O requests a property or class x *explicitly* from an r-ERDF ontology O' then it should hold that $O' \in \mathcal{R}$ and O' is willing to export x to O. Assume now that it exists $r \in P_O$ s.t. $Cond(r)$ refers to an ERDF formula F of an r-ERDF ontology O'. Constraint (3.i) expresses that it should hold $O' \in \mathcal{R}$. Constraint (3.ii) expresses that if r refers to $p(s, o)$ of O', where $p \neq rdf{:}type$, then O' should be willing to export property p to O. Additionally, constraint (3.iii) expresses that if O refers to $rdf{:}type(x, c)$ of O' then O' should be willing to either (a) export to O the property $rdf{:}type$, expressing that all classes of O' are exported to O, or (b) export to O just the class c (if $c \in URI$).

Example 4. Modular rule base $\mathcal{R}$ of Example 1 is valid. □

In this work, we consider valid modular ERDF ontologies, only. Additionally, by $\mathcal{R}$, we will denote a valid modular ERDF ontology.

3   Modular ERDF and Herbrand Interpretations

In this section, we define the modular ERDF interpretations of an r-ERDF ontology w.r.t. a modular ERDF ontology. Additionally, we define satisfaction of an r-ERDF formula by such a modular ERDF interpretation and an r-ERDF ontology. Further, we define the modular Herbrand interpretations of an r-ERDF ontology w.r.t. a modular ERDF ontology.

Below we review the definition of a partial interpretation of a vocabulary V [1], which is an extension of the definition of a simple interpretation of V [6], such that each property is associated not only with a truth extension but also with a falsity extension, allowing for partial properties.

Definition 6 (Partial interpretation of a vocabulary). A *partial interpretation* I of a vocabulary V_I consists of:

- A non-empty set of resources Res_I, called the *domain* or *universe* of I.
- A set of properties $Prop_I$.
- A vocabulary interpretation mapping $I_v : V_I \cap URI \to Res_I \cup Prop_I$.
- A property-truth extension mapping $PT_I : Prop_I \to \mathcal{P}(Res_I \times Res_I)$.
- A property-falsity extension mapping $PF_I : Prop_I \to \mathcal{P}(Res_I \times Res_I)$.
- A mapping $IL_I : V_I \cap \mathcal{TL} \to Res_I$.
- A set of literal values $LV_I \subseteq Res_I$, which contains $V \cap \mathcal{PL}$.

We define the mapping: $I : V_I \to Res_I \cup Prop_I$ such that: (i) $I(x) = I_v(x)$, $\forall x \in V_I \cap URI$, (ii) $I(x) = x$, $\forall\ x \in V_I \cap \mathcal{PL}$, and (iii) $I(x) = IL_I(x)$, $\forall\ x \in V_I \cap \mathcal{TL}$. □

A partial interpretation I is *coherent* iff for all $x \in Prop_I$, $PT_I(x) \cap PF_I(x) = \emptyset$.

Let $O \in \mathcal{R}$. Below we define the dependencies of O w.r.t. $\mathcal{R}$.

Definition 7 (Dependencies of an r-ERDF ontology w.r.t. a MEO)
Let $O \in \mathcal{R}$. The *dependencies* of O w.r.t. $\mathcal{R}$, denoted by $D_O^{\mathcal{R}}$, is the minimum set of r-ERDF ontologies s.t.: (i) $O \in D_O^{\mathcal{R}}$, (ii) if $O' \in D_O^{\mathcal{R}}$ and it exists $x \in Imported_{O'}^{\mathtt{t}}$, for $\mathtt{t} \in \{\mathtt{pr}, \mathtt{cl}\}$, s.t. $Nam_{O''} \in Import_{O',\mathcal{R}}^{\mathtt{t}}(x)$ then $O'' \in D_O^{\mathcal{R}}$, and (iii) if $O' \in D_O^{\mathcal{R}}$, $r \in P_{O'}$, and it exists a qualified ERDF formula $F@Nam_{O''}$ in $Cond(r)$ then $O'' \in D_O^{\mathcal{R}}$.   □

Example 5. Consider the modular ERDF ontology $\mathcal{R}$ of Example 1. It holds: $D_{O_1}^{\mathcal{R}} = \{O_1\}$, $D_{O_2}^{\mathcal{R}} = \{O_2, O_1\}$, and $D_{O_5}^{\mathcal{R}} = \{O_5, O_1, O_2, O_3, O_4\}$.   □

The vocabulary of RDF, $\mathcal{V}_{RDF}$, is a set of *URI* references in the *rdf:* namespace [6], and the vocabulary of RDFS, $\mathcal{V}_{RDFS}$, is a set of *URI* references in the *rdfs:* namespace [6]. Let $n \in I\!N$. We define $\mathcal{V}_{RDF}^{\#n} = \mathcal{V}_{RDF} - \{rdf{:}\_i \mid i > n\}$. The *vocabulary of ERDF* is defined as $\mathcal{V}_{ERDF} = \{erdf{:}TotalClass, erdf{:}TotalProperty\}$.

Let $O \in \mathcal{R}$. We define: (i) $n_O = 0$, if $(V_{G_O} \cup V_{P_O}) \cap \{rdf{:}\_i \mid i \geq 1\} = \emptyset$, and (ii) $n_O = max(\{i \in I\!N \mid rdf{:}\_i \in V_{G_O} \cup V_{P_O}\})$, otherwise. Further, we define: $n_{\mathcal{R}} = max(\{n_O \mid O \in \mathcal{R}\} \cup \{1\})$. Intuitively, $n_{\mathcal{R}}$ is the largest i ($i \in I\!N$) such that $rdf{:}\_i$ appears in an $O \in \mathcal{R}$. In the case that no such an $rdf{:}\_i$ exists then $n_{\mathcal{R}} = 1$. Recall that the $rdf{:}\_i$ properties are used in RDF(S) [6] to express members of containers (i.e. bags, sequences, and alternatives), which are in practice finitely limited.

Let $O \in \mathcal{R}$, and let $n \in I\!N$. The *n#-vocabulary* of O is defined as: $V_O^{\#n} = V_{sk(G_O)} \cup V_{P_O} \cup Exported_O^{\mathtt{pr}} \cup Exported_O^{\mathtt{cl}} \cup Imported_O^{\mathtt{pr}} \cup Imported_O^{\mathtt{cl}} \cup V_{RDF}^{\#n} \cup \mathcal{V}_{RDFS} \cup \mathcal{V}_{ERDF}$. The *vocabulary* of O w.r.t. $\mathcal{R}$ is defined as: $V_{O,\mathcal{R}} = \cup\{V_{O'}^{\#n_{\mathcal{R}}} \mid O' \in D_O^{\mathcal{R}}\}$. Intuitively, $V_{O,\mathcal{R}}$ corresponds to the local domain of O w.r.t. $\mathcal{R}$.

Let $n \in I\!N$. Below we define the modular ERDF interpretations of an r-ERDF ontology w.r.t. a modular ERDF ontology. In this definition, we use the definition of an *ERDF #n-interpretation* over a vocabulary V (see [2]), not reviewed here due to space limitations. Intuitively, an *ERDF #n-interpretation* I of a vocabulary V is a partial interpretation of $V_I = V \cup V_{RDF}^{\#n} \cup \mathcal{V}_{RDFS} \cup \mathcal{V}_{ERDF}$ that assigns truth and falsity extensions to the classes[5] and properties in V_I, satisfying: (i) all semantic conditions of an RDFS interpretation [6] of V, except these referring to $\{rdf{:}\_i \mid i > n\}$ terms, as well as (ii) new semantic conditions, particular to ERDF.

Definition 8 (Modular ERDF interpretation). Let $O \in \mathcal{R}$. A *modular ERDF interpretation* of O w.r.t. $\mathcal{R}$ is a set $I = \{I_{O'} \mid O' \in D_O^{\mathcal{R}}\}$, where $I_{O'}$ is an ERDF $\#n_{\mathcal{R}}$-interpretation of $V_{O',\mathcal{R}}$ and it holds:

1. If $O' \in D_O^{\mathcal{R}}$, $p \in Imported_{O',\mathcal{R}}^{\mathtt{pr}}$, and $Nam_{O''} \in Import_{O',\mathcal{R}}^{\mathtt{pr}}(p)$ then $PT_{I_{O'}}(I_{O'}(p)) \supseteq PT_{I_{O''}}(I_{O''}(p))$, and $PF_{I_{O'}}(I_{O'}(p)) \supseteq PF_{I_{O''}}(I_{O''}(p))$, and
2. If $O' \in D_O^{\mathcal{R}}$, $c \in Imported_{O',\mathcal{R}}^{\mathtt{cl}}$, and $Nam_{O''} \in Import_{O',\mathcal{R}}^{\mathtt{cl}}(c)$ then $CT_{I_{O'}}(I_{O'}(c)) \supseteq CT_{I_{O''}}(I_{O''}(c))$, and $CF_{I_{O'}}(I_{O'}(c)) \supseteq CF_{I_{O''}}(I_{O''}(c))$.   □

---

[5] The *truth* and *falsity extension* of a class $c \in V_I$ is indicated by $CT_I(I(c))$ and $CF_I(I(c))$, respectively. It holds: (i) $x \in CT_I(y)$ iff $\langle x, y \rangle \in PT_I(I(rdf{:}type))$, and (ii) $x \in CF_I(y)$ iff $\langle x, y \rangle \in PF_I(I(rdf{:}type))$.

Below, we define satisfaction of an r-ERDF formula w.r.t. a modular ERDF interpretation, an r-ERDF ontology, and a valuation. First, we provide an auxiliary definition. Let I be a partial interpretation of a vocabulary V_I, let Res be a set, and let v be a partial function $v : Var \to Res$ (called *valuation*). If $x \in Var$, we define $[I + v](x) = v(x)$. If $x \in V_I$, we define $[I + v](x) = I(x)$.

Definition 9 (Satisfaction of an r-ERDF formula w.r.t. a modular ERDF interpretation, an r-ERDF ontology, and a valuation). Let $O \in \mathcal{R}$. Let $\mathsf{I} = \{I_{O'} \mid O' \in D_O^{\mathcal{R}}\}$ be a modular ERDF interpretation of O w.r.t. $\mathcal{R}$. Additionally, let F, G be r-ERDF formulas over $\{Nam_{O'} \mid O' \in D_O^{\mathcal{R}}\}$. For each $O', O'' \in D_O^{\mathcal{R}}$ and for each mapping $v : Var(F) \to Res_{I_{O'}}$:

- If $F = p(s, o)$ then $\langle \mathsf{I}, O', v \rangle \models F$ iff $p \in V_{I_{O'}} \cap URI$, $s, o \in V_{I_{O'}} \cup Var$, $I_{O'}(p) \in Prop_{I_{O'}}$, and $\langle [I_{O'} + v](s), [I_{O'} + v](o) \rangle \in PT_{I_{O'}}(I_{O'}(p))$.
- If $F = \neg p(s, o)$ then $\langle \mathsf{I}, O', v \rangle \models F$ iff $p \in V_{I_{O'}} \cap URI$, $s, o \in V_{I_{O'}} \cup Var$, $I_{O'}(p) \in Prop_{I_{O'}}$, and $\langle [I_{O'} + v](s), [I_{O'} + v](o) \rangle \in PF_{I_{O'}}(I_{O'}(p))$.
- If $F = \sim G$ then $\langle \mathsf{I}, O', v \rangle \models F$ iff $V_G \subseteq V_{I_{O'}}$ and $\langle \mathsf{I}, O', v \rangle \not\models G$.
- If $F = F_1 \wedge F_2$ then $\langle \mathsf{I}, O', v \rangle \models F$ iff $\langle \mathsf{I}, O', v \rangle \models F_1$ and $\langle \mathsf{I}, O', v \rangle \models F_2$.
- If $F = F_1 \vee F_2$ then $\langle \mathsf{I}, O', v \rangle \models F$ iff $\langle \mathsf{I}, O', v \rangle \models F_1$ or $\langle \mathsf{I}, O', v \rangle \models F_2$.
- If $F = F_1 \supset F_2$ then $\langle \mathsf{I}, O', v \rangle \models F$ iff $\langle \mathsf{I}, O', v \rangle \models \sim F_1 \vee F_2$.
- If $F = \exists x \, G$ then $\langle \mathsf{I}, O', v \rangle \models F$ iff there exists a mapping $u : Var(G) \to Res_{I_{O'}}$ such that $u(y) = v(y)$, $\forall y \in Var(G) - \{x\}$, and $\langle \mathsf{I}, O', u \rangle \models G$.
- If $F = \forall x \, G$ then $\langle \mathsf{I}, O', v \rangle \models F$ iff for all mappings $u : Var(G) \to Res_{I_{O'}}$ such that $u(y) = v(y)$, $\forall y \in Var(G) - \{x\}$, it holds $\langle \mathsf{I}, O', u \rangle \models G$.
- If $F = G@Nam_{O''}$ then $\langle \mathsf{I}, O', v \rangle \models F$ iff $\langle \mathsf{I}, O'', v \rangle \models G$. $\qquad\square$

Let I be a modular ERDF interpretation of O w.r.t. $\mathcal{R}$ and let F be an r-ERDF formula. We define: $\langle \mathsf{I}, O' \rangle \models F$ iff for each mapping $v : Var(F) \to Res_{I_{O'}}$, it holds that $\langle \mathsf{I}, O', v \rangle \models F$. Additionally, let G be an ERDF graph. We define: $\langle \mathsf{I}, O' \rangle \models G$ iff $\langle \mathsf{I}, O' \rangle \models formula(G)$. We assume that for every function $v : Var \to Res_{I_{O'}}$, it holds that $\langle \mathsf{I}, O', v \rangle \models \texttt{true}$ and $\langle \mathsf{I}, O', v \rangle \not\models \texttt{false}$.

Below, we define the modular models of an r-ERDF ontology w.r.t. a modular ERDF ontology.

Definition 10 (Modular ERDF model). Let $O \in \mathcal{R}$. Let $\mathsf{I} = \{I_{O'} \mid O' \in D_O^{\mathcal{R}}\}$ be a modular ERDF interpretation of O w.r.t. $\mathcal{R}$ and let $O' \in D_O^{\mathcal{R}}$:

- We say that $\langle \mathsf{I}, O' \rangle$ *satisfies* an r-ERDF rule r, denoted by $\langle \mathsf{I}, O' \rangle \models r$, iff it holds: For all mappings $v : Var(r) \to Res_{I_{O'}}$, if $\langle \mathsf{I}, O', v \rangle \models Cond(r)$ then $\langle \mathsf{I}, O', v \rangle \models Concl(r)$.
- We say that I is a *modular (ERDF) model* of O w.r.t. $\mathcal{R}$, denoted by $\mathsf{I} \models_{\mathcal{R}} O$, iff for all $O' \in D_O^{\mathcal{R}}$, $\langle \mathsf{I}, O' \rangle \models G_{O'}$ and $\langle \mathsf{I}, O' \rangle \models r$, $\forall r \in P_{O'}$. $\qquad\square$

Let $O \in \mathcal{R}$. We denote by $Res_{O,\mathcal{R}}^{\mathrm{H}}$ the union of $V_{O,\mathcal{R}}$ and the set of XML values of the well-typed XML literals in $V_{O,\mathcal{R}}$ minus the well-typed XML literals. Below we define the modular Herbrand interpretations of O w.r.t. $\mathcal{R}$, extending the definition of a Herbrand interpretation of an ERDF ontology [1].

Definition 11 (Modular ERDF Herbrand interpretation). Let $O \in \mathcal{R}$. Let $\mathsf{I} = \{I_{O'} \mid O' \in D_O^{\mathcal{R}}\}$ be a modular ERDF interpretation of O w.r.t. $\mathcal{R}$. We say that I is a *modular (ERDF) Herbrand interpretation* of O w.r.t. $\mathcal{R}$ iff for each $O' \in D_O^{\mathcal{R}}$:

- $Res_{I_{O'}} = Res^{\text{II}}_{O',\mathcal{R}}$.
- $I_{O'_V}(x) = x$, for all $x \in V_{O',\mathcal{R}} \cap URI$.
- $IL_{I_{O'}}(x) = x$, if x is a typed literal in $V_{O',\mathcal{R}}$ other than a well-typed XML literal, and $IL_{I_{O'}}(x)$ is the XML value of x, if x is a well-typed XML literal in $V_{O',\mathcal{R}}$.

We denote the set of modular Herbrand interpretations of O w.r.t. $\mathcal{R}$ by $\mathcal{I}^{\text{H}}_{O,\mathcal{R}}$.

□

Let $O \in \mathcal{R}$. Let $\mathsf{I} = \{I_{O'}\,|\,O' \in D^{\mathcal{R}}_O\}$ be a modular Herbrand interpretation of O w.r.t. $\mathcal{R}$. We say that I is a *modular (ERDF) Herbrand model* of O w.r.t. $\mathcal{R}$ iff for all $O' \in D^{\mathcal{R}}_O$, (i) $\langle \mathsf{I}, O' \rangle \models sk(G_{O'})$ and (ii) for all $r \in P_{O'}$, $\langle \mathsf{I}, O' \rangle \models r$. We denote the set of modular Herbrand models of O w.r.t. $\mathcal{R}$ by $\mathcal{M}^{\text{H}}_{O,\mathcal{R}}$.

It holds that: if M is a modular Herbrand model of O w.r.t. $\mathcal{R}$ then M is a modular model of O w.r.t. $\mathcal{R}$.

4   Modular Stable Models and Complexity Results

In this Section, we define the modular stable models of an r-ERDF ontology w.r.t. a modular ERDF ontology, and provide some of their properties. Additionally, we provide several complexity results.

Let $O \in \mathcal{R}$. We proceed by defining a partial ordering on the modular Herbrand interpretations of O w.r.t. $\mathcal{R}$.

Definition 12 (Modular Herbrand interpretation ordering). Let $O \in \mathcal{R}$. Let $\mathsf{I}, \mathsf{J} \in \mathcal{I}^{\text{H}}_{O,\mathcal{R}}$. We say that J *extends* I, denoted by $\mathsf{I} \leq \mathsf{J}$ (or $\mathsf{J} \geq \mathsf{I}$) iff: For all $O' \in D^{\mathcal{R}}_O$, it holds that (i) $Prop_{I_{O'}} \subseteq Prop_{J_{O'}}$, and for all $p \in Prop_{I_{O'}}$, it holds $PT_{I_{O'}}(p) \subseteq PT_{J_{O'}}(p)$ and $PF_{I_{O'}}(p) \subseteq PF_{J_{O'}}(p)$.   □

Let $O \in \mathcal{R}$. The intuition behind Definition 12 is that by extending a modular Herbrand interpretation of O w.r.t. $\mathcal{R}$, we extend both the truth and falsity extension for all properties of $O' \in D^{\mathcal{R}}_O$, and thus (since *rdf:type* is a property), for all classes.

Let $\mathcal{I} \subseteq \mathcal{I}^{\text{H}}_{O,\mathcal{R}}$. We define $minimal(\mathcal{I}) = \{\mathsf{I} \in \mathcal{I} \mid \not\exists \mathsf{J} \in \mathcal{I} : \mathsf{J} \neq \mathsf{I} \text{ and } \mathsf{J} \leq \mathsf{I}\}$. Let $\mathsf{I}, \mathsf{J} \in \mathcal{I}^{\text{H}}_{O,\mathcal{R}}$. We define $[\mathsf{I}, \mathsf{J}]_{O,\mathcal{R}} = \{\mathsf{I}' \in \mathcal{I}^{\text{H}}_{O,\mathcal{R}}, \mathsf{I} \leq \mathsf{I}' \leq \mathsf{J}\}$.

Let V be a vocabulary and let r be an r-ERDF rule. We denote by $[r]_V$ the set of rules that result from r if we replace each variable $x \in FVar(r)$ by $v(x)$, for all mappings $v : FVar(r) \to V$. Let P be an r-ERDF program. We define $[P]_V = \bigcup_{r \in P}[r]_V$.

In [2], we defined the #n-stable models of an ERDF ontology (for an $n \in I\!N$), based on the coherent stable models of partial logic [7] (which, on ELPs, are equivalent [7] to Answer Sets [5]). Here, we extend this definition to modular stable models of an r-ERDF ontology w.r.t. a modular ERDF ontology.

Definition 13 (Modular ERDF stable model). Let $O \in \mathcal{R}$, and let $\mathsf{M} \in \mathcal{I}^{\text{H}}_{O,\mathcal{R}}$. We say that M is a *modular (ERDF) stable model* of O w.r.t. $\mathcal{R}$ iff there is a chain of modular Herbrand interpretations of O w.r.t. $\mathcal{R}$, $\mathsf{I}_0 \leq ... \leq \mathsf{I}_k$ such that $\mathsf{I}_{k-1} = \mathsf{I}_k = \mathsf{M}$ and:

1. $l_0 \in minimal(\{I \in \mathcal{I}^{\mathsf{H}}_{O,\mathcal{R}} \mid \forall O' \in D^{\mathcal{R}}_O, \ \langle I, O' \rangle \models sk(G_{O'})\})$.
2. For $0 < \alpha \leq k$:
   $l_\alpha \in minimal(\{I \in \mathcal{I}^{\mathsf{H}}_{O,\mathcal{R}} \mid I \geq l_{\alpha-1}$ and $\forall O' \in D^{\mathcal{R}}_O$, it holds that:
   if $r \in [P_{O'}]_{V_{O',\mathcal{R}}}$ s.t. $\langle J, O' \rangle \models Cond(r)$, $\forall J \in [l_{\alpha-1}, \mathsf{M}]_{O,\mathcal{R}}$, then $\langle I, O' \rangle \models Concl(r)\})$.

The set of modular stable models of O w.r.t. $\mathcal{R}$ is denoted by $\mathcal{M}^{\mathsf{st}}_{O,\mathcal{R}}$. $\qquad\square$

Example 6. Consider the modular ERDF ontology $\mathcal{R}$ of Example 1. For every $\mathsf{M} \in \mathcal{M}^{\mathsf{st}}_{O_5,\mathcal{R}}$, it holds $\langle \mathsf{M}, O_1 \rangle \models \neg \ \mathtt{rdf:type(geo:Egypt,geo:Europ\_}$ $\mathtt{Country)}$. This is due to the local CWA in P_{O_1}. Now, since O_5 imports class $\mathtt{geo:Europ\_Country}$ from O_1, for every $\mathsf{M} \in \mathcal{M}^{\mathsf{st}}_{O_5,\mathcal{R}}$, it holds $\langle \mathsf{M}, O_5 \rangle \models \neg$ $\mathtt{rdf:type(geo:Egypt,geo:Europ\_Country)}$. Therefore, due to the 2nd rule of P_{O_5}, for every $\mathsf{M} \in \mathcal{M}^{\mathsf{st}}_{O_5,\mathcal{R}}$, $\langle \mathsf{M}, O_5 \rangle \models \mathtt{ann:choose\_trav\_package(pyr:}$ $\mathtt{package1,geo:Egypt)}$.

Note the ERDF triple $\mathtt{rdf:type(eu:CountryEU,erdf:TotalClass)}$ in G_{O_2}, expressing a local OWA. Therefore, in some $\mathsf{M} \in \mathcal{M}^{\mathsf{st}}_{O_5,\mathcal{R}}$, it holds $\langle \mathsf{M}, O_2 \rangle \models \neg$ $\mathtt{rdf:type(geo:Croatia,eu:CountryEU)}$, while in the rest $\mathsf{M}' \in \mathcal{M}^{\mathsf{st}}_{O_5,\mathcal{R}}$, it holds $\langle \mathsf{M}', O_2 \rangle \models \mathtt{rdf:type(geo:Croatia,eu:CountryEU)}$. Now note that O_5 imports class $\mathtt{eu:CountryEU}$ from O_2. Thus, in some $\mathsf{M} \in \mathcal{M}^{\mathsf{st}}_{O_5,\mathcal{R}}$, it holds $\langle \mathsf{M}, O_5 \rangle \models \neg$ $\mathtt{rdf:type(geo:Croatia,eu:CountryEU)}$, while in the rest $\mathsf{M}' \in \mathcal{M}^{\mathsf{st}}_{O_5,\mathcal{R}}$, it holds $\langle \mathsf{M}', O_5 \rangle \models \mathtt{rdf:type(geo:Croatia,eu:CountryEU)}$. Reasoning now by cases and due to the 3rd and 4th rule of P_{O_5}, for every $\mathsf{M} \in \mathcal{M}^{\mathsf{st}}_{O_5,\mathcal{R}}$, it holds $\langle \mathsf{M}, O_5 \rangle \models$ $\mathtt{ann:choose\_trav\_package(trav:package2,geo:Croatia)}$. $\qquad\square$

The following proposition shows that any modular stable model of O w.r.t. $\mathcal{R}$ is a modular Herbrand model of O w.r.t. $\mathcal{R}$.

Proposition 1. Let $O \in \mathcal{R}$. If $\mathsf{M} \in \mathcal{M}^{\mathsf{st}}_{O,\mathcal{R}}$ then $\mathsf{M} \in \mathcal{M}^{\mathsf{H}}_{O,\mathcal{R}}$.

On the other hand, if all properties of $O' \in D^{\mathcal{R}}_O$ are total, a modular Herbrand model M of O w.r.t. $\mathcal{R}$ is a modular stable model of O w.r.t. $\mathcal{R}$.

Proposition 2. Let $O \in \mathcal{R}$. If $rdfs{:}subclass(\,rdf{:}Property, erdf{:}TotalProperty) \in G_{O'}$, for all $O' \in D^{\mathcal{R}}_O$, then $\mathcal{M}^{\mathsf{st}}_{O,\mathcal{R}} = \mathcal{M}^{\mathsf{H}}_{O,\mathcal{R}}$.

The following proposition relates the modular stable models of different r-ERDF ontologies w.r.t. a modular ERDF ontology.

Proposition 3. Let $O \in \mathcal{R}$ and let $O' \in D^{\mathcal{R}}_O$. Let $\mathsf{M} \in \mathcal{I}^{\mathsf{H}}_{O,\mathcal{R}}$ and let $\mathsf{M}' = \{M_{O''} \in \mathsf{M} \mid O'' \in D^{\mathcal{R}}_{O'}\}$. It holds that: If $\mathsf{M} \in \mathcal{M}^{\mathsf{st}}_{O,\mathcal{R}}$ then $\mathsf{M}' \in \mathcal{M}^{\mathsf{st}}_{O',\mathcal{R}}$. $\qquad\square$

Let $O \in \mathcal{R}$. We say that O is *inconsistent* under the modular stable model semantics w.r.t. $\mathcal{R}$ iff $\mathcal{M}^{\mathsf{st}}_{O,\mathcal{R}} = \{\}$.

Let $O \in \mathcal{R}$, and let $O' \in D^{\mathcal{R}}_O$. It follows directly from Proposition 3 that if O' is inconsistent under the modular stable model semantics w.r.t. $\mathcal{R}$ then O is also inconsistent under the modular stable model semantics w.r.t. $\mathcal{R}$. Obviously, due to the definition of $D^{\mathcal{R}}_O$ in Definition 7, all other r-ERDF ontologies in $\mathcal{R}$ remain unaffected from the local inconsistency.

Definition 14 (Modular ERDF stable model entailment). Let $O \in \mathcal{R}$. Additionally, let F be an r-ERDF formula. We say that O *entails* F w.r.t. $\mathcal{R}$ under the *modular ERDF stable model semantics*, denoted by $O \models^{st}_{\mathcal{R}} F$ iff for all $M \in \mathcal{M}^{st}_{O,\mathcal{R}}$, $\langle M, O \rangle \models F$. □

The following proposition shows that modular stable model entailment extends #n-stable model entailment from ERDF ontologies to modular ERDF ontologies.

Proposition 4. Let $O = \langle G, P \rangle$ be an ERDF ontology and let F be an ERDF formula. Additionally, let O' be an r-ERDF ontology such that $G_{O'} = G$, $P_{O'} = P$, and $Int_{O'} = \{\}$. It holds: $O \models^{st\#n_{\mathcal{R}}} F$ iff $O' \models^{st}_{\mathcal{R}} F$, where $\mathcal{R} = \{O'\}$.

The following corollary follows directly from the above proposition and Proposition 3 in [2], and it shows that modular stable model entailment extends RDFS entailment from RDF graphs to modular ERDF ontologies.

Corollary 1. Let G, G' be RDF graphs such that $\mathcal{V}_G \cap \mathcal{V}_{ERDF} = \emptyset$, $\mathcal{V}_{G'} \cap \mathcal{V}_{ERDF} = \emptyset$, and $\mathcal{V}_{G'} \cap sk_G(Var(G)) = \emptyset$. Let O be an r-ERDF ontology with $G_O = G$, $P_O = \{\}$, and $Int_O = \{\}$. If $max(\{i \in \mathbb{N} \mid rdf\text{:\_}i \in \mathcal{V}_{G'}\}) \leq n_{\mathcal{R}}$ then: $G \models^{RDFS} G'$ iff $O \models^{st}_{\mathcal{R}} G'$, where $\mathcal{R} = \{O\}$.

Let $O \in \mathcal{R}$ and let F be an r-ERDF formula. The *modular stable answers* of F w.r.t. O and $\mathcal{R}$ are defined as follows[6]:

$$Ans^{st}_{O,\mathcal{R}}(F) = \begin{cases} \text{"yes"} & \text{if } FVar(F) = \emptyset \text{ and } O \models^{st}_{\mathcal{R}} F \\ \text{"no"} & \text{if } FVar(F) = \emptyset \text{ and } O \not\models^{st}_{\mathcal{R}} F \\ \{v : FVar(F) \rightarrow V_{O,\mathcal{R}} \mid O \models^{st}_{\mathcal{R}} v(F)\}, & \text{if } FVar(F) \neq \emptyset \end{cases}$$

Example 7. Consider the modular ERDF ontology $\mathcal{R}$ of Example 1. Then:
$Ans^{st}_{O_5,\mathcal{R}}(\text{ann:choose\_trav\_package}(?x, ?y)) = \{\langle ?x = \text{pyr:package1}, ?y = \text{geo:Egypt}\rangle,$
$\langle ?x = \text{trav:package2}, ?y = \text{geo:Croatia}\rangle\}$. □

An r-ERDF formula is called *simple*, if it has the form: $t_1 \wedge ... \wedge t_k \wedge \sim t_{k+1} \wedge ... \wedge \sim t_n \wedge (\sim t'_1)@Nam_{O_1} \wedge ... \wedge (\sim t'_m)@Nam_{O_m}$, where (i) each t_i is a normal or qualified ERDF triple (positive or negative), (ii) each t'_i is a normal ERDF triple (positive or negative), and (iii) each O_i is an r-ERDF ontology. An r-ERDF program P is called *simple* if the body of each rule in P is simple, or true. Let $\mathcal{R}$ be a modular ERDF ontology and let $O \in \mathcal{R}$. O is called *simple* w.r.t. $\mathcal{R}$, if for each $O' \in D^{\mathcal{R}}_O$, $P_{O'}$ is simple.

Let $O \in \mathcal{R}$ s.t. O is simple w.r.t. $\mathcal{R}$. We can show that the modular stable answers of a simple r-ERDF formula F w.r.t. O and $\mathcal{R}$ can be computed through Answer Set Programming [5] on an ELP $\Pi_{O,\mathcal{R}}$.

Below, we state several complexity results of the modular ERDF stable model semantics. We define $size\_inst(O, \mathcal{R}) = \sum\{\text{size of } (G_{O'} \cup [P_{O'}]_{V_{O',\mathcal{R}}}) \mid O' \in D^{\mathcal{R}}_O\}$.

Proposition 5. Let $O \in \mathcal{R}$, and let F be an r-ERDF formula. Additionally, let v be (i) one of $\{\text{"yes"}, \text{"no"}\}$, if $Var(F) = \emptyset$, or (ii) a mapping $v : Var(F) \rightarrow V_{O,\mathcal{R}}$, if $Var(F) \neq \emptyset$.

---

[6] $v(F)$ is the formula F after replacing all the free variables x in F by $v(x)$.

1. If O is a simple r-ERDF ontology w.r.t. $\mathcal{R}$ then: (i) the problem of establishing whether O has a modular stable model w.r.t. $\mathcal{R}$ is *NP*-complete w.r.t. $size\_inst(O, \mathcal{R})$, and (ii) the problem of establishing whether $v \in Ans^{\text{st}}_{O,\mathcal{R}}(F)$ is *co-NP*-complete w.r.t. $size\_inst(O, \mathcal{R})$.
2. If for all $O' \in D^{\mathcal{R}}_O$, no quantifies $\forall, \exists$ appear in $P_{O'}$ then: (i) the problem of establishing whether O has a modular stable model w.r.t. $\mathcal{R}$ is $\Sigma^P_2 = NP^{NP}$-complete w.r.t. $size\_inst(O, \mathcal{R})$, and (ii) the problem of establishing whether $v \in Ans^{\text{st}}_{O,\mathcal{R}}(F)$ is $\Pi^P_2 = co\text{-}NP^{NP}$-complete w.r.t. $size\_inst(O, \mathcal{R})$.
3. In the *general* case, (i) the problem of establishing whether O has a modular stable model w.r.t. $\mathcal{R}$ is *PSPACE*-complete w.r.t. $size\_inst(O, \mathcal{R})$, and (ii) the problem of establishing whether $v \in Ans^{\text{st}}_{O,\mathcal{R}}(F)$ is *PSPACE*-complete w.r.t. $size\_inst(O, \mathcal{R})$.

5   Conclusions and Related Work

In this paper, we extended ERDF ontologies [1], and thus RDF graphs to r-ERDF ontologies. In particular, an r-ERDF ontology is an ERDF ontology that (i) is associated with a set of export and import statements, and (ii) interacts with other r-ERDF ontologies (through qualified ERDF formulas in the program rules). Further, we defined a modular ERDF ontology as a set of r-ERDF ontologies and defined its modular stable model semantics, model-theoretically, based on partial logic [7]. We showed that modular stable model entailment on modular ERDF ontologies extends #n-stable model entailment on ERDF ontologies [2], and thus it also extends RDFS entailment on RDF graphs [6]. Future work concerns (i) the extension of the modular stable model semantics such that meaning is assigned to inconsistent r-ERDF ontologies of a modular ERDF ontology, and (ii) the implementation of the modular ERDF framework.

N3Logic [3] allows rules to be integrated with RDF. Indeed, part of the RDFS semantics is represented by program rules. Yet, the supported form of negation as failure, expressed through the built-in `log:notincludes`, is limited. Additionally, N3Logic does not have a model-theoretic semantics that faithfully extends RDFS semantics [6], does not support explicit negation and general formulas in the body of the rules, and ignores visibility issues.

A modularity framework for RDF rule bases (without blank nodes) is proposed in [11]. There, RDFS semantics are partially represented through a normal logic program, associated with a special context/module c_{RDFS}. The *contextually closed* AS and *contextually closed* WFS semantics of such a modular RDF rule base $\mathcal{R}$ are defined, through the AS [5] and WFS [4] semantics of a normal logic program $\mathcal{R}_{CC}$, generated from $\mathcal{R}$, respectively. Yet, this framework does not have a model-theoretic semantics that faithfully extends RDFS semantics [6], does not support explicit negation and general formulas in the body of the rules, and ignores visibility issues.

TRIPLE [12] is a rule language for the Semantic Web that supports modules (called, *models* there), qualified literals, and dynamic module transformation. Arbitrary formulas can be used in the body of a rule, handled through the Lloyd-Topor transformations [9]. Part of the semantics of the RDF(S) vocabulary is represented as pre-defined rules (and not as semantic conditions on interpretations), which are grouped together in a module. The semantics of a modular rule base is defined, based on the *well-founded semantics* (WFS) [4] of an equivalent logic program. Yet, the model-theoretic semantics of TRIPLE [13] does not faithfully extend RDFS semantics [6] and is not, in general, equivalent to its transformational semantics. Additionally, TRIPLE does not support explicit negation and ignores visibility issues.

References

1. Analyti, A., Antoniou, G., Damásio, C.V., Wagner, G.: Extended RDF as a Semantic Foundation of Rule Markup Languages. Journal of Artificial Intelligence Research (JAIR) 32, 37–94 (2008)
2. Analyti, A., Antoniou, G., Damásio, C.V., Wagner, G.: On the Computability and Complexity Issues of Extended RDF. In: Ho, T.-B., Zhou, Z.-H. (eds.) PRICAI 2008. LNCS (LNAI), vol. 5351, pp. 5–16. Springer, Heidelberg (2008)
3. Berners-Lee, T., Connolly, D., Kagal, L., Scharf, Y., Hendler, J.: N3Logic: A Logical Framework For the World Wide Web. TPLP 8(3), 249–269 (2008)
4. Gelder, A.V., Ross, K.A., Schlipf, J.S.: The Well-Founded Semantics for General Logic Programs. Journal of the ACM 38(3), 620–650 (1991)
5. Gelfond, M., Lifschitz, V.: Logic programs with Classical Negation. In: 7th International Conference on Logic Programming, pp. 579–597 (1990)
6. Hayes, P.: RDF Semantics. W3C Recommendation, February 10 (2004),
   http://www.w3.org/TR/2004/REC-rdf-mt-20040210/
7. Herre, H., Jaspars, J., Wagner, G.: Partial Logics with Two Kinds of Negation as a Foundation of Knowledge-Based Reasoning. In: Gabbay, D.M., Wansing, H. (eds.) What Is Negation? Kluwer Academic Publishers, Dordrecht (1999)
8. Klyne, G., Carroll, J.J.: Resource Description Framework (RDF): Concepts and Abstract Syntax. W3C Recommendation, February 10 (2004),
   http://www.w3.org/TR/2004/REC-rdf-concepts-20040210/
9. Lloyd, J.W., Topor, R.W.: Making Prolog more Expressive. Journal of Logic Programming 1(3), 225–240 (1984)
10. Oikarinen, E., Janhunen, T.: Achieving compositionality of the stable model semantics for smodels programs. TPLP 8(5–6), 717–761 (2008)
11. Polleres, A., Feier, C., Harth, A.: Rules with Contextually Scoped Negation. In: Sure, Y., Domingue, J. (eds.) ESWC 2006. LNCS, vol. 4011, pp. 332–347. Springer, Heidelberg (2006)
12. Sintek, M., Decker, S.: TRIPLE - A Query, Inference, and Transformation Language for the Semantic Web. In: Horrocks, I., Hendler, J. (eds.) ISWC 2002. LNCS, vol. 2342, pp. 364–378. Springer, Heidelberg (2002)
13. Stefan Decker, W.N., Sintek, M.: The Model-Theoretic Semantics of TRIPLE. Technical Report (2002)
14. Wagner, G., Giurca, A., Diaconescu, I.-M., Antoniou, G., Analyti, A., Damasio, C.: Reasoning on the Web with Open and Closed Predicates. In: 3rd International Workshop on Applications of Logic Programming to the (Semantic) Web and Web Services (ALPSWS 2008), in conjunction with ICLP 2008, pp. 71–84 (2008)

The Perfect Match: RPL and RDF Rule Languages

François Bry, Tim Furche, and Benedikt Linse

University of Munich
http://www.pms.ifi.lmu.de

Abstract. Path query languages have been previously shown to complement RDF rule languages in a natural way and have been used as a means to implement the RDFS derivation rules. RPL is a novel path query language specifically designed to be incorporated with RDF rules and comes in three flavors: *Node-*, *edge-* and *path*-flavored expressions allow to express conditional regular expressions over the nodes, edges, or nodes *and* edges appearing on paths within RDF graphs. Providing regular *string* expressions and negation, *RPL* is more expressive than other RDF path languages that have been proposed. We give a compositional semantics for *RPL* and show that it can be evaluated efficiently, while several possible extensions of it cannot.

Graph traversal operators play a crucial role in rule and query languages for semi-structured data and for RDF rule languages in particular. This need bas been acknowledged by the development of languages like Versa [14] SPARQLeR [10] and nested regular expressions (NREs) [15] and underlined in [3]. Moreover, the need for traversal of semi-structured data in general, and XML in particular is underscored by the huge success of XPath, arguably the most prominent XML query language. In [15] it has been shown that SPARQL augmented with conditional (in the sense of [12]) regular path expressions is expressive enough to query RDF graphs under the RDFS semantics without computing the closure of the graph under the RDFS entailment rules.

Most path query languages proposed up until now are unfit for clean integration with RDF rule languages for the following reasons: (i) their use of variables interferes with the use of logical variables already present in rule languages, (ii) they do not always evaluate to pairs of nodes and thus cannot be safely used at the place of RDF predicates in query patterns and (iii) they lack negation, regular string expressions and often also conditional operators.

We propose the RDF path language RPL, that is designed for easy integration with RDF rule languages such as SPARQL [19], XCERPT[RDF] [7,16] and RDFLog [5,4]. RPL is an orthogonal extension to RDF rule languages in that it sets out to extend RDF rule languages by features they lack, and in that it tries to avoid duplication of features they already provide. RPL expressions always evaluate to pairs of nodes within an RDF graph, and can thus be safely used at the place of predicates within the body of RDF rules. Despite of this restriction,

A. Polleres and T. Swift (Eds.): RR 2009, LNCS 5837, pp. 227–241, 2009.

SPARQL extended with RPL predicates is capable, just as NREs, to query RDF graphs under the RDFS semantics without computing the closure of the queried graphs under the RDFS entailment rules. RPL is more expressive than previously proposed RDF query languages in that it provides regular string expressions and negation.

RDF Path Expressions (RPEs) come in three flavors: *node-restricting*, *edge-restricting* and *path-restricting*, identified by the keywords NODES, EDGES, PATH, respectively. *Node-restricting* (*edge-restricting*) RPEs only place restrictions on the nodes (edges) appearing within a path. *Path-restricting* expressions may place restrictions on both, nodes and edges. RPEs evaluate to sets of pairs of nodes – i.e. binary relations over the set N of nodes of an RDF graph. The three unrestrictive RPEs [PATH (_ _)*], [EDGES _*] and [NODES _*] evaluate to $N \times N$.

This paper is organized as follows: Section 1 informally introduces the semantics of RPL by example, before its syntax and semantics is formally defined in Sections 2 and 3. Section 5 compares RPL to related path query languages and comes up with first complexity results. Section 6 shows the tractability of RPL as a whole, and the intractability of node and edge flavored path RPL expressions augmented with *unordered* paths.

The contributions of this paper are as follows: (i) We formalize the syntax and semantics of RPL expressions, and (ii) show that RPL can express all relevant RDFS queries. (iii) We show that RPL can be evaluated efficiently, and (iv) that also NREs could be extended by regular string expressions and negation without sacrificing tractability. (v) Finally we show that extensions of RPL and NREs to unordered paths results in the loss of the tractability of both languages.

1   RPL by Example

Before introducing RPL, we define the notions of RDF triples, graphs, and paths in RDF graphs.

Definition 1 (RDF triple, graph). *Let U, B, L be three disjoint sets of URIs, blank node identifiers and RDF literals. Then $t = (s, p, o) \in U \cup B \times U \times U \cup B \cup L$ is an RDF triple, and $t_g \in U \times U \times U \cup L$ is a ground RDF triple. s, p, o are the subject, predicate' and object of t, respectively. A (ground) RDF graph is a set of (ground) RDF triples. The set of nodes N of an RDF graph G are all elements in $U \cup B \cup L$ that appear in subject or object position of a triple in G.*

Definition 2 (Path in an RDF graph). *Let G be an RDF graph. The sequence $n_1, \ldots, n_k$ is a path in G, iff the triples (n_1, n_2, n_3), (n_3, n_4, n_5), $\ldots$, (n_{k-2}, n_{k-1}, n_k) are in G.*

Example 1. [PATH (_ eg:/.*/)* rdf:type]: All pairs (n_1, n_2) of nodes connected over intermediate nodes of the namespace eg. Additionally, the last edge on the connecting path must correspond to the qualified name rdf:type. This first example demonstrates the following points:

- RPEs start with an opening square bracket followed by one of the keywords PATH, EDGES and NODES specifying the flavor of the path expression, and end with a closing square bracket.
- As in SPARQL, XPath, XQuery, XSLT and XCERPT^{RDF}, URIs may be abbreviated by qualified names.
- Wildcards (_) and regular expressions (e.g. /.*/) play an important role within RPEs. Together with qualified names, URIs and literals, they constitute the atomic building blocks of RPEs, called *atomic RPEs*.
- From atomic RPEs, *compound* RPEs can be built via *sequencing* (denoted by whitespace), *alternation* (|), *Kleene closure* (* and +), *optionality* (?), and *negation* (not(...)).

Example 2. The expression [PATH (>eg:p ^_[not(PATH eg:p1])])* eg:p] collects all pairs of nodes connected over a path with at least one predicate with URI eg:p. All intermediate nodes must not have an outgoing eg:p1 edge.

This second example introduces *path directions* and *path predicates* and demonstrates the following points:

- URIs, regular expressions or qualified names within RPEs may be modified by one of the directions '>' (*forward predicate*), '<' (*reverse predicate*) or '^' (*node*). If an atomic RPE is prefixed with '<' ('^') then it must match with a reverse edge (node) on the path connecting the nodes n_1 and n_2. If an atomic RPE is undirected or prefixed by '>', then it must match a *forward* edge on the path connecting n_1 and n_2.
- Path expressions may be nested via *path predicates*, which roughly correspond to XPath predicates. While URIs, qualified names or regular expressions within RPEs represent *local restrictions* only, predicates allow the specification of *non-local* restrictions, i.e., restrictions that are not directly enforced on nodes or edges on the path, but on nodes or edges connected via a nested path expression.

Example 3. The edge-flavored query [EDGES rdf:type (rdfs:subClassOf)*] evaluates to all pairs of nodes connected via one rdf:type edge and zero or more rdfs:subClassOf edges (in this order).

This query determines the direct or indirect class membership of resources under the RDFS semantics. Note that also for many other RDF queries, only the edges along a path are relevant. The reverse relation is obtained by the query [EDGES (<rdfs:subClassOf)* <rdf:type].

Example 4. The node-flavored expression [NODES (eg:a eg:b)] finds all pairs of nodes that are connected over nodes eg:a and eg:b (in this order), with arbitrary predicates on the path. The query [NODES (eg:/.*/ | foaf:/.*/)*] on the other hand, finds all pairs of nodes connected over a path of length zero or more which contains only intermediate nodes belonging to the namespaces eg or foaf. The predicates on the path are irrelevant, as indicated by the keyword NODES.

Example 5 (RDFS querying with RDFLog augmented by RPL). This example shows how RDF rule languages can be augmented by RPL path expressions to immitate the RDFS semantics.

Due to its simplicity, we choose RDFLog [5] as the rule language to be extended. But similar embeddings can be given for most RDF rule languages, including the various SPARQL extensions with rules [17,18,8]. The RDFLog rule[1]

$$\forall x \; p \; y \; p_1 \; z \; . \; (x \; p \; y) \leftarrow (x \; p_1 \; z), (p_1 \; \texttt{[EDGES sp*]} \; y) \tag{1}$$

can be used to materialize the extension of the predicate p under the RDFS semantics. In a backward chaining evaluation of an RDFLog program, materialization is only carried out *on demand*, and is thus more efficient than computing the RDFS closure of the queried graph. If only single rules or queries are allowed (such as in SPARQL), then the body of Equation 1 can simply be used in the query at the place of p.

The extension of predicates with a special semantics under the RDFS model theory deserve special treatment. E.g the extension of rdf:type is computed by the following RDFLog rules with RPL predicates:

$$\forall x \; y \; . \; (x \; type \; y) \leftarrow (x \; \texttt{[EDGES type sc*]} \; y)$$
$$\forall x \; y \; p_1 \; z. \; (x \; type \; y) \leftarrow (x \; p_1 \; z), (p_1 \; \texttt{[EDGES sp* dom sc*]} \; y)$$
$$\forall x \; y \; p_1 \; z. \; (x \; type \; y) \leftarrow (z \; p_1 \; x), (p_1 \; \texttt{[EDGES sp* range sc*]} \; y)$$

It can be shown that also extensions of the remaining RDFS predicates subclassOf, subPropertyOf, domain and range can be encoded as RDFLog or SPARQL rule bodies augmented with RPL. The encoding is analogous to the one presented in [15] and is omitted here for the sake of brevity.

2   RPL Syntax

Definition 3 (Abstract syntax of RPEs). *The abstract syntax of RPL is recursively defined as follows:*

- *A URI u, regular expression re, qualified name q, literal l and wild card _ is an atomic RPE. Moreover, a qualified name $prefix{:}localpart$ where localpart is a regular expression, is an atomic RPE.*
- *If p is an atomic path expression, then p, $< p$, $> p$ and $\hat{}p$ are directed path expressions.*
- *if p_1 is an atomic RPE, and $q_1, \ldots q_n$ are RPL predicates (see below), then p_1 and $p_1[q_1] \ldots [q_n]$ are predicated RPEs.*
- *If p is a predicated or concatenated RPE, then p, $p*$, $p+$ and $p?$ are adorned RPEs.*

---

[1] rdf:type, rdfs:subClassOf, rdfs:subPropertyOf, rdfs:range, and rdfs:domain are abbreviated by type, sc, sp, range and dom, respectively.

- If $p_1, \ldots p_n$ are adorned or disjunctive (see below) RPEs, then $(p_1 \ldots p_n)$ with $n \geq 1$ is a concatenated RPE.
- If $p_1, \ldots p_n$ are concatenated RPEs, then $(p_1 \mid \ldots \mid p_n)$ with $n \geq 1$ is a disjunctive RPE.
- If p is a concatenated RPE, $PATH\ p$, $EDGES\ p$, $NODES\ p$ are flavored RPEs. They are called path-restricting, edge-restricting and node-restricting expressions, respectively.
- If p is a flavored RPE, then p and $not(p)$ are RPL predicates.

Figure 1 summarizes the relationships between the different types of subexpressions in RPL. An arrow labeled with 1, $+$ or $*$ from type A to type B means that expressions of type B are made up of exactly one, at least one, or zero or more expression of type A, respectively. It holds that any atomic RPL expression is also a directed subexpression, which are in turn also predicated subexpressions, which are in turn adorned subexpressions. As in XQuery, a concatenated expression (called sequence in XQuery) of one element is equivalent to the element itself. Also a disjunctive RPE of one element is equivalent to the element itself.

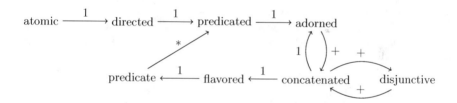

Fig. 1. Relationships among subexpressions of RPEs

The following remarks clarify Definition 3.

- Atomic RPEs correspond to the building blocks of ground RDF graphs with the following exceptions: (i) qualified names are allowed as shorthand notations for URIs, (ii) regular expressions are allowed as a means for matching URIs and Literals[2], (iii) the local part of a qualified name may be expressed by a regular expression, (iv) wildcards can be used to match any blank node, URI or literal.
- RPEs do not provide any means for selecting RDF literals based on their types or based on their language tags, other than using a regular expression for this purpose.
- Just as with ordinary regular (string) expressions, parentheses are used to influence operator precedence. The operators Kleene star ($*$), Kleene plus ($+$), optionality ($?$) are mutually exclusive and have precedence over all other operators. The concatenation operator (denoted by whitespace) binds stronger

---

[2] Matching blank nodes with regular expressions is not allowed, since this would mean *syntactic* matching of RDF graphs, i.e. the semantics of an RPE would be dependent on the syntactic representation of the RDF graph that is being queried.

than the disjunctive operator |, i.e. $a\ b\ |\ c$ is equivalent to $(a\ b)\ |\ c$. Parentheses may be omitted, if they do not alter operator precedence.

3   RPL Compositional Semantics

The intuitive presentation of the RPEs is now formalized by a compositional semantics, which is given by the function $[\![\cdot]\!]$ and its four helper functions $[\![\cdot]\!]^P$ for path-restricting expressions, $[\![\cdot]\!]^E$ for edge-restricting expressions, $[\![\cdot]\!]^N$ for node-restricting expressions and $[\![a]\!]^V$ for atomic expressions a that are evaluated in *vertex position*. While the functions $[\![\cdot]\!]$, $[\![\cdot]\!]^P$, $[\![\cdot]\!]^E$ and $[\![\cdot]\!]^N$ evaluate to subsets of $N \times N$, i.e. binary relations on the set N of nodes of the queried RDF graph, the function $[\![\cdot]\!]^V$ evaluates to subsets of N.

In order to present the semantics in an easily digestible manner, we split the entire definition according to the flavor of the RPE to be formalized. Definition 4 gives the semantics for *edge-restricting* RPEs, Definitions 5, 6 and 7 add the necessary equations for *node-restricting*, *path-restricting* and arbitrary RPEs, respectively. The three flavors of RPEs differ in the way subexpressions are concatenated. In contrast, most equations for evaluating atomic RPEs, alternatives and Kleene closures are independent of the flavor and are only given once. A more detailed natural-language description of the semantics is given in the online version of this contribution [6].

In the following, let G be an RDF graph over the vocabulary $U \cup B \cup L$, u a URI, l an RDF Literal, re a regular expression, a an atomic RPE, pe a predicated RPE, $f_1, \ldots f_k$ flavored RPEs, and $e, e_1, \ldots, e_k$ arbitrary RPEs.

Definition 4 (Semantics of edge-restricting RPEs). *The semantics of edge-restricting RPEs is given by the function $[\![\cdot]\!]^E$ defined as follows:*

$$[\![u]\!]^{E,P} = \{(n_1, n_2) \mid (n_1, u, n_2) \in G\} \tag{2}$$

$$[\![\_]\!]^{E,P} = \{(n_1, n_2) \mid \exists p \ . \ (n_1, p, n_2) \in G\} \tag{3}$$

$$[\![/re/]\!]^{E,P} = \{(n_1, n_2) \mid \exists p \in \mathcal{L}(re) \ . \ (n_1, p, n_2) \in G\} \tag{4}$$

$$[\![>pe]\!]^X = [\![pe]\!]^X \ for \ X \in \{E, P\} \tag{5}$$

$$[\![<pe]\!]^X = \{(n_2, n_1) \mid (n_1, n_2) \in [\![>pe]\!]^X\} \ for \ X \in \{E, P\} \tag{6}$$

$$[\![e_1 \ \ldots \ e_k]\!]^E = \{(n_1, n_{k-1}) \mid \tag{7}$$
$$\exists n_2, \ldots n_k \ . \ \forall 1 \le i \le k \ ((n_i, n_{i+1}) \in [\![e_i]\!]^E)\}$$

$$[\![(e_1 \mid \ldots \mid e_k)]\!]^X = [\![e_1]\!]^X \cup \ldots \cup [\![e_k]\!]^X \ for \ X \in \{P, E, N\} \tag{8}$$

$$[\![a[f_1] \ldots [f_k]]\!]^E = \bigcup_{a' \in [\![a[f_1] \ldots [f_k]]\!]^V} [\![a']\!]^E \tag{9}$$

$$[\![\epsilon]\!]^{E,P} = \{(n, n) \mid n \in N\} \tag{10}$$

$$[\![e+]\!]^X = [\![e]\!]^X \cup [\![e\ e+]\!]^X \ for \ X \in \{P, E, N\} \tag{11}$$

$$[\![e*]\!]^X = [\![\epsilon]\!] \cup [\![e+]\!]^X \ for \ X \in \{P, E, N\} \tag{12}$$

$$[\![e?]\!]^X = [\![\epsilon]\!] \cup [\![e]\!]^X \ for \ X \in \{P, E, N\} \tag{13}$$

Definition 5 (Semantics of node-restricting RPEs). *The semantics for node-restricting RPEs is defined as follows:*

$$\llbracket \_ \rrbracket^V = N \tag{14}$$

$$\llbracket /re/ \rrbracket^V = N \cap \mathcal{L}(re) \tag{15}$$

$$\llbracket u \rrbracket^V = \{u\} \cap N \tag{16}$$

$$\llbracket l \rrbracket^V = \{l\} \cap N \tag{17}$$

$$\llbracket pe \rrbracket^N = \{(n,n) \mid n \in \llbracket pe \rrbracket^V\} \tag{18}$$

$$\llbracket a[f_1] \ldots [f_k] \rrbracket^V = \llbracket a \rrbracket^V \cap \{n_1 \mid \exists n_2 \ . \ (n_1, n_2) \in \llbracket f_1 \rrbracket\} \cap \tag{19}$$

$$\ldots \cap \{n_1 \mid \exists n_2 \ . \ (n_1, n_2) \in \llbracket f_k \rrbracket\} \tag{20}$$

$$\llbracket e_1 \ \ldots \ e_k \rrbracket^N = \{(n_1, n_{2k}) \mid \exists n_2, \ldots n_{2k-1}, p_1, \ldots, p_{k-1} \ . \tag{21}$$

$$\forall 1 \le i \le k \ ((n_{2i-1}, n_{2i}) \in \llbracket e_i \rrbracket^N) \wedge$$

$$\forall 1 \le i \le k-1 \ ((n_{2i}, p_i, n_{2i+1}) \in G)\}$$

Definition 6 (Semantics of path-restricting RPEs). *The semantics of path-restricting RPEs is defined as follows:*

$$\llbracket \hat{} a \rrbracket = \llbracket a \rrbracket^V \tag{22}$$

$$\llbracket e_1 \ \ldots \ e_k \rrbracket^P = \{(n_1, n_j) \mid \exists n_2, \ldots, n_{j-1} \ . \ (n_1, n_2) \in \llbracket e_1 \rrbracket^P \tag{23}$$

$$\wedge \ n_2 \in \llbracket e_2 \rrbracket^V \wedge \ldots \wedge n_{j-1} \in \llbracket e_{k-1} \rrbracket^V \wedge (n_{j-1}, n_j) \in \llbracket e_k \rrbracket^P\}$$

$$\llbracket \hat{} pe \ e \rrbracket^P = \{(n_1, n_2) \in \llbracket e \rrbracket^P \mid n_1 \in \llbracket pe \rrbracket^V\} \tag{24}$$

$$\llbracket e \ \hat{} pe \rrbracket^P = \{(n_1, n_2) \in \llbracket e \rrbracket^P \mid n_1 \in \llbracket pe \rrbracket^V\} \tag{25}$$

Definition 7 (Semantics of flavored RPEs)

$$\llbracket PATH \ e \rrbracket = \llbracket e \rrbracket^P \tag{26}$$

$$\llbracket EDGES \ e \rrbracket = \llbracket e \rrbracket^E \tag{27}$$

$$\llbracket NODES \ e \rrbracket = \{(n_1, n_4) \mid \exists n_2, n_3, p_1, p_2 \ (\tag{28}$$

$$(n_2, n_3) \in \llbracket e \rrbracket^N \wedge (n_1, p_1, n_2), (n_3, p_2, n_4) \in G \ )\}$$

$$\llbracket not(u) \rrbracket = \llbracket \_ \rrbracket \setminus \llbracket u \rrbracket \tag{29}$$

4   RPL Restrictions and Extensions

In order to compare RPL to other regular path languages over ordinary graphs and RDF graphs, and to study the complexity of RPL fragments, we introduce the following set of sublanguages:

Definition 8 (RPL sublanguages). *Besides the operators +, ? and *, RPL makes use of the following features:*

- *regular string expressions (denoted by RSE)*
- *the EDGE keyword (denoted by →)*

- *the NODE keyword (denoted by $\cup$)*
- *the PATH keyword (denoted by $\dashrightarrow$)*
- *predicates (denoted by $[]$)*
- *concatenation (denoted by $/$)*
- *disjunction (denoted by $|$)*
- *predicate negation (denoted by $\neg$)*
- *direction modifiers (denoted by μ)*

$RPL^{f_1,\ldots,f_k}$ *with* $f_1,\ldots,f_k \in \{RSE, \rightarrow, \circ, \dashrightarrow, [], /, |, \neg, \mu\}$ *denotes the sublanguage of RPL making use of the operators* +, ?, *and* * *and the features* $f_1,\ldots,f_n$ *only.*

Languages such as XPath and Xcerpt allow queries to be incompletely specified in depth, or with respect to order. Incompleteness in depth is specified via the descendant axis in XPath and via the desc keyword in Xcerpt. Incompleteness with respect to order is the default querying mode in XPath and can be overridden by using the << operator; in Xcerpt it is specified via curly braces.

An obvious extension of RPL is thus to introduce *unordered* and *incomplete* paths. While the order in Xcerpt query terms is enforced/relaxed with respect to the *sibling* axis of an XML document, the order in RPEs may be relaxed with respect to the paths traversed, i.e. the *descendant* axis. Also the concept of *incomplete* specification of *siblings* in Xcerpt query terms may be transferred to the *descendant* axis by allowing double brackets within RPL. We denote the extensions of the sublanguages of RPL by unordered paths, incomplete paths and both by adding the symbols {}, [[]] or both to the feature list of the sublanguage. The RPL expression Nodes { x y z } thus evaluates to all pairs of nodes that are connected by a path containing only the intermediate nodes x, y, and z in an arbitrary order. The RPL expression Nodes [[x y]] on the other hand evaluates to all pairs of nodes that are connected via a path that contains the nodes x and y with x appearing before y, and an arbitrary number of nodes before x, between x and y and following y.

The semantics of {} is formalized by the functions $[\![\cdot]\!]^{UN}$, $[\![\cdot]\!]^{UE}$, and $[\![\cdot]\!]^{UP}$ for unordered node-flavored, edge-flavored and path-flavored expressions, respectively. The semantics of [[]] is given by the functions $[\![\cdot]\!]^{IN}$, $[\![\cdot]\!]^{IE}$, and $[\![\cdot]\!]^{IP}$.

Definition 9 (Semantics of unordered and incomplete RPEs)

$$[\![e]\!]^{UN} = \bigcup_{p \in Perm(e)} [\![p]\!]^N \qquad\qquad [\![e]\!]^{IN} = \bigcup_{c \in Comp(e)} [\![c]\!]^N$$

$$[\![e]\!]^{UE} = \bigcup_{p \in Perm(e)} [\![p]\!]^E \qquad\qquad [\![e]\!]^{IE} = \bigcup_{c \in Comp(e)} [\![c]\!]^E$$

$$[\![e]\!]^{UP} = \bigcup_{p \in Perm(e)} [\![p]\!]^P \qquad\qquad [\![e]\!]^{IP} = \bigcup_{c \in Comp(e)} [\![c]\!]^P$$

A completion *of a sequence* $e := e_1,\ldots,e_n$ *is a sequence* c *that contains all elements of* e *plus an arbitrary number of wildcards. A completion of* e *is called* order-respecting, *iff for $e_i, e_j \in e$ with $i < j$, e_i appears in c before e_j. Perm(e) and Comp(e) denotes the set of all permutations and order respecting completions of e, respectively.*

Both extensions of RPL – to unordered paths and to incomplete paths – are mere syntactic sugar. The RPE Nodes { x y } can be rewritten to the equivalent RPE Nodes (x y) | (y x) and the RPE Nodes [[x y]] can be rewritten to Nodes _* x _* y _*. Observe that whereas the rewriting of incomplete path expressions is linear in the size of the original expression, the rewriting of unordered paths is exponential in the size of the original expression. We chose not to include incomplete RPEs in standard RPL, since one can easily do without them. On the other hand we chose not to include unordered RPEs in standard RPL, because it would make evaluation of RPL NP-hard as shown in Section 6.

The semantics of RPEs that are both unordered and incomplete (denoted by {{ }}) is easily defined at the aid of non-order-respecting permutations. For the sake of brevity, we omit this extension of RPL.

5   RPL Compared to Lorel, SPARQLeR and Nested Regular Expressions

[2] extends SPARQL by regular expression patterns which may occur at the place of predicates in RDF graphs. These regular expression patterns include amongst others kleene closure, disjunction, concatenation, but not predicate negation and regular string expressions. Moreover, node labels are are not considered part of the path to be matched by the regular expression pattern.

The Lorel query language [1] is an offspring of the XML database system Lore, but can be used to query all kinds of semi-structured data. It has received considerable attention in the research community, partially due to its incorporation of regular path expressions.

RPEs compare to Lorel path expressions as follows:

- The data model of Lorel is an edge-labeled graph, without node labels. Therefore Lorel does not distinguish the three flavors of RPEs.
- Both languages provide the unary operators Kleene plus (+), Kleene star (*) and optionality (?), and the binary operators concatenation (denoted by '.' in Lorel), and alternative.
- Lorel allows the use of the character '%' to match 0 or more characters within a label. RPL on the other hand allows regular string expressions. Wildcards for entire labels are denoted by '#' in Lorel and '_' in RPL.
- Lorel allows the extraction of values from traversed paths by so-called path variables. RPEs do not use variables since they may be embedded in RDF query language such as SPARQL or XCERPT[RDF], that provide themselves variables.
- RPEs allow the restriction of paths based on path predicates, Lorel does not. Hence query 2 is not expressible in Lorel.

In [13] the evaluation of regular expressions over the alphabet σ of an edge-labeled graph g is studied. Compared to RPEs, [13] considers the labels of edges to be atomic, i.e. they do not consider regular string expressions on node or edge-level. Moreover, non-local restrictions on paths (i.e. predicates) and traversal in reverse direction are not expressible. Since nodes in the queried graphs are unlabelled, only the edge labels are relevant, i.e. the path expressions in [13] correspond to a subset of edge-flavored RPEs.

[13] considers the problems *Regular Simple Path*, *Fixed Regular Path (R)*, and *Regular Path*. The problem Regular Simple Path takes a regular expression e, a graph g over the same alphabet Σ, and a pair of nodes (x, y) as input, and returns true iff g contains a directed simple path from x to y that satisfies e. A path is called *simple*, if it does not contain the same vertex twice. The problem Fixed Regular Path is the same as regular simple path, but e is not considered as input. Regular Path is the same as Regular Simple Path, but the path is not required to be simple.

[13] show that Fixed Regular Simple Path is NP-complete and Regular Simple Path is NP-hard by a simple reduction from the problems Even Path and Disjoint Paths treated in [11] and [9], respectively. Regular Path, however, is decidable in time $O(|E|\,|D|)$, where $|E|$ is the size of the regular path expression and $|D|$ is the size of the data – shown by the construction of a product automaton of the NFA of a regular path expression and the database graph interpreted as a NFA. In RPL we choose to accept arbitrary paths, including non-simple paths as possible connections among two nodes. RPEs are more expressive than the regular path expressions of [13] in three respects: (i) They allow the specification of predicates on nodes, (ii) regular expressions for matching edge and node labels, and (iii) in that they take into account also the labels of *nodes*. Therefore, the results of [13] leave the question, if there is a polynomial time algorithm for the evaluation problem of RPEs, open. The following result for the complexity of $RPL^{\rightarrow,/,|,\mu}$ expressions is a direct consequence of the complexity *Regular Path*.

Corollary 1. $RPL^{\rightarrow,/,|,\mu}$ *can be evaluated in time* $O(|E|\,|G|)$, *where* $|E|$ *is the size of the path expression and* $|G|$ *is the size of the queried RDF graph.*

[15] propose the regular path language nested regular expressions (NRE) with the following syntax:

$$exp := \text{axis} \mid \text{axis::}a \ (a \in U) \mid \text{axis::}[exp] \mid exp/exp \mid exp|exp \mid exp^* \quad (30)$$

where $axis \in \{self, next, next^{-1}, node, node^{-1}, edge, edge^{-1}\}$ and U denotes the set of URIs. The axes *next*, *edge* and *node* are used to navigate from one node in an RDF graph to an adjacent one, from a node to one of its outgoing edges and from an edge to its sink. If the starting node is left unspecified, *next*, *edge* and *node* can be interpreted as binary relations over an RDF graph G. Node tests following the axes *next*, *edge* and *node* constrain the label of a traversed edge, the object of an arc, and the subject, respectively. The semantics of the predicates [], alternatives |, Kleene star *, and concatenation / are as expected.

In this section we briefly give an intuitive semantics of NRE by translating Examples 1, 2, 3 and 4 to NRE.

We abbreviate URIs in NRE by qualified names to shorten the examples.

Example 6 (Nested regular expressions)

- Example 1 is *contained* in the NRE (next)*/next::rdf:type. An exact translation is not possible due to the absence of regular string expressions for matching nodes or edges of RDF graphs.
- Example 2 is *contained* in the NRE (next::eg:p)$^+$. An exact translation is not possible due to the absence of negation in NRE predicates.
- Example 3 is *equivalent* to next::rdf:type/(next::rdfs:subClassOf)*.
- The first RPL expression in Example 4 is *equivalent* to the following NRE.

$$next/\text{self::eg:a}/next/\text{self::eg:b}$$

- The NRE

$$\text{next::a}/(\text{next::[next::a/self::b]})*/(\text{next::[node::b]} \mid \text{next::a})^+ \qquad (31)$$

  from [15] is *contained* in the RPE [EDGES a(_[PATH a b]) * _]. An exact translation to an RPE is not possible, since RPEs always evaluate to pairs of *nodes* of an RDF graph. In contrast, NREs may also evaluate to pairs of edges and nodes, as the subexpression "node::b" of Expression 6 does. Expression 6 can, however, be translated to an equivalent XCERPT^{RDF} query term or SPARQL query pattern that makes use of a single RPE.

Given an NRE *exp*, an RDF graph G, and a pair of nodes (n_1, n_2), the problem whether there is a path from n_1 to n_2 matching *exp* within G, can be decided in $O(|G| \cdot |exp|)$.

Corollaries 2 and 3 shed light on the expressive relationship between fragments of RPL and NRE. An immediate consequence of Corollary 2 is Corollary 4. The proofs for Corollaries 2 and 3 are contained in the online version of this paper [6].

Corollary 2. *Any RPE* $r \in RPL^{\rightarrow,\circ,--\rightarrow,[],/,|,\mu}$ *can be translated to an equivalent NRE of length* $\mathcal{O}(|p_c|)$.

Corollary 3. *Any NRE* p_n *excluding the axes node, node^{-1}, edge, and edge^{-1} can be translated to an equivalent RPE* p_c *of length* $\mathcal{O}(|p_n|)$.

Corollary 4. *A RPE* p_c *in* $RPL^{\rightarrow,\circ,--\rightarrow,[],/,|,\mu}$ *can be evaluated in* $O(|G| \cdot |p_c|)$.

6   Further Complexity Results

The comparison of *RPL* to related path query languages in the last section has already brought up some complexity results for sublanguages of *RPL*. In this section we establish the tractability of *RPL* as a whole and the intractability of *RPL* with unordered paths.

Theorem 1 (Tractability of RPL and $NRE^{RSE,\neg}$). *RPL and the extension of NRE by regular string expressions and predicate negation (denoted by $NRE^{RSE,\neg}$) can be evaluated in time $O(|exp| \cdot |G|)$.*

Proof. (Sketch) Theorem 1 builds upon Corollary 4, that establishes that the evaluation of $RPL^{\rightarrow,\circ,--\rightarrow,[],/,|,\mu}$ is in $O(|exp| \cdot |G|)$. The only features missing in $RPL^{\rightarrow,\circ,--\rightarrow,[],/,|,\mu}$ when compared to full RPL are predicate negation ($\neg$) and regular string expressions (RSE). The evaluation of regular string expressions is linear. Thus, defining the size of an RDF graph as the total length of the characters appearing within its nodes and edges, the complexity remains in $O(|exp| \cdot |G|)$ when regular string expressions are added to the language.

Showing that predicate negation has no effect on evaluation complexity is a little more tricky: Consider the proof of the tractability of NRE in [15]. It involves the construction of product automata $G \times \mathcal{A}_p$ for each predicate p appearing in the expression exp to be evaluated. We can extend NRE to $NRE^\neg$ by allowing predicate negation in the same way as RPL allows predicate negation. A RPE p_c with predicate negation can then be translated to an $NRE^\neg$ expression p_n in linear time, such that the size of p_n remains linear in the size of p_c.

It remains to be shown that $NRE^\neg$ is in $O(|exp| \cdot |G|)$. For this end, we adapt the algorithm $LABEL(G, exp)$ from [15] to label both positive and negative predicates appearing in exp. For each negative predicate $not(p)$ we introduce the label not_p which is attached to each node n in G *not* matching with p. Then, for each negative predicate $not(p)$ in exp, we substitute $not(p)$ in not_p, thereby obtaining an ordinary NRE expression exp^+. exp^+ evaluates to over G with the adapted labeling algorithm if and only if exp evaluates to true over G with the original labeling algorithm.

Theorem 2 (NP-Completeness of $RPL^{\circ,/,\{\}}$). *The evaluation problem of $RPL^{\circ,/,\{\}}$ is NP-complete.*

Proof. Obviously the evaluation problem for $RPL^{\circ,/,\{\}}$ is in NP. We show its NP-hardness by a reduction from the directed Hamiltonian path problem. Let G be an arbitrary RDF graph with nodes $\{n_1, \ldots, n_k\}$. Then G has a directed Hamiltonian path if and only if the RPE { NODES $n_1, \ldots n_k$ } has a non-empty solution over G.

Theorem 3. *The evaluation problem for $RPL^{\rightarrow,/,\{\}}$ is in $O(n \cdot \sigma^w \cdot e)$ where n is the number of nodes of the RDF graph, e the number of edges, σ the number of edge labels, and w is the length of the path expression.*

Corollary 5. *The evaluation problem for $RPL^{\rightarrow,/}$ is in $O(e \cdot w)$ where w is the length of the regular path expression and e is the number of edges in the RDF graph.*

Proof. Theorem 3 only gives an upper bound for the evaluation of $RPL^{\rightarrow,/,\{\}}$, therefore it suffices to give an algorithm that runs in $O(n \cdot \sigma^w \cdot e)$ time.

Let G be an RDF graph, and $p \in RPL^{\rightarrow,/,\{\}}$. The idea of the algorithm is to view G as a non-deterministic finite automaton, and p as a word to be checked by the automaton. p is checked from the first element to the last, and the set of valid states in the automaton is remembered in each step, starting out from the set of all nodes in the RDF graph. For $RPL^{\rightarrow,/}$ (i.e. only ordered edge-flavored expressions), this view gives us an algorithm in $O(e \cdot w)$, where e is the number of edges in G, and w is the length of p (Corollary 5).

For unordered edge-flavored path expressions, a naive implementation would compute all possible permutations, and check the RDF graph for correspondence with each of these permuations. Since there $w!$ permutations for a path of length w, this procedure has a complexity of $O(w! \cdot e)$. The following algorithm is more efficient:

Again, the RDF graph G is viewed as a finite automaton, which is traversed using symbols occurring in the path expression p. In step i of the computation, each node n in G is labeled with all paths p of length i such that n is reachable over p from some other node m in G. Initially, all nodes are labeled with the empty path ϵ. After w steps (or earlier), the algorithm terminates and exactly the set of labeled nodes in G is reachable over p. In Listing 1.1 we use set notation to represent paths, since the order of traversal is irrelevant; however we must think of paths as multisets, because the same edge label may occur multiple times in p. For this reason, the set difference operator $\setminus$ and the union operator $\cup$ in Listing 1.1 are the set difference and the union operator for *multisets*, not *sets*, respectively.

Listing 1.1. Evaluation algorithm for expressions in $RPL^{\rightarrow,/}$

```
for each node n in G do labels(n) = {ε} end
for i = 1 to w do            // w is the length of path p
   for each e in E do        // follow every edge
      for each l in labels(source(e)) do
         if label(e) is in p \ l then
            labels(sink(e)).add({l} ∪ label(e))
         end
      end
   end
   remove all labels of length i − 1
end
```

In the i-th iteration of the outermost loop of Listing 1.1, the set of labels for the nodes in G is bounded by $\sigma^i \cdot |n|$. Thus, the number of edge traversals in step i is bounded by $\sigma^i \cdot |n|$. The total number of edge traversals is thus $\sigma^{w+1} \cdot |n|$ (geometric series).

Theorem 4 (NP-Completeness of $RPL^{\rightarrow,/,\{\}}$). *The evaluation problem of* $RPL^{\rightarrow,/,\{\}}$ *is NP-complete.*

Proof. For the proof of Theorem 4 we use a reduction from the Hamiltonian Cycle Problem. The idea of the proof is illustrated in Figure 2. Let $G = (V, E)$ be a directed labeled graph with nodes $\{1, \ldots, k\}$. G has a Hamiltonian Cycle if and only if the RPE { EDGES $1_{in}, 1_{out}, \ldots, k_{in}, k_{out}$ } has a non-empty solution over the *edge expansion graph* of G, which is defined as follows:

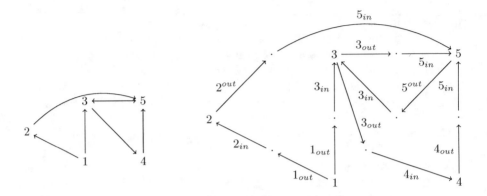

Fig. 2. Reduction from the Hamilton Cycle Problem to $RPL^{\rightarrow, /, \{\}}$ evaluation

Definition 10 (Edge expansion graph). *Let $G = (V, E)$ with $V = 1, \ldots, k$ be a graph. The* edge expansion graph *$F = (V', E', \mu)$ of G is an edge labeled graph with the following properties:*

- *$V \subseteq V'$*
- *For each edge $(u, v) \in E$ there is some node n in V' and edges $(u, n), (n, v) \in E'$ with $\mu(u, n) = u_{out}$ and $\mu(n, v) = v_{in}$. There are no other edges in E' involving n.*
- *These are all nodes and edges in F.*

The edge expansion graph F of a given Graph G with v vertices and e edges contains $v + e$ vertices and $2 \cdot e$ edges. Obviously, F can be constructed from G in polynomial time.

7   Conclusion and Future Work

This paper describes the novel RDF path language *RPL*. *RPL* is one of the few RDF path languages with a formal semantics. Compared to other query languages it omits features that are rarely used, but includes features such as regular string expressions, direction modifiers, and predicate negation, that may turn out to be extremely useful for query authors. *RPL* can be evaluated efficiently, but extensions of *RPL* with unordered paths or variables cannot. *RPL* is currently being implemented. Future work includes the experimental affirmation of the tractability of *RPE* evaluation.

References

1. Abiteboul, S., Quass, D., McHugh, J., Widom, J., Wiener, J.L.: The Lorel query language for semistructured data. International Journal on Digital Libraries 1(1), 68–88 (1997)

2. Alkhateeba, F., Baget, J.-F., Euzenat, J.: Extending SPARQL with regular expression patterns. Journal of Web Semantics (2009)
3. Angles, R., Gutierrez, C., Hayes, J.: RDF query languages need support for graph properties. Technical report (2004)
4. Bry, F., Furche, T., Ley, C., Linse, B.: RDFLog—taming existence - a logic-based query language for RDF (2007)
5. Bry, F., Furche, T., Ley, C., Linse, B., Marnette, B.: RDFLog: It's like datalog for RDF. In: Proceedings of 22nd Workshop on (Constraint) Logic Programming, Dresden, September 30 –October 1 (2008)
6. Bry, F., Furche, T., Linse, B.: Online version,
   http://www.pms.ifi.lmu.de/mitarbeiter/linse/RPL-full.pdf
7. Bry, F., Furche, T., Linse, B., Pohl, A.: XcerptRDF: A pattern-based answer to the versatile web challenge. In: Proceedings of 22nd Workshop on (Constraint) Logic Programming, Dresden, Germany, September 30 –October 1, pp. 27–36 (2008)
8. Bry, F., Furche, T., Ley, C., Linse, B., Marnette, B., Poppe, O.: SPARQLog: SPARQL with rules and quantification. In: Virgilio, R.D., Giunchiglia, F., Tanca, L. (eds.) Semantic Web Information Management: A Model-based Perspective, ch. 12. Springer, Heidelberg (2009)
9. Fortune, S., Hopcroft, J.E., Wyllie, J.C.: The Directed Subgraph Homeomorphism Problem (1978)
10. Kochut, K., Janik, M.: SPARQLeR: Extended SPARQL for semantic association discovery. In: Franconi, E., Kifer, M., May, W. (eds.) ESWC 2007. LNCS, vol. 4519, pp. 145–159. Springer, Heidelberg (2007)
11. La Paugh, A., Papadimitrou, C.: The even-path problem for graphs and digraphs. Networks 14(4), 507–513 (1984)
12. Marx, M.: Conditional XPath. ACM Transactions on Database Systems (TODS) 30(4), 929–959 (2005)
13. Mendelzon, A.O., Wood, P.T.: Finding Regular Simple Paths in Graph Databases. SIAM Journal on Computing 24, 1235 (1995)
14. Ogbuji, C.: Versa: Path-based RDF query language (2005)
15. Pérez, J., Arenas, M., Gutierrez, C.: nSPARQL: A navigational language for RDF. In: Sheth, A.P., Staab, S., Dean, M., Paolucci, M., Maynard, D., Finin, T., Thirunarayan, K. (eds.) ISWC 2008. LNCS, vol. 5318, pp. 66–81. Springer, Heidelberg (2008)
16. Pohl, A.: RDF Querying in Xcerpt: Language Constructs and Implementation. Deliverable I4-Dx2, REWERSE (2008)
17. Polleres, A.: From SPARQL to rules (and back). In: Williamson, C.L., Zurko, M.E., Patel-Schneider, P.F., Shenoy, P.J. (eds.) WWW, pp. 787–796. ACM, New York (2007)
18. Schenk, S., Staab, S.: Networked graphs: A declarative mechanism for SPARQL rules, SPARQL views and RDF data integration on the Web. In: Proceedings of the 17th International World Wide Web Conference, Bejing, China (April 2008)
19. Seaborne, A., Prud'hommeaux, E.: SPARQL query language for RDF. W3C recommendation, W3C (January 2008),
   http://www.w3.org/TR/2008/REC-rdf-sparql-query-20080115/

A Hybrid Architecture for a Preoperative Decision Support System Using a Rule Engine and a Reasoner on a Clinical Ontology

Matt-Mouley Bouamrane[1,2], Alan Rector[1], and Martin Hurrell[2]

[1] School of Computer Science
Manchester University, UK
{mBouamrane,Rector}@cs.man.ac.uk
[2] CIS Informatics, Glasgow, UK
martin.hurrell@informatics.co.uk

Abstract. We report on a preventive care software system for preoperative risk assessment of patient undergoing elective surgery. The system combines a rule engine and a reasoner which uses a decision support ontology developed with a logic based knowledge representation formalism. We specifically discuss our experience of using a representation of a patient's medical history in OWL, combined with a reasoning tool to suggest appropriate preoperative tests based on an implementation of preoperative assessment guidelines. We illustrate the reasoning functionalities of the system with a number of practical examples.

1 Introduction

The primary goal of clinical guidelines is to improve the quality and efficiency of healthcare delivery, while providing an efficient, cost-effective and consistent service across healthcare institutions. Clinical guidelines are expressed as statements, rules, recommendations, management protocols, etc. They suggest a certain course of action given a specific medical *context*. Clinical guidance provide support to health professionals without overruling their clinical judgment and ability to use their own discretion in order to make decisions appropriate to the individual circumstances of the patients and the broader medical context. Clinical guidelines are generally designed after a lengthy and thorough process, involving clinical topic selection, consultation with stakeholders, establishing an expert panel which will examine the best evidence available on the topic, the generation of draft recommendations, typically by consensus, submission to stakeholders or peer-review, and finally, issuance of final guidelines. The guidelines may be used for training health professionals and supporting them in the intellectually demanding and knowledge-intensive environment of the health services. In addition, guidelines may lead to a significant improvement in service efficiency and a substantial reduction in costs. The study by Ferrando et al. on 702 patients undergoing preoperative assessment estimated that applying preoperative guidelines would reduce the cost of preoperative tests by 63% and in excess of

A. Polleres and T. Swift (Eds.): RR 2009, LNCS 5837, pp. 242–253, 2009.
© Springer-Verlag Berlin Heidelberg 2009

40% for hospital stays per patient [1]. In addition, tests performed on patients, which were not deemed relevant by the guidelines, did not provided additional clinical information of interest.

While the benefits of clinical guidelines may be confirmed by a number of studies, the problem remains as ever of how to apply these effectively on the ground and integrate them in clinical workflows and work practices, amidst the huge volume of medical information available to health professionals, a proliferation of guidelines, as well as rapidly changing health policies [2]. The difficulty is thus to deliver the guidelines in a format which make them readily usable by health professionals. In addition, the considerable effort in producing and implementing the guidelines is another incentive to make these guidelines sharable and interoperable across services and institutions. Thus, the use of Information Technology in the health services, and in particular Clinical Decision Support Systems (CDSS), can potentially efficiently support health professionals in their duties. Tasks which may routinely be performed by CDSS include: efficiently managing patient clinical information, handling clinical guidelines, providing risk assessment, preventing errors of omission through alerts and reminders, extracting information of interest and making recommendations appropriate to the circumstances of the patients, such as treatment protocols [3,4,5]. Indeed, the potential to exploit computer systems to implement and share clinical guidelines has long been recognised, as evidenced by the development of the GuideLine Interchange Format (GLIF) [6], and efforts to develop guidelines models and guidelines based decision support systems [7,8].

We here report on our work combining a preventive care software system for preoperative risk assessment of patient using a rule engine with a decision support ontology developed with a logic based knowledge representation formalism. We specifically discuss our experience of using a representation of a patient's medical history in OWL, combined with a reasoning tool to suggest appropriate preoperative tests based on an implementation of the NICE preoperative assessment guidelines[1]. The paper is structured as follows: first we present the research motivation and background information on the preoperative decision support system. We provide an overview of the system decision support ontology and reasoning functionalities. We then discuss in details our implementation of the NICE preoperative assessment guidelines. We conclude with a discussion on this approach and directions for future work.

2   Background

This work is part of an ongoing project to introduce *semantic* technology into a preoperative risk assessment system software called Synopsis. The overall architecture of the preoperative assessment system is illustrated in Figure 1. The use of knowledge representation and reasoning both completes functionalities and overcomes a number of limitations of the existing system. While numeric risk

---

[1] NHS, National Institute for Clinical Excellence
http://www.nice.org.uk/Guidance/CG3

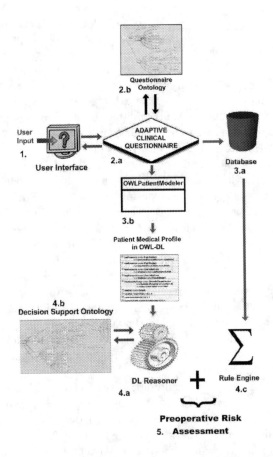

Fig. 1. Hybrid architecture of the rule-engine / clinical knowledge-base preoperative risk assessment system

score calculation is currently easily done by the system using an open source java-based rule engine, JBoss Rules[2], other tasks including categorisation, classification and logical inference were beyond the capacity of the system prior to introduction of semantic technology.

Functionalities introduced in the system as the result of this project include the development of an adaptive questionnaire, whereby clinical content of the information collection process adapts automatically to information specific to a patient clinical profile. Another enhancement to the system includes the ability to generate a high level semantic representation of a patient clinical history in the web ontology language OWL-DL. This is subsequently used to provide decision support functionalities via the use of automated reasoning tools. The web ontology language OWL is supported by an active research community, readily

---

[2] http://www.jboss.com/products/rules

and freely available development and reasoning tools and a well maintained application programming interface [9,10,11].

High level medical history representation is done automatically for new patients entered in the system and we have proposed a methodology for the semi-automatic generation of this medical history from legacy clinical databases in [12]. Note that prior to introduction to semantic technology to the system, the preoperative software was composed only of the following elements: user input (step 1), clinical data storage (3.a) and rule engine (4.c). Therefore, the preoperative risk assessment (5) was almost entirely based on the calculation of numeric scores. Thus, the introduction of semantic based technology in the system enables adaptive information collection (2a and 2b), high level semantic patient modelling (3.b) and decision support based on classification (4.a and 4.b) rather than numeric rules only. This provides for a significant enhancement to the functionalities and capabilities of the system. We refer the interested reader to [13,14,15,16] for a detailed description of the system features and functionalities.

3   Decision Support Ontology and Reasoning Functionalities

Once the system has completed the adaptive information collection phase and has generated a patient medical history semantic profile (3.b), reasoning on the decision support ontology can then be performed. The purpose of the decision support is to provide advice to clinicians and flag potential risks of complications so appropriate preventive measures can be taken accordingly. The decision support ontology is divided into three domains with distinct purposes. The first one consists of a Risk Assessment ontology and its purpose is to highlight potential intra-operative and post-operative complications given a patient medical profile and the planned surgical procedure. The second one is the Recommended Test ontology, which purpose is to suggest certain preoperative tests, which may help to decide whether it is safe to proceed with a planned surgery or whether further actions need to be taken (e.g. referral to a specialist, optimisation of patient's health prior to surgery, etc.) The Recommended Test ontology we are currently using is based on the preoperative tests clinical guidelines issued by the British National Health System (NHS) National Institute for Clinical Excellence (NICE). This will be discussed further in the next section. Finally, the last domain of the Decision Support ontology is the Precaution Protocol ontology, which may suggest a management or treatment protocol given a specific medical context.

In the system, decision support is usually provided in a 2 step process. The first step typically calculates risk scores or derives risk grades (ASA grades, surgical risk grades, etc.) using numerical formulas such as the Goldman and Detsky cardiac risk index, the Physiological and Operative Severity Score for the enUmeration of Mortality and Morbidity (POSSUM) [17], etc. Once the risk grades and categories have been derived from the first risk calculation step, the system can then perform decision support using the open-source java-based

PELLET reasoner [18] to reason on the decision support ontology given a patient OWL medical history profile.

4   Recommended Preoperative Investigations

4.1   Description of Investigation Guidelines

The purpose of the NICE guidelines recommendations for preoperative investigations is both to avoid patients undergoing unnecessary investigations, which can be detrimental to their health, as well as more efficiently managing limited resources in the public health services. The implementation of the NICE guidelines in the decision support system is both: (i) a pragmatic and useful functionality provided to health professionals and (ii) a good example of how the use of a clinical ontology and reasoner can provide functionalities beyond the capabilities of a traditional rule engine. Regarding the first point, the table in Figure 2 illustrates the format of the NICE guidelines recommendation for preoperative investigations. We here describe the guidelines in more details:

Type of investigations: the guidelines include 9 potential investigations: Chest X-Ray, ECG (Electrocardiogram), Full Blood Count, Haemostasis, Renal Function, Random glucose, Urine analysis, Blood gases and Lung Function.

Type of recommendations: there are currently 3 types of recommendations for each test: "test recommended", "test not recommended" and "consider test".

Factors Influencing recommendations: There are 5 factors taken into consideration in order to find the relevant recommendations: the (i) age of the patient, (ii) his ASA grade, (iii) the type of comorbidities the patient has (e.g. respiratory, cardiovascular, renal) (iv) the type of surgery (e.g. cardiovascular surgery, neurosurgery, etc.) (v) the risk grade of the surgery (from 1 to 4).

Number of cases in the guidelines: the guidelines are summarised for preoperative health assessors into 36 tables such as the one illustrated in Figure 2. There are different tables for different combinations of the 5 factors previously described, including different tables for children under 16 years old and adults over 16 years old. In total, there are at least 1242 different possible cases.

Perhaps not surprisingly, we found that in practice, preoperative health assessors faced considerable difficulties in using the guidelines. The important number of factors to take into consideration in order to find the correct table and then the specific case within this table, combined with the significant number of tables meant that too much time was being spent by preoperative health assessors trying to refer to the correct case. In addition, the preoperative health assessors would need to be able to categorise (i) the type of comorbidities (ii) their severity (e.g. for determining the patient's ASA grade) (iii) the type of surgical procedures and (iv) their surgical risk grades, all of this before being able to refer to

Grade 1 Surgery - Adults with ASA 2 with
comorbidity from Renal Disease

Test	age in years			
	16 to 40	40 to 60	60 to 80	over 80
Chest X-ray	NO	NO	NO	
ECG	NO			YES
Full Blood Count				
Haemostasis	NO	NO	NO	NO
Renal Function	YES	YES	YES	YES
Random Glucose	NO	NO	NO	NO
Urine Analysis				
Blood gases	NO	NO	NO	NO
Lung function	NO	NO	NO	NO

Fig. 2. Adapted from the NICE preoperative guidelines: investigations are recommended based on: patient (i) age, (ii) ASA, (iii) comorbidities (iv) type of surgical procedure and (v) risk grade of surgical procedure. There are 3 types of result for each test: "test recommended", "test not recommended" and "consider test".

the correct table. All of these tasks are obviously highly knowledge intensive as well as being intellectually demanding. In addition, preoperative health assessors typically see dozens of different patients a day, each with a wide variety of health conditions and scheduled for various types of surgical procedures. In practice, the consequences are that, if in doubt, preoperative investigations would probably be requested regardless of the guideline (i.e. *better safe than sorry*) , thus defeating the purpose of the guidelines in managing efficiently the allocation of preoperative investigation in the health services.

4.2   Investigation Rule Axioms Generation

We combined the use of an ontology and reasoner in the preoperative decision support system in order to automatically make recommendations regarding the suitability of tests based on the NICE guidelines. The first step consisted into transforming the NICE tables into rules. This enabled to considerably reduce overlap and redundant information in the current format of the guidelines. See Table 1 as an illustration of the NICE guidelines as rules. Figure 3 illustrates the same rule seen through the Protégé-OWL development tool [10]. Thus, the 1242 different possible cases currently covered by the NICE guidelines were reduced to around a hundred rules of the type illustrated in Table 1.

As the amount of redundant information in the guidelines is substantial, there are a number of different ways the rule axioms can be expressed: according to comorbidity, surgery grade, ASA grade, etc. Thus we have use a number of heuristics to develop the rules:

Table 1. NICE guidelines as a rule: refers to the first line (Chest X-ray) of the table in Figure 2

Rule Id:	NiceGuidelineRule_ChestXRay_NOT-RECOMMENDED05_N0007
Class name:	ChestXRay_NOT_RECOMMENDED_NT+SurgeryGrade1+ ASA2+Renal_Comorb+Between16_and_80Years
Location	Subclass of: ChestXRay_NOT_RECOMMENDED_NICE_Test
Meaning:	*"Adults between the age of 16 years to 60 years old* *of ASA 2, with renal comorbidities* *undergoing surgery of risk grade 1* *should NOT undergo a chest X-ray investigation* *unless undergoing cardiovascular surgery"*

OWL Axiom	PatientClassificationEntity
	and ChestXRayClassificationEntity
	and (has some (Age
	    and (hasTemporalUnit some Year)
	    and (hasIntNumericValue some int[≥"16"integer,<"60"integer])))
	and (hasASA-gradeValue value "2"integer)
	and (hasAssociatedComorbidity some RenalComorbidity)
	and (hasPlannedSurgicalProcedure some (SurgicalProcedure
	    and (hasSurgicalGrade some SurgicalGrade1)))
	and not (hasPlannedSurgicalProcedure some CardioVascularSurgery)

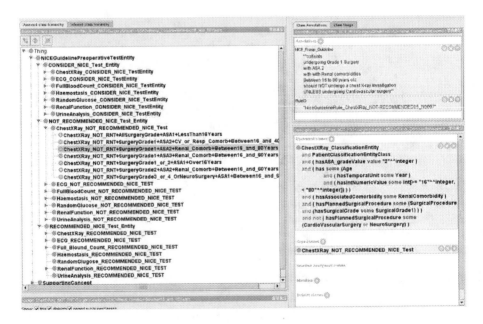

Fig. 3. The NICE Guidelines as OWL Rules as viewed through the Protégé-OWL User Interface

General rules: where possible, we have tried to use as generic rules as possible (e.g. *"All* Adults aged over 80 years old should undergo an ECGTest *regardless* of any other factorsĔ (ASA or surgery gradeĔ))"

Consistency in Rule Generation: Rules can often be generated according to different combinations of factors. For example, the same guidelines could be based on different combinations of factors based on patient ASA, or surgery grade, age, etc. We generally chose to express rules along an interpretable factor whenever possible, such as comorbidities (e.g. *"all patients with a renal comorbidity..."*)

Trade-off between number of rules and complexity of rules: There is a trade-off between an optimum (i.e. minimum) number of rules and their complexity. It is possible to minimise the number of rules by including conjunction, disjunction and exclusion clauses. However, ones needs to ensure that the rules remain interpretable and as close as possible to the original guidelines format. This is mainly for explanation purposes as the decision support tool should be able to provide an understandable explanation of results to end users. Practicaly, this involved chosing to split complex rules into simpler rules when too many factors in the rule made it difficult to interpret. Typically, this involved limiting the number and occurrences of disjunction and exclusion clauses.

The main advantage of modelling the preoperative investigation guidelines as OWL axioms is that the preoperative decision support system can now (i) use third party clinical taxonomies in order to allocate a surgical risk grade to a specific surgical procedure and (ii) use a third party clinical ontology to infer patient comorbidities. Thus, using the OWL patient medical history profile generated at the step 3.b in Figure 1, we can now automatically infer which investigations a patient should have based on his specific medical history.

Figure 4 provides an example of preoperative test recommendation based on reasoning on the decision support ontology. Mark is a 67 year old patient, with arrhythmia, his ASA status has been estimated to be 3 and he is to undergo an open excision of lesion of duodenum. Reasoning on the recommended tests decision support ontology returns that a chest X-Ray, ECG test, full blood count and a renal function tests are all recommended, a haemostasis test may be considered and a lung function test is not recommended. The recommendations are made by the system based on the following reasons:

chest X-Ray: patient is over 60 and of ASA 3, in addition he has arrhythmia which is classified in the decision support ontology as a cardiovascular comorbidity and he is to undergo an open excision of lesion of duodenum, a surgical procedure of grade 4 in the ontology. All this criteria mean that this patient falls within one of 2 categories of patients which are recommended for a chest X-ray investigation.

ECG: test recommended as the patient has a cardiovascular comorbidity so he should undergo an ECG test regardless of all other factors.

Full Blood Count: test recommended as the patient is over 16 years old and is undergoing either grade 3 or 4 surgery.

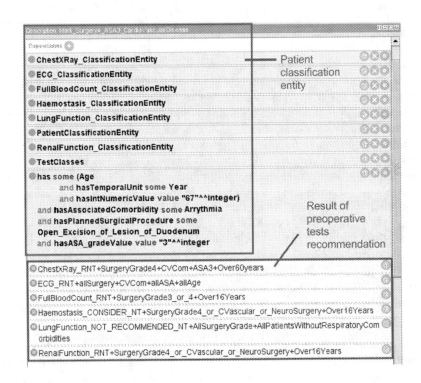

Fig. 4. Example of preoperative test recommendation based on reasoning on the decision support ontology

Renal function: test recommended as patient is over 16 and undergoing grade 4 surgery

Haemostasis: test may be considered as patient is over 16 and undergoing grade 4 surgery

Lung Function: this test is not recommended as the patient does not have any respiratory comorbidities

4.3   Dealing with Multiple Comorbidities

The preoperative guidelines do not explicitly deal with the issue of multiple comorbidities and this is an other area where the decision support tool can provide additional functionalities.

Duplication of test recommendations: In the case of a patient with multiple comorbidities, it is possible that a test may be recommended for multiple reasons. As an example, a patient of ASA 2, over 60 years old, undergoing grade 2 surgery could be recommended an ECG test twice if he has renal comorbidities and cardiovascular comorbidities. In this case, the system can issue a *strong* recommendation alongside relevant explanation.

Conflicting test recommendations: the guidelines are not mutually exclusive and especially not in the case of multiple comorbidities. It is possible
for example for a patient of ASA 2, less than 40 years old, undergoing grade
2 surgery to be recommended an ECG test if he has cardiovascular comorbidities but not if he has respiratory comorbidities. The contradiction is only
apparent: what is not necessary for a patient with *only* respiratory comorbidities obviously becomes necessary if the patient *also* has cardiovascular
comorbidities. Thus, one instance of a "recommended test" within a batch
of test results lead to a positive test recommendation regardless of all other
test recommendations. According to the same principle, if the system returns
instances of "consider test" along "test not recommended" instances, then the
system issues a "consider test" recommendation. Finally, the system issues
a "test not recommended" advice only if all instances returned are negative
for this specific test.

5   Discussion

The work on developing the GuideLine Interchange Format by Ohno-Machado
et al. [6] contains a very interesting discussion on the motivation behind the
development of GLIF and the process and issues encountered while translating
paper-based guidelines into computer-based guidelines. The authors emphasise
that this transcription process will usually reveal inconsistencies or flaws in the
guidelines. We agree that this is essentially a virtuous cycle as inconsistencies
in the guidelines may be highlighted and addressed. Although we found a very
small number of inconsistencies in the preoperative guidelines, this did not prove
much of a serious issue in our case. A recurring remark about the preoperative
guidelines is that they are not exhaustive: they do not cover all possible combinations of surgery, comorbidities and ASA cases. Our understanding is that
the guidelines cover a large number and possibly the majority of cases (although
we can not confirm this at the time of writing) of patients presenting for elective surgery. Most serious cases are not covered by the guidelines, but this is
likely to be when they would be the least helpful, as in these situations, health
professionals will use their own clinical judgement to request all relevant preoperative tests or even perhaps seek an alternative to surgery. Our experience would
suggest that the major obstacle to effective use of the guidelines in the health
services in the format in which they are represented, as they remain both intellectually demanding and knowledge-intensive. This is where a computer-based
decision support system can make a genuine difference in practice.

Getting the right format is without doubt a very difficult challenge for the is-
suers of the guidelines. To be useful, the guidelines need to be comprehensive in
covering a large number of cases, as well as being systematic in the presentation
of the results, while guaranteeing a rigorous path to a given recommendation for
the safety of the patients. Somehow, the format may then take precedence over
the underlying *meaning* behind the guidelines. Effectively, since the preoperative
guidelines use 5 types of information items as input (surgery grade, surgery type,

ASA grade, type of comorbidities and age), these factors are always present in the formulation of the guidelines, even if they are not necessarily relevant in every context. As an example: if a patient has a renal comorbidity, the guidelines recommend a renal function test regardless of any other factors (age, other comorbidities, surgery, etc.) However, this information is not necessarily clear, as it is both repeated and dispersed across multiple tables. Thus, the current format of the guidelines does not convey this information in an efficient manner. An OWL representation of the guidelines make these relations explicit, and is thus perhaps a closer representation of the *intended* meaning behind the guidelines.

6   Conclusion and Future Work

We discussed our experience of using a representation of a patient's medical history in OWL, combined with a reasoning tool to suggest appropriate preoperative tests based on an implementation of preoperative assessment guidelines. We illustrated the reasoning functionalities of the system with a number of practical examples. The system described can reuse third parties ontologies and taxonomies in order to provide advanced decision support functionalities. Future work will include deploying the system in selected pilot sites and observing and analysing how health professionals use the recommendations made by the system. Of particular interest will be to see if health professionals show an interest in the system underlying knowledge representation models and whether these relate to their personal understanding of the guidelines. In other words, we will observe whether the knowledge representation formalisms promote the use and understanding of the guidelines or whether they are of little interest to the health professionals who may be more concerned with the applicability and reliability of the results of the decision support system.

References

1. Ferrando, A., Ivaldi, C., Buttiglieri, A., Pagano, E., Bonetto, C., Arione, R., Scaglione, L., Gelormino, E., Merletti, F., Ciccone, G.: Guidelines for preoperative assessment: impact on clinical practice and costs. International Journal for Quality in Health Care 17(4), 323–329 (2005)
2. Audet, A., Greenfield, S., Field, M.: Medical practice guidelines: current activities and future directions. Annals of Internal Medicine 113, 709–714 (1990)
3. Hunt, D.L., Haynes, R.B., Hanna, S.E., Smith, K.: Effects of computer-based clinical decision support systems on physician performance and patient outcomes. JAMA, Journal of American Medical Association 280(15), 1339–1346 (1998)
4. Amit, G., Neill, A., McDonald, H., Rosas-Arellano, M., Devereaux, P., Beyene, J., Sam, J., Haynes, R.: Effects of computerized clinical decision support systems on practitioner performance and patient outcomes, a systematic review. Journal of the American Medical Association, JAMA 293(10), 1223–1238 (2005)
5. Leong, T.Y., Kaiser, K., Miksch, S.: Free and open source enabling technologies for patient-centric, guideline-based clinical decision support: A survey. IMIA Yearbook of Medical Informatics, Methods of Information in Medicine 46(1), 74–86 (2007)

6. Ohno-Machado, L., Gennari, J.H., Murphy, S.N., Jain, N.L., Tu, S.W., Oliver, D.E., Pattison-Gordon, E., Greenes, R.A., Shortliffe, E.H., Barnett, G.O.: The GuideLine Interchange Format, a model for representing guidelines. Journal of American Medical Informatics Association, JAMIA 5, 357–372 (1998)
7. Johnson, P.D., Tu, S.W., Musen, M.A., Purves, I.: A virtual medical record for guideline-based decision support. In: Proceedings of the 25th Symposium of the American Medical Informatics Association (AMIA), Washington, DC, US, pp. 294–298 (2001)
8. de Clercq, P.A., Blom, J.A., Korsten, H.H., Hasman, A.: Approaches for creating computer-interpretable guidelines that facilitate decision support. Journal of Artificial Intelligence in Medicine 31(1), 1–27 (2004)
9. OWL, Web Ontology Language (2004), http://www.w3.org/2004/OWL
10. Knublauch, H., Fergerson, R.W., Noy, N.F., Musen, M.A.: The Protégé OWL plugin: an open development environment for semantic web applications. In: McIlraith, S.A., Plexousakis, D., van Harmelen, F. (eds.) ISWC 2004. LNCS, vol. 3298, pp. 229–243. Springer, Heidelberg (2004)
11. Horridge, M., Bechhofer, S., Noppens, O.: Igniting the OWL 1.1 touch paper: The OWL API. In: Proceedings of the third International Workshop of OWL Experiences and Directions, OWLED 2007, Innsbruck, Austria (2007)
12. Bouamrane, M.M., Rector, A., Hurrell, M.: Semi-automatic generation of a patient preoperative knowledge-base from a legacy clinical database. In: Proceedings of the 8th International Conference on Ontologies, DataBases, and Applications of Semantics, ODBASE 2009, OTM 2009 Internet Systems. LNCS. Springer, Heidelberg (to appear, 2009)
13. Bouamrane, M.M., Rector, A., Hurrell, M.: Gathering precise patient medical history with an ontology-driven adaptive questionnaire. In: Proceedings of the 21st IEEE International Symposium on Computer-Based Medical Systems, CBMS 2008, Jyväskylä, Finland, pp. 539–541. IEEE Computer Society, Los Alamitos (2008)
14. Bouamrane, M.M., Rector, A., Hurrell, M.: Ontology-driven adaptive medical information collection system. In: An, A., Matwin, S., Raś, Z.W., Ślęzak, D. (eds.) Foundations of Intelligent Systems. LNCS (LNAI), vol. 4994, pp. 574–584. Springer, Heidelberg (2008)
15. Bouamrane, M.M., Rector, A.L., Hurrell, M.: Using ontologies for an intelligent patient modelling, adaptation and management system. In: Meersman, R., Tari, Z. (eds.) OTM 2008, Part II. LNCS, vol. 5332, pp. 1458–1470. Springer, Heidelberg (2008)
16. Bouamrane, M.M., Rector, A., Hurrell, M.: Development of an ontology of preoperative risk assessment for a clinical decision support system. In: Proceedings of the 22nd IEEE International Symposium on Computer-Based Medical Systems, CBMS 2009, Albuquerque, US. IEEE Computer Society, Los Alamitos (to appear, 2009)
17. Copeland, G., Jones, D., Walters, M.: Possum: a scoring system for surgical audit. British Journal of Surgery 78(3), 355–360 (1991)
18. Sirin, E., Parsia, B., Grau, B.C., Kalyanpur, A., Katz, Y.: Pellet: A practical OWL-DL reasoner. Journal of Web Semantic 5(2), 51–53 (2007)

A Logic Based Approach to the Static Analysis of Production Systems

Jos de Bruijn and Martín Rezk

KRDB Research Center
Free University of Bozen-Bolzano, Italy
{debruijn,rezk}@inf.unibz.it

Abstract. In this paper we present an embedding of propositional production systems into μ-calculus, and first-order production systems into fixed-point logic, with the aim of using these logics for the static analysis of production systems with varying working memories. We encode properties such as termination and confluence in these logics, and briefly discuss which ones cannot be expressed, depending on the expressivity of the logic. We show how the embeddings can be used for reasoning over the production system, and use known results to obtain upper bounds for special cases. The strong correspondence between the structure of the models of the encodings and the runs of the production systems enables the straightforward modeling of properties of the system in the logic.

1 Introduction

Production systems (PS) are one of the oldest knowledge representation paradigms in artificial intelligence, and are still widely used today[1]. Such a system consists of a set of rules r of the form "if $condition_r$ then $action_r$", a working memory, which contains the current state of knowledge, and a rule interpreter, which executes the rules and makes changes in the working memory, based on the actions in the rules.

In general rule-based systems are administered and executed in a distributed environment where the rules are interchanged using standardized rule languages, e.g. RIF, RuleML, SWRL. The new system obtained from adding (or removing) the interchanged rules need to be consistent, and some properties be preserved, e.g. termination. In this work we address the static analysis of such production systems, which means deciding properties like termination and confluence. We propose using logics and their reasoning techniques from the area of software specification and verification, in particular μ-calculus [1] and fixed-point logic (FPL) [2].

In this work we consider rules in which conditions are first-order logic (FOL) formulas with free variables and the actions are *additions* and *removals* of atomic formulas. We also consider the special case of variable-free, i.e., propositional rules. We note here that in case a limited number of constant symbols is available and the rules are quantifier-free, the first-order case can be reduced to the propositional case through grounding, i.e., replacing each rule with all possible ground variable substitutions.

---

[1] http://www.jessrules.com/   http://clipsrules.sourceforge.net/
http://www.ilog.com/products/jrules/

A. Polleres and T. Swift (Eds.): RR 2009, LNCS 5837, pp. 254–268, 2009.
© Springer-Verlag Berlin Heidelberg 2009

The working memory of a production system is a set of facts, i.e., ground atomic formulas. Given a working memory, the rule interpreter applies rules in three steps: (1) *pattern matching*, (2) *conflict resolution*, and (3) *rule execution*. In the first step, the interpreter decides – nowadays typically using the RETE algorithm [3] – for each rule r_i and for each variable substitution σ_j whether r_i can be applied in the working memory using σ_j, i.e., whether the working memory satisfies $\sigma_j(condition_{r_i})$. This step returns all pairs (r_i, σ_j) such that r_i can be applied using σ_j; this set is called the conflict resolution set. In step (2), the interpreter non deterministically chooses a pair from the conflict resolution set; in case the set is empty the system terminates. In the last step, the working memory is updated following the additions and removals in the action part of the selected rule. The interpreter then starts again with step (1).

We note here that the choice of conflict resolution strategy affects certain properties of the production system – for example, if one chooses a conflict resolution strategy that allows each rule to be executed at most once, the system is guaranteed to terminate. Therefore, the conflict resolution strategy of a production system must be known when performing static analysis. In the present paper we assume a simple conflict resolution strategy: the rule interpreter arbitrarily selects one pair (r_i, σ_j) from the resolution sets such that applying $\sigma_j(action_{r_i})$ to the working memory yields an updated working memory that is different from the current one. We leave consideration of further conflict resolution strategies for future work.

The operational semantics of production systems makes it difficult to analyze their behavior. Therefore, it is desirable to use a formalism with a declarative semantics for static analysis. In addition, we are interested in deciding properties of production systems for which the initial working memory varies – e.g., we want to be able to decide whether the execution of a set of rules terminates: no matter what the start state is. We use two well-known logics that are frequently used in the area of software verification. For the analysis of propositional systems we use μ-calculus, which is a modal logic extended with the least and greatest fixpoint operators, and for which common reasoning tasks, such as entailment, are decidable in exponential time. For the first-order systems we use fixed-point logic (FPL), an extension of FOL with least and greatest fixpoint operators. Even though reasoning with FPL is not decidable in the general case, there are decidable subsets [4].

Our main contributions with this paper are as follows. We present an embedding of propositional production systems into μ-calculus and show how this embedding can be used for the static analysis of production systems. We then present an embedding of first-order production systems in fixed-point logic, show how the embedding can be used for reasoning over the production system, and discuss two decidable cases.

We use properties of these logics to derive (un)decidability and complexity results for deciding properties such as termination and confluence of production systems. The embedding of first-order production systems into FPL serves as a starting point for investigating further decidable subsets (e.g., based on the guarded fragment [4]), in particular when considering further strategies that limit the choice in the conflict resolution step (2) of the rule application – for example, such strategies may guarantee termination, and thus finite models of the embedding.

The papers is further structured as follows. We review related work in Section 2 and give preliminary definitions in Section 3. We present our embedding of propositional production systems in μ-calculus and show how it can be used for static analysis in Section 4. In Section 5, we present our embedding of first-order systems in FPL. We conclude and discuss future work in Section 6.

2   Related Work

We consider two streams of related work: *action languages and planning* and *rules in active databases*.

The situation calculus [5] is one of the classic formalisms for representing action and change in artificial intelligence. One might thus consider it an alternative formalism that may also be used for capturing production systems. A distinguishing feature between situation calculus, on the one hand, and μ-calculus and our use of FPL, on the other, is that in the former there is a notion of situation – essentially a term capturing the history of the actions – and in the latter there is a notion of state – essentially a collection of all the facts that hold at a given point in time. Arguably, the latter are conceptually a better match with the notion of working memory in production systems. Nonetheless, the situation calculus has been used in the context of production systems by [6]. Specifically, [6] used logic programs with the stable model semantics and situation calculus notation for characterizing production systems, and deciding properties such as confluence and termination. A notable difference between our approach and the one base on stable model semantics, is that, in general, we allow the initial working memory to vary – we can do that since μ-calculus and FPL allow reasoning over all possible models.

In the planning domain and STRIP-like languages, one assumes a set of operators with preconditions and actions, an initial database and a goal. The planning problem is to find a sequence of operations that updates the initial database so that the goal holds. One might reuse planning results for the static analysis of production systems by considering properties of systems as goals and finding a plan involving rule applications as atomic actions (for example $G = p$ meaning that p eventually holds). This problem has been approached from the logic point of view in many ways, many of them (that we are aware of) started with [7] approaching planning as a satisfiability problem. Reiter [8] allows first order sentences for the goal, which makes the problem undecidable. In [9,10], the authors address the propositional case with LTL. The main problem in trying to apply these works to PS, is that in the planning problem, they need to find *one* sequence which satisfies the goal, or the *absence* of a sequence. This lead the research to linear temporal logic, where basic properties like confluence can be expressed in a first order extension (but maybe encoded in the propositional case) and some other properties, like **[PE5]** in Section 4.2 can not be neither expressed nor encoded. On the other hand, if we consider the operators as the PS's rules, the present work can be used to solve to planning problem using **[PE4]** (in Section 4.2) and replacing p by the Goal.

In [11], the authors present a new logic ($\mathcal{DIFR}$) which is an extension of PDL that can encode propositional situation calculus. They present a formal framework for modeling, and reasoning about actions. Consequently, each particular problem has to be modeled ad-hoc. In the present work we model not just the conditions and effect of an action, but several specific features of Production systems like strategies, constrains, and the behavior of the system in time. We provide an axiomatization of PS (Section 4), and a formal proof of the correspondence with the set of runs of a PS, and the models of our axiomatization (Theorem 1). This link is required to do formal verification of properties of PS, using the models of the axiomatization. The choice of μ-calculus over $\mathcal{DIFR}$ for modeling has been based on two points: First, certain properties of interest, like finiteness of runs (among others), cannot be expressed in $\mathcal{DIFR}$, while they can be expressed in μ-calculus (see Section 4.2). Second, in Section 5 we extend the propositional case, and we model PS with variables, First Order Production Systems (FO-PS). This model is in Fix Point Logic, which can be seen as a first order extension of μ-calculus, therefore the choice of μ-calculus makes the path from the propositional PS to FO-PS more understandable.

Rules in active databases are strongly related to production rules. While production rules are condition-action rules, active databases contain event-condition-action rules; there are external events that may trigger firing of rules. The techniques we are aware of that have been proposed for the static analysis of such rules ([12,13,14]) are based on checking properties of graphs, where nodes are rules, and an edge between r_1 and r_2 means that the action of r_1 can trigger the firing of r_2. The general problem, where conditions are arbitrary SQL queries, is (unsurprisingly) known to be undecidable [15]; an analogous result for our case is stated in Theorem 4. In [12], [13], and [14] the authors study sufficient conditions for deciding termination and confluence. In contrast, our embeddings in μ-calculus and FPL are used to find sufficient and *necessary* conditions for deciding these and other properties for classes of production systems.

3 Preliminaries

μ-Calculus. Let Var be a (infinite) set of variable names, typically written $Y, Z \ldots$ and let $Prop$ be a set of atomic propositions, typically written p, q. μ-calculus extends propositional logic with the modal operator $\Diamond$ and with formulas of the form $\mu.Z.\phi(Z)$, where $\phi(Z)$ is a μ-calculus formula in which the variable Z occurs positively, i.e., under an even number of negations.

As usual, $\Box\phi$ is short for $\neg\Diamond\neg\phi$ and $\nu.Z.\phi(Z)$ is short for $\neg\mu.Z.\neg\phi(\neg Z)$.

A *Kripke structure* is a tuple $K = (S, R, V)$, where S is a non-empty (possibly infinite) set of states, $R \subseteq S \times S$ a binary relation over S, and $V : S \to 2^{Prop}$ assigns to each proposition $p \in Prop$ a (possibly empty) set of states.

A valuation $\mathcal{V} : S \to 2^{Var}$ assigns to each variable a set of states. For a valuation $\mathcal{V}$, a variable Y and a set S, we denote by $\mathcal{V}[Y \leftarrow S]$ the valuation obtained from $\mathcal{V}$ by assigning S to Y.

Given a Kripke structure $K = (S, R, V)$, we define the set of states satisfying a formula ϕ, relative to a valuation $\mathcal{V}$, denoted $\phi^K(\mathcal{V})$, as follows:
- $p^K(\mathcal{V}) = \{s \in S \mid p \in V(s)\}$ for propositions p, $Y^K(\mathcal{V}) = \mathcal{V}(Y)$ for variables Y,
- $(\phi_1 \wedge \phi_2)^K(\mathcal{V}) = (\phi_1)^K(\mathcal{V}) \cap (\phi_2)^K(\mathcal{V})$,
- $(\neg\phi)^K(\mathcal{V}) = S \backslash (\phi)^K(\mathcal{V})$,
- $(\Diamond\phi)^K(\mathcal{V}) = \{s \in S \mid \exists s' \in \phi^K(\mathcal{V}).(s, s') \in R$, and
- $(\mu.Z.\phi(Z))^K(\mathcal{V}) = \bigcap\{S' \subseteq S \mid \phi^K(\mathcal{V}[Z \leftarrow S']) \subseteq S'\}$.

A μ-calculus formula ϕ is *satisfiable* iff there exists a structure K s.t. $(\phi)^K(\mathcal{V}) \neq \emptyset$, for every valuation $\mathcal{V}$; in this case K is a *model* of ϕ. A formula ϕ *entails* a formula ψ iff $(\phi \wedge \neg\psi)$ is not satisfiable.

Fixed Point Logic. Fixed Point Logics FPL [4]) extend standard first order logic (with equality and without function symbols) with least fixed-point formulas of the form $[\mu W.\boldsymbol{x}.\psi(W, \boldsymbol{x})](\boldsymbol{x})$, where W is a k-ary relation symbol (a second order variable), $\psi(W, \boldsymbol{x})$ contains only positive occurrence of W and its free first-order variables are in $\boldsymbol{x}$. As usual, the greatest fixed-point formula $[\nu W.\boldsymbol{x}.\psi(W, \boldsymbol{x})](\boldsymbol{x})$ is short for $\neg[\mu W.\boldsymbol{x}.\neg\psi(\neg W, \boldsymbol{x})](\boldsymbol{x})$.

In order to obtain the necessary correspondence with the constants employed in the production system, we assume *standard names*. Let C be the set of constants. Then, all interpretations are of the form $\mathcal{M} = \langle C, \cdot^{\mathcal{M}} \rangle$ and we have that $c^{\mathcal{M}} = c$, for every $c \in C$.

Given a structure $\mathcal{M} = \langle \Delta, \cdot^{\mathcal{M}} \rangle$ providing interpretations for all the free second order variables in ψ, except W, the formula $\psi(W, \boldsymbol{x})$ defines an operator on k-ary relations $W \subseteq A^K$:

$$\psi^{\mathcal{M}} : W \mapsto \psi^{\mathcal{M}}(W) := \{\boldsymbol{a} \in \Delta^k : \mathcal{M} \models \psi(W, \boldsymbol{a})\}$$

Since W occurs only positively in ψ, this operator is monotone and therefore has a least fixed point $LFP(\psi^{\mathcal{M}})$. We then define

$$\mathcal{M}, B \models [\mu W.\boldsymbol{x}.\psi(W, \boldsymbol{x})](\boldsymbol{x}) \text{ iff } B(\boldsymbol{x}) \in LFP(\psi^{\mathcal{M}})$$

for interpretation $\mathcal{M}$ and first-order variable assignment B.

Satisfaction and entailment are then defined in the usual way.

4   Propositional Production Systems

An intensively investigated area of temporal logic is automatic verification of propositional temporal logic properties of finite state systems. As is well known, semantic properties of programs written in Turing-complete languages are undecidable, thus a full verification of such a program intrinsically includes hand work. However, there are many applications from components of microprocessors to communication protocols, which work with finite data domains, and which can be represented by finite state transition systems. Even if a system is by nature infinite, sometimes the possibility exists to abstract from infinitary aspects and obtain a finitary representation of the relevant aspects of the behavior. Propositional logic succeeds to specify the static properties of finite state systems, i.e. the properties of the states. Moreover, if we have finite state systems on the one hand, and propositional temporal logics, i.e. propositional logics amalgamated with temporal constructs, on the other hand, then automatic verification is possible. This means, there exist programs taking a description of a finite state system and a propositional temporal specification as input and return as result, whether the system satisfies the temporal property.

We first present formal definitions of propositional production systems.

Definition 1. *A* Generic Production System *(GPS) is a tuple* $PS = (Prop, L, R)$, *where*

- $Prop$ *is a finite set of propositions, representing the set of potential facts,*
- L *is a set of rule labels, and*
- R *is a set of* rules, *which are statements of the form*

$$r : \text{ if } \phi_r \text{ then } \psi_r$$

where $r \in L$, ϕ_r *is a propositional formula, and* $\psi_r = a_1 \wedge \cdots \wedge a_k \wedge \neg b_1 \wedge \cdots \wedge \neg b_l$, *with* $a_i \neq b_j$ $(a_i, b_j \in Prop)$ *signifying the propositions added, respectively removed by the rule, such that every rule has a distinct label and* $L \cap Prop = \emptyset$. *We define* $\phi_r^{add} = \{a_1, \ldots, a_k\}$ *and* $\phi_r^{remove} = \{b_1, \ldots, b_l\}$.

In the following, let $PS = (Prop, L, R)$ be a production system. A *Working Memory* $WM \subseteq Prop$ for PS is a set of propositions. As an abuse of notation, we use WM for both the working memory and the propositional valuation induced by it (i.e., $WM \models p$ iff $p \in WM$).

A rule r is *fireable* in a working memory WM if $WM \models \phi_r$ and $WM' = WM \cup \psi_r^{add} \setminus \psi_r^{remove} \neq WM$.[2]

A *concrete production system* (CPS) is a pair (PS, WM_0), where WM_0 is a working memory. For a set of symbols Σ, a Σ-*labeled tree* is a pair (T, V), where $T \subseteq \mathbb{N}^+$ is such that if $x.c \in T$, then also $x \in T$ and $V : T \to 2^\Sigma$ maps each node to a set of symbols in Σ.

Definition 2. *A* computation tree $CT_{WM_0}^{PS}$ *for a CPS* (PS, WM_0) *is a* $(Prop \cup L)$-*labeled tree* (T, V) *such that the root of* T *is* 0, $V(0) = WM_0$, *and for each node* $n \in T$ *and every rule* r *that is fireable in the working memory* $WM = V(n) \cap Prop$, *there is a child node* $n' \in T$ *of* n *such that* $V(n') = WM' \cup \{r\}$, *with* $WM' = WM \cup \psi_r^{add} \setminus \psi_r^{remove}$. *There are no other nodes in* $CT_{WM_0}^{PS}$.

Note that the rule label r in the label of each node represents the rule that has been fired to obtain the current node from the parent. Note also that $CT_{WM_0}^{PS}$ is unique up to isomorphisms, and so we may speak about *the* computation tree of a CPS. A *run* of a CPS is a branch in $CT_{WM_0}^{PS}$. A run is *terminating* if it is finite.

In the remainder of this section we present an axiomatization of production systems in μ-calculus and show how this can be used for static analysis.

4.1 Axiomatization

The existence of a formal description of any language is a prerequisite to any rigorous method of proof, validation, or verification. Here we present an axiomatization of production systems in μ-calculus. In the following subsections we will show how this axiomatization can be used for reasoning about production systems.

In the following, let $PS = (Prop, L, R)$ be a generic production system and let WM_0 be a working memory. We first define the necessary components of the formula

---

[2] Note that we assume here a simple conflict resolution strategy, namely a rule is only fired if it brings about a change in the working memory.

comprising the axiomatization. These components encode the constrains and requirements in the relation between one state and its successors depending if it is an intermediate state in the execution of the PS, or a state representing the end of a run. A
greatest fix point composed of these components restricts the models to the ones which
are bisimilar to a computation tree (CT). The states in the models of the axiomatization can be seen as nodes in the computation tree. We assume that b does not appear in
$Prop \cup L$.

Root. The current state represents the root of the CT.

$$b \wedge \bigwedge_{r \in L} \neg r$$

RApp. Rule application.

$$(\bigwedge_{r \in L} (r \to \psi_r))$$

Appl. If a rule is applied, it must be applicable.

$$(\bigwedge_{r \in L} \Diamond r \to \phi_r \wedge \neg \psi_r)$$

Frame. Frame axiom: if q holds, it holds in the next state unless q is removed and if
$\neg q$ holds, $\neg q$ holds, unless q is added.

$$(\bigwedge_{q \in Prop} (q \to \Box(q \vee (\bigvee_{r \in L.q \in \psi_r^{remove}} r))) \wedge$$
$$(\neg q \to \Box(\neg q \vee (\bigvee_{r \in L.q \in \psi_r^{add}} r)))))$$

NoFireable. No rule is fireable and there is no successor.

$$[(\bigwedge_{r \in L} (\phi_r \to \psi_r)) \wedge (\Box \bot]$$

Fireable. At least one rule is fireable and there is a successor.

$$(\bigvee_{r \in L} \phi_r \wedge \neg(\psi_r)) \wedge \Diamond \top$$

Complete. If a rule is fireable, it is applied in some successor states.

$$(\bigwedge_{r \in L} (\phi_i \wedge \neg(\psi_r)) \to \Diamond r)$$

1Rule. Exactly one rule is applied.

$$(\bigvee_{r \in L} r \wedge (\bigwedge_{r \in L} (r \to \neg \bigvee_{r' \in L \& r' \neq r} r')))$$

WM. (Optional) The initial working memory holds.

$$\bigwedge_{q \in WM_0} q \wedge \bigwedge_{q \in Prop \setminus WM_0} \neg q$$

The axiom **Root.** captures the root of the computation tree. We now define the axioms
which capture intermediate and end (i.e., leaf) nodes.

Intermediate = **RApp.** $\wedge$ **1Rule.** $\wedge$ **Appl.** $\wedge$ **Frame.** $\wedge$ **Fireable.** $\wedge$ **Complete.** $\wedge \neg b$
End = **RApp.** $\wedge$ **1Rule.** $\wedge$ **Frame.** $\wedge$ **NoFireable.** $\wedge \neg b$

We now define the μ-calculus formula that captures the production system PS:

$$\Phi_{PS} = [(\textbf{Root.} \wedge \textbf{NoFireable.}) \vee (\textbf{Root.} \wedge \textbf{Appl.} \wedge \textbf{Frame.} \wedge$$
$$\textbf{Complete.} \wedge \textbf{Fireable.} \wedge \Box(\nu.X.(\textbf{Intermediate} \vee \textbf{End}) \wedge \Box X)))]$$

We now proceed to prove bisimilarity between the models of Φ_{PS} and the computation trees of PS. We will exploit this result later for reasoning about PS.

A *bisimulation* between two pointed Kripke structures, $K = ((S, R, V), s_0)$ and $K' = ((S', R', V'), t_0')$ is a relation $Z \subseteq S \times S'$ such that:
- $(s_0, t_0') \in Z$
- if $(s_i, t_i') \in Z$, then $p \in V(s_i)$ iff $p \in V'(t_i')$, for every proposition p,
- if $(s_i, t_i') \in Z$ and $(s_i, s') \in R$ implies that there is a $t' \in S'$ such that $(t_i', t') \in R'$ and $(s', t') \in Z$
- if $(s_i, t_i') \in Z$ and $(t_i, t') \in R'$ implies that there is a $s' \in S$ such that $(s_i, s') \in R$ and $(s', t') \in Z$

We view a computation tree (T, V), with 0 being the root, also as a Kripke structure $K = (T, R, V')$, where $V'(0) = V(0) \cup \{b\}$, $V'(n) = V(n)$ for $n \neq 0$, and $(n, n') \in R$ iff $n.n' \in T$.

Theorem 1. *Given a Production system $PS = (Prop, L, R)$, a starting working memory WM_0, and the formula Φ_{PS}.*

1. *A Kripke structure $K = (S, R, V)$ is a model of Φ_{PS} iff there is a working memory WM for PS such that there is an $s \in S$ and (K, s) is bisimilar to $(CT_{WM}^{PS}, 0)$*
2. *A Kripke structure $K = (S, R, V)$ is a model of $\Phi_{PS} \wedge \textbf{WM.}$ iff there is an $s \in S$ such that (K, s) is bisimilar to $(CT_{WM_0}^{PS}, 0)$.*

Proof (Sketch). We start with 2. If $K = (S, R, V)$ is a model of $\Phi_{PS} \wedge \textbf{WM.}$ then there is at least a node s_0 s.t. $K, s_0 \models \textbf{Root.} \wedge \textbf{WM.}$. We start defining the bisimulation Z: $(s_0, 0) \in Z$. Now we have to define bisimulation in such a way that $(s_i, xi) \in Z$ iff $(s_j, x) \in Z$ and s_i successor of s_j, $x.i$ is the successor of x, and $s_i, x.i$ agree on the proposition constants. Let's take a node s_i successor of s_j such that $(s_j, x) \in Z$ and $V(s_j) = V(x)$. By **Complete.** we know that there is at least one successor for each applicable rule in s_j, and by **RApp.** and **Frame.**, we know that the label of the successor is just the result of the rule application. By **Fireable.** we know that the precondition of every successor hold, therefore, for each successor of s_j in K, we have a successor of x in CT_{ps} (note it can be many-to-one) with the same label, so $(s_i, xj) \in Z$ for every s_i and some xj which is determined by the rule label. The other direction is analogous.

Now, the proof of 1 is straightforward: if we have a model K of Φ_{PS} where some state s_0 (as defined above) is the set of proposition WM, we know that PS with WM as the initial working memory is bisimilar to K by point 2. The converse is analogous.

4.2 Deciding Properties of Production Systems

Typical properties of production systems one would like to check are termination and confluence of the system. However, one could imagine additional properties of interest, e.g., redundancy of rules (useful in the design of the system). In this section we showcase a number of properties that we feel might be of interest, and that can be reduced to μ-calculus satisfiability or entailment checking, using the axiomatization of the previous section.

The properties stated below are defined for both generic and concrete production systems. **PEi** are the properties that can be decided by checking entailment . With ϕ_{PEi} we denote the formula associated with **PEi**.

PE1. All runs are finite (i.e., Termination)

$$(\mu.X.\Box X)$$

PE2. All runs terminate with the same working memory (Confluence)

$$\bigwedge_{q_i \in Prop}(\mu.X.(\Box \perp \wedge q_i) \vee \Diamond X) \to (\nu.X.(\Box \perp \to q_i) \wedge \Box X)$$

PE3. There is a fireable rule in the initial working memory

$$(\bigvee_{r \in R} \phi_r)$$

PE4. A proposition p eventually holds in some run.

$$(\mu.X.(p \wedge (\nu.Y.p \wedge \Diamond Y)) \vee \Diamond X)$$

PE5. A proposition p eventually holds for ever in every run.

$$(\mu.X.(p \wedge (\nu.Y.p \wedge \Box Y)) \vee \Box X)$$

PE6. Some rule r is never applied

$$\neg(\mu.X.\Diamond X \vee r)$$

PE7. All rules are applied in every run

$$\bigwedge_{r_i \in R}(\mu.Z.r_i \vee \Box Z \wedge \Diamond \top)$$

We now show how deciding the above properties can be reduced to μ-calculus entailment checking, by exploiting Theorem 1.

Theorem 2. *A property PEi, for $i \in \{1, \ldots, 7\}$ holds for a generic production system PS iff Φ_{PS} entails PEi and PEi holds for a concrete production system (PS, WM_0) iff WM. $\wedge \Phi_{PS}$ entails ϕ_{PEi}.*

Note that when considering concrete production systems, some of the mentioned properties (e.g., **PE3**) can be decided by simply running the system. However, certain other properties (e.g., termination) cannot.

From the fact that Φ_{PS} is polynomial in the size of PS and the fact that μ-calculus entailment can be decided in exponential time, we immediately obtained the following complexity results.

Proposition 1. *The properties PE1-7 can be decided in exponential time, both on generic and concrete production systems.*

5 First Order Production Systems

We now consider the case of production systems with variables.

Definition 3. *A* Generic FO-Production System *is a tuple* $PS = (\tau, L, R)$, *where*
- $\tau = (P, C)$ *is a first-order signature, with P a set of predicate symbols, each with an associated nonnegative arity, and C a nonempty (possibly infinite) set of constant symbols,*
- L *is a set of rule labels, and*
- R *is a set of rules, which are statements of the form*

$$r : \text{ if } \phi_r(\boldsymbol{x}) \text{ then } \psi_r(\boldsymbol{x})$$

where $r \in L$, ϕ_r is an FO formula with free variables $\boldsymbol{x}$ and $\psi_r(\boldsymbol{x}) = (a_1 \wedge \cdots \wedge a_k \wedge \neg b_1 \wedge \cdots \wedge \neg b_l)$, where $a_1, \ldots, a_k, b_1, \ldots, b_l$ are atomic formulas with free variables among $\boldsymbol{x}$, such that no a_i and b_j share the same predicate symbol, each rule has a distinct label and $L \cap Prop = \emptyset$. We define $\phi_r^{add} = \{a_1, \ldots, a_k\}$ and $\phi_r^{remove} = \{b_1, \ldots, b_l\}$.

In the following, let $PS = (\tau, L, R)$ be an FO-production system. With AT we denote the set of equality-free ground atomic formulas (atoms) of τ. A working memory WM for PS is a subset of AT. As an abuse of notation, we use WM to denote both the working memory and the first-order structure induced by the working memory, i.e., the domain of WM is C, $c^{WM} = c$, for any $c \in C$, and $\boldsymbol{c} \in p^{WM}$ iff $p(\boldsymbol{c}) \in WM$ for any $p(\boldsymbol{c}) \in AT$.

A variable substitution $\mathcal{S}$ is a mapping from variables to constants in C. The application of a variable substitution to a term or formula φ, written $\mathcal{S}(\varphi)$, is defined in the usual way. A rule is *fireable* in a working memory WM using a substitution $\mathcal{S}$ if $WM \models \mathcal{S}(\phi_r)$ and $WM' = WM \cup \mathcal{S}(\psi_r^{add}) \setminus \mathcal{S}(\psi_r^{remove}) \neq WM$.

A *concrete FO-production system* is a pair (PS, WM_0), where WM_0 is a working memory. We view rule labels $r \in L$ also as n-ary predicates, where n is the number of free variables in the condition ϕ_r; with AL we denote the set of ground atoms constructed from the predicate symbols in L and the constants in C.

Definition 4. *A computation tree $CT^{PS}_{WM_0}$ for a (PS, WM_0) is an $(AT \cup AL)$-labeled tree (T, V) such that the root of T is 0, $V(0) = WM_0$, and for each node $n \in T$, every rule r, and every variable substitution $\mathcal{S}$ such that r is fireable in the working memory $WM = V(n) \cap Prop$ using $\mathcal{S}$, there is a child node $n' \in T$ of n such that $V(n') = WM' \cup \{\mathcal{S}(r(\boldsymbol{x}))\}$, with $WM' = WM \cup \mathcal{S}(\psi_r^{add}) \setminus \mathcal{S}(\psi_r^{remove})$. There are no other nodes in $CT^{PS}_{WM_0}$.*

A *run* of (PS, WM_0) is a branch of $CT^{PS}_{WM_0}$. A run is *terminating* if it is finite.

In the remainder of this section we discuss special cases of FO production systems that can be reduced to propositional production systems, we present an axiomatization of general FO production systems in fixed-point logic (FPL), and show how static analysis can we reduced to reasoning with FPL. This axiomatization will be a starting point for future investigation of decidable fragments.

5.1   Grounding FO Production Systems

The *grounding* of an FO production system $PS = (\tau, L, R)$, denoted $gr(PS)$, is obtained from PS by replacing each rule r : if $\phi_r(\boldsymbol{x})$ then $\psi_r(\boldsymbol{x})$ with a set of rules $\mathcal{S}(r(\boldsymbol{x}))$: if $\mathcal{S}(\phi_r(\boldsymbol{x}))$ then $\mathcal{S}(\psi_r(\boldsymbol{x}))$, for every substitution $\mathcal{S}$ of variables with constants in C.

Clearly, for any working memory WM, the computation trees of (PS, WM) and $(gr(PS), WM)$ are the same. Also, if the rules in PS are quantifier-free, $gr(PS)$ can be seen as a propositional generic production system – in the absence of variables, atomic formulas are essentially propositions. This allows us to apply some of the results for the propositional case to FO production systems.

We first exploit the fact that if the set of constants C is finite, the grounding $gr(PS)$ is finite, and its size exponential in the size of PS.

Proposition 2. *Let* $PS = (\tau, L, R)$ *be an FO production system such that R is quantifier-free and C is finite, and let WM be a working memory.[3] Then, the properties **PE1-7** can be decided in double exponential time, on both PS and (PS, WM).*

When considering concrete FO production systems, i.e., the initial working memory is given, we can also exploit grounding, provided the conditions in the rules are *domain-independent* (cf. [16]): a first-order formula with n free variables $\phi(\boldsymbol{x})$ is domain-independent iff whenever $\mathcal{M} = \langle \Delta, \cdot^{\mathcal{M}} \rangle$ and $\mathcal{M}' = \langle \Delta, \cdot^{\mathcal{M}'} \rangle$ are structures, $\mathcal{M}$ is a substructure of $\mathcal{M}'$, and the interpretations functions are identical ($\cdot^{\mathcal{M}'} = \cdot^{\mathcal{M}}$), then for all $a_1 \in \mathcal{M}', \ldots, a_n \in \mathcal{M}'$:

$$\mathcal{M}' \models \phi(a_1 \ldots a_n) \leftrightarrow a_1 \in \mathcal{M} \wedge \cdots \wedge a_n \in \mathcal{M} \wedge \mathcal{M} \models \phi(a_1 \ldots a_n)$$

If all conditions are domain-independent and the initial working memory WM_0 is given, one only needs to consider grounding with the constants appearing in (PS, WM_0). Examples of domain-independent formulas are conjunctions of literals such that each variable occurs in a positive literal.

Proposition 3. *Let* $PS = (\tau, L, R)$ *be an FO production system such that R is quantifier-free and for every rule $r \in R$ holds that $\phi_r(\boldsymbol{x})$ is domain-independent, and let WM be a working memory. Then, the properties **PE1-7** can be decided in double exponential time, on (PS, WM).*

5.2   Axiomatizing FO Production Systems

In the remainder we assume that the signature of each of the production systems contains a countably infinite set of constant symbols. In our μ-calculus axiomatization for the propositional case the structure of the computation tree was reflected in the accessibility relation of the Kripke models. Interpretations in FPL are first-order structures; therefore, we capture the structure of the computation tree using the binary predicate R,

---

[3] Note that if C is finite, the existential quantifier could be replaced with a disjunction of all possible ground variable substitutions; analogous for universal quantifier. In this case, the grounding would be double exponential.

and we divide the domain into two parts: the nodes of the tree, i.e., the states (A), and the objects in the working memories (U). The arity of the predicates in $P \cup L$ is increased by one, and the first argument of each predicates will signify the state; $p(y, x_1, \ldots, x_n)$ intuitively means that $p(x_1, \ldots, x_n)$ holds in state y.

In the remainder, let $PS = (\tau, L, R)$ be a generic FO production system and let WM_0 be a working memory. We first define the *foundational axioms*, which encode the basic structure of the models and the tree shape of R.

We assume that the unary predicates B (signifying the start state), U and A and the binary predicate R are not in $P \cup L$.

Structure. Partitioning of the domain.

$$\forall x : A(x) \leftrightarrow \neg U(x) \wedge (\bigwedge_{p \in P \cup L \cup \{B\}} \forall y, \boldsymbol{x} : p(y, \boldsymbol{x}) \rightarrow A(y) \wedge U(x_1) \wedge \cdots \wedge$$
$$U(x_n)) \wedge (\forall x, y : R(x, y) \rightarrow A(x) \wedge A(y))$$

Tree. The predicate R encodes a tree.

$$\forall x \exists^{\leq 1} y : A(x) \rightarrow R(y, x) \wedge \exists^{\leq 1} x : \forall y : A(y) \rightarrow (\neg R(y, x) \wedge$$
$$(\mu.W.x, y.R(x, y) \rightarrow W(x, y) \wedge (\exists z : W(x, y) \wedge R(y, z) \rightarrow W(x, z)))(x, y))$$

We denote the set of foundational axioms with $\Sigma_{found} = \{\textbf{Structure., Tree.}\}$. We now turn to the axioms that encode the behavior of the production system. We omit explanations of axioms that are simply extensions of the propositional case.

Root. $B(y) \wedge \bigwedge_{r \in L} \forall x : \neg r(y, x)$

RApp. $(\bigwedge_{r \in L} \forall \boldsymbol{x} : r(y, \boldsymbol{x}) \rightarrow \psi_r(\boldsymbol{x}))$

Appl. $(\bigwedge_{r \in L} \forall \boldsymbol{x} : \exists w(R(y, w) \wedge r_i(w, \boldsymbol{x})) \rightarrow \phi_r(y, \boldsymbol{x}) \wedge \neg \psi_r(y, \boldsymbol{x}))$

Frame. $\bigwedge_{p \in P} \forall x_1, \ldots, x_n([p(y, x_1, \ldots, x_n) \rightarrow (\forall w : R(y, w) \rightarrow$
$p(w, x_1, \ldots, x_n) \vee (\bigvee_{r \in L. \psi_r(z) = \ldots \neg p(t_1, \ldots, t_n) \wedge \ldots} \exists z : r(y, z) \wedge x_1 = t_1 \wedge \cdots \wedge x_n = t_n))] \wedge [\neg p(y, x_1, \ldots, x_n) \rightarrow (\forall w : R(y, w) \rightarrow \neg p(w, x_1, \ldots, x_n) \vee$
$(\bigvee_{r \in L. \psi_r(z) = \ldots p(t_1, \ldots, t_n) \wedge \ldots} \exists z. r(y, z) \wedge x_1 = t_1 \wedge \cdots \wedge x_n = t_n))])$

NoFirable. $(\bigwedge_{r \in L} \forall \boldsymbol{x} : \phi_r(y, \boldsymbol{x}) \rightarrow \psi_r(y, \boldsymbol{x})) \wedge (\forall w : \neg R(y, w)))$

Firable. $(\bigvee_{i=1}^{n} \exists \boldsymbol{x} : \phi_{r_i}(y, \boldsymbol{x})) \wedge \exists w : R(y, w)$

Complete. If a rule is fireable, it is applied once.
$\bigwedge_{r \in L} \forall \boldsymbol{x} (\phi_r(y, \boldsymbol{x}) \wedge \neg \psi_r(y, \boldsymbol{x}) \rightarrow \exists^{=1} w(R(y, w) \wedge r(w, \boldsymbol{x})))$

1Rule. $(\bigvee_{r \in L} \exists \boldsymbol{x} : r(y, \boldsymbol{x})) \wedge (\bigwedge_{r \in L} (\exists z : r(y, z) \rightarrow \neg \bigvee_{r' \in L \& r' \neq r} \exists \boldsymbol{x} : r'(y, \boldsymbol{x})))$

WM. $\bigwedge_{p \in P} \forall x_1 \ldots x_n (p(y, x_1 \ldots x_n) \leftrightarrow$
$\bigvee \{x_1 = c_1 \wedge \cdots \wedge x_n = c_n \mid p(c_1, \ldots, c_n) \in WM_0\})$

Only. A rule can not be applied twice in the same state.
$(\bigwedge_{r \in L} \forall \boldsymbol{x} : r(y, \boldsymbol{x}) \rightarrow \exists^{=1} z : (r(y, z)))$

Intermediate $=$ **RApp.** $\wedge$ **1Rule.** $\wedge$ **Only.** $\wedge$ **Appl.** $\wedge$ **Frame.** $\wedge$ **Firable.** $\wedge$ **Complete.** $\wedge$
$\quad \neg B(y)$
End $=$ **RApp.** $\wedge$ **1Rule.** $\wedge$ **Only.** $\wedge$ **Frame.** $\wedge$ **NoFirable.** $\wedge \neg B(y)$

Analogous to the propositional case, we defined a formula that captures the behavior of PS:

$$\Phi_{PS} = (\exists y : (\textbf{Root.} \wedge \textbf{NoFirable.}) \vee (\textbf{Root.} \wedge \textbf{Appl.} \wedge \textbf{Complete.} \wedge \textbf{Firable.} \wedge$$
$$\forall w (R(y,w) \rightarrow (\nu . X . y . (\textbf{Intermediate} \vee \textbf{End}) \wedge \forall w (R(y,w) \rightarrow X(w)))(w))))$$

The most notable difference with the propositional axiomatization is in the **Complete.** axiom. In the propositional case, we could require that a fireable rule is applied at least once, but it could be applied several times. In the first-order case, we can require a fireable rule to be applied exactly once. We can therefore obtain a stronger correspondence between computation trees and Kripke models: they are essentially isomorphic.

Definition 5. *Let PS be an production system, WM_0 the working memory, $\mathcal{M}$ a model of Σ_{found}, and $CT_{WM_0}^{PS} = (T,V)$ the computation tree of (PS, WM_0). Then, we say that $CT_{WM_0}^{PS}$ and $\mathcal{M} = (\Delta, \cdot^{\mathcal{M}})$ are isomorphic if there is a bijective function $f : V \mapsto A^{\mathcal{M}}$ such that:*

1. *$x.i \in T^{CT}$ iff $(f(x), f(x.i)) \in (R)^{\mathcal{M}}$,*
2. *for every $x \in T$, every n-ary $p \in P \cup L$, and every $\mathbf{c} \in C^n$, $p(c_1 \ldots c_n) \in V(x)$ iff $(f(x), c_1 \ldots c_n) \in p^{\mathcal{M}}$,*
3. *$(z,w) \in (R)^{\mathcal{M}}$ iff $f^{-1}(w).f^{-1}(z) \in T^{CT}$, and*
4. *for every $x \in A^{\mathcal{M}}$, every n-ary $p \in P \cup L$, and every $\mathbf{c} \in C^n$, $(x, c_1 \ldots c_n) \in p^{\mathcal{M}}$ iff $p(c_1 \ldots c_n) \in V(f^{-1}(x))$.*

Theorem 3. *Given an FO production system $PS = (\tau, L, R)$, a starting working memory WM_0, and the formula Φ_{PS},*

1. *a model $\mathcal{M}$ of Σ_{found} is a model of Φ_{PS} iff there is a working memory WM for PS s.t. $\mathcal{M}$ is isomorphic to CT_{WM}^{PS}, and*
2. *a model $\mathcal{M}$ of Σ_{found} is a model of $\Phi_{PS} \wedge$ **WM.** iff $\mathcal{M}$ is isomorphic to $CT_{WM_0}^{PS}$.*

Proof (Sketch). We start with 2. We construct a mapping f; one can verify that it satisfies conditions 1–4 from Definition 5.

We take $y_0 \in C$ s.t. **Root.**$(y_0) \wedge$ **WM.**(y_0) holds. By **WM.** we know that $0 \in T$ and $y_0^{\mathcal{M}} \in A^{\mathcal{M}}$ "share" the same predicates. Therefore we can define $f(0) = y_0$. We proceed by induction.

For every $x.i$ in $CT_{WM_0}^{PS}$, s.t. $f(x) = y$, (recall that x is a predecessor of $x.i$ by definition) for some $y \in A^{\mathcal{M}}$, the node x and the state y share the same predicates in the sense of definition 5. We have that $r(\mathbf{c}) \in V(x.i)$ for some $r \in L$. We define $f(x.i) = z$, where $(y, z) \in R^{\mathcal{M}}$ and $(z, \mathbf{c}) \in r^{\mathcal{M}}$. There is such a unique z, by satisfaction of **Complete.**, **Firable.**, and **Appl.**. This establishes satisfaction of condition 1. Satisfaction of condition 3 is established analogously.

Satisfaction of conditions 2 and 4 is established by induction and satisfaction of **1Rule.** and **Only.** (for $p \in L$) and by satisfaction of **RApp.** and **Frame.** (for $p \in P$). Satisfaction of condition 4 is established analogously. The first part of the theorem is proved analogously.

The following result follows immediately from the undecidability of first-order logic and the fact that ϕ_r is an arbitrary first-order formula.

Theorem 4. *The satisfiability problem for ϕ_{PS} under Σ_{found} is undecidable.*

Using Theorem 3 it is straightforward to verify that finiteness of all runs can be reduced to checking entailment of

$$\forall y : A(y) \rightarrow (\mu.X.\forall w(R(y,w) \rightarrow X(w)))(y)$$

and confluence can be reduced to checking entailment of

$$\bigwedge_{p \in P} (\exists y, \boldsymbol{x} : A(y) \wedge (\forall z : A(z) \rightarrow \neg R(y,z)) \wedge p(y,\boldsymbol{x}) \rightarrow (\forall w : A(w) \wedge (\forall z : A(z) \rightarrow$$
$$\neg R(w,z)) \rightarrow p(w,\boldsymbol{x}))$$

Even in the very expressive logics we consider in this paper, there are properties that might be of interest, but cannot be expressed. For example: every run of the system has the same length. This particular property cannot be expressed in FPL, or even in monadic second-order logic over countable trees [17], for that matter.

6 Conclusions and Future Work

In this paper we presented an embedding of propositional production systems into μ-calculus, and first-order production systems into fixed-point logic. We exploited the fixpoint operator in both logics to encode properties of the system over time. One of the advantages of our encodings is the strong correspondence between the structure of the models and the runs of the production systems, which enables straightforward modeling of properties of the system in the logic.

We have illustrated the versatility of our approach by encoding a number of properties discussed in the literature [12,14], as well as a number of other properties that have not been previously considered. Another possible application of our encodings is the optimization of production systems. We have already shown how one can check that a particular rule is never applied (cf. property **PE6**), and thus may be discarded. Deciding equivalence of production systems can be reduced to entailment in μ-calculus and FPL. Equivalence can be exploited for optimization by replacing a production system with an equivalent system that is potentially easier to execute.

We plan to extend the work presented in this paper in a number of directions. We plan to extend both the propositional and first-order case with additional conflict resolution strategies, e.g., based on rule priorities. We plan to extend the first-order case with object invention, i.e., the rules may assert information about new (anonymous) objects; this is strongly related to existential quantification in logic. Another topic we plan to address are new decidable fragments of our first-order encoding, in particular restricting the conditions and possibly the working memory, and conflict resolution strategies in order to exploit the guarded fragment of FPL [4], as well as translations to monadic second-order logic over trees; both fragments are known to be decidable. Finally, we plan to investigate the combination of production systems with languages for describing background knowledge, in the form of description logic ontologies.

Acknowledgements. We thank Sergio Tessaris and Enrico Franconi for valuable discussions and the anonymous reviewers for useful comments and feedback. The work presented in this paper was partially supported by the European Commission under the project ONTORULE (IST-2009-231875).

References

1. Kozen, D.: Results on the propositional μ-calculus. In: Proceedings of the 9th Colloquium on Automata, Languages and Programming, London, UK, pp. 348–359. Springer, Heidelberg (1982)
2. Gurevich, Y., Shelah, S.: Fixed-point extensions of first-order logic. In: Symposium on Foundations of Computer Science, pp. 346–353 (1985)
3. Forgy, C.: Rete: A fast algorithm for the many patterns/many objects match problem. Artif. Intell. 19(1), 17–37 (1982)
4. Grädel, E.: Guarded fixed point logics and the monadic theory of countable trees. Theor. Comput. Sci. 288(1), 129–152 (2002)
5. McCarthy, J., Hayes, P.: Some philosophical problems from the standpoint of artificial intelligence. In: Meltzer, B., Michie, D. (eds.) Machine Intelligence, vol. 4, pp. 463–502. Edinburgh University press, Edinburgh (1969)
6. Baral, C., Lobo, J.: Characterizing production systems using logic programming and situation calculus,
   `http://www.public.asu.edu/~cbaral/papers/char-prod-systems.ps`
7. Kautz, H., Selman, B.: Planning as satisfiability. In: ECAI 1992: Proceedings of the 10th European Conference on Artificial Intelligence, New York, NY, USA, pp. 359–363. John Wiley & Sons, Inc., Chichester (1992)
8. Reiter, R.: Knowledge in Action. Logical Foundations for Specifying and Implementing Dynamical Systems. MIT Press, Cambridge (2001)
9. Mattmller, R., Rintanen, J.: Planning for temporally extended goals as propositional satisfiability. In: Veloso, M. (ed.) Proceedings of the 20th International Joint Conference on Artificial Intelligence, Hyderabad, India, January 2007, pp. 1966–1971. AAAI Press, Menlo Park (2007)
10. Cerrito, S., Mayer, M.C.: Using linear temporal logic to model and solve planning problems. In: Giunchiglia, F. (ed.) AIMSA 1998. LNCS (LNAI), vol. 1480, p. 141. Springer, Heidelberg (1998)
11. De Giacomo, G., Lenzerini, M.: Pdl-based framework for reasoning about actions. In: AI*IA 1995: Proceedings of the 4th Congress of the Italian Association for Artificial Intelligence on Topics in Artificial Intelligence, London, UK, pp. 103–114. Springer, Heidelberg (1995)
12. Aiken, A., Hellerstein, J.M., Widom, J.: Static analysis techniques for predicting the behavior of active database rules. ACM Transactions on Database Systems 20, 3–41 (1995)
13. Baralis, E., Ceri, S., Paraboschi, S.: Compile-time and runtime analysis of active behaviors. IEEE Trans. on Knowl. and Data Eng. 10(3), 353–370 (1998)
14. Baralis, E., Torino, P.D., Widom, J., Widom, N.J.: An algebraic approach to static analysis of active database rules. ACM TODS 25, 269–332 (2000)
15. Bailey, J., Dong, G., Ramamohanarao, K.: Decidability and undecidability results for the termination problem of active database rules. In: Proceedings of the 17th ACM SIGMOD-SIGACT-SIGART Symposium on Principles of Database Systems, pp. 264–273 (1998)
16. Abiteboul, S., Hull, R., Vianu, V.: Foundations of Databases. Addison-Wesley, Reading (1995)
17. Courcelle, B.: The expression of graph properties and graph transformations in monadic second-order logic, pp. 313–400. World Scientific, Singapore (1997)

Author Index